Hybrid Computation

Hybrid Computation

George A. Bekey
*Department of Electrical Engineering
University of Southern California*

Walter J. Karplus
*Department of Engineering
University of California, Los Angeles*

John Wiley & Sons, Inc.
New York London Sydney Toronto

Copyright © 1968
by John Wiley & Sons, Inc.

All rights reserved.
No part of this book may
be reproduced by any
means, nor transmitted, nor
translated into a machine
language without the written
permission of the publisher.

Library of Congress Catalog Card
Number: 68-8103 SBN 471 06355 X

2 3 4 5 6 7 8 9 10

Printed in the United States of America

Preface

In a broad sense the field of hybrid computation includes all computing techniques combining some of the features of digital computations with some of the features of analog computations. However, for a variety of technical and historical reasons, the term "hybrid computer" has required a more restricted meaning and is now used primarily to characterize computer systems involving linkages between general-purpose digital computers and electronic differential analyzers. The potential advantages of this type of hybridization were recognized as soon as electronic digital computers appeared on the scene in the early 1950's.

Initial efforts to interconnect analog and digital computers suffered from a variety of technical and economic handicaps. Foremost among these were the inadequacies of the vacuum-tube linkage elements, the unsuitability of available digital computers for on-line operation, and the lack of adequate software. The introduction of high-quality, fully transistorized, 100-volt analog computers in the early 1960's permitted for the first time the construction of reliable analog computer systems containing many hundreds of amplifiers. This, in turn, made it possible for users to contemplate an entirely new and more ambitious series of applications. At approximately the same time, relatively small on-line digital computers became available. As a result, hybrid computers changed from novel experiments to useful and efficient computing systems. Gradually, as the reliability and the capability of hybrid systems for the solution of large problems increased, larger and larger digital computers were connected to analog computers. By 1968, analog/hybrid computers had attained a high level of maturity. Numerous hybrid computer systems using an excess of 500 analog amplifiers were in use. At the same time many installations successfully employed more modest hybrid computers in which 50- to 100-amplifier analog machines were connected to relatively small digital computers. The time therefore appears ripe to summarize the theoretical foundations underlying the hybrid computer field as well as to survey the variety of applications in which hybrid computers have been successfully used.

It is the purpose of the present text to provide a comprehensive perspective of the theory, the mechanization, and the application of hybrid computers. For a number of years this material has been presented in first-year graduate courses in computer applications at the University of California at Los Angeles and at the University of Southern California. In these courses the students are expected to have completed introductory course work in both analog and digital computing, but no electronic or electrical engineering background is required. This book is also intended for use as a reference by academic and industrial users of hybrid computing techniques and by those interested in the possible application of hybrid methods to their computational problems. To this end the text contains some introductory material concerning numerical analysis, sampled-data theory, and the formulation of typical problems in the various application areas.

This is definitely not a programming text. Digital computer flow charts and analog schematics are used wherever appropriate to illustrate and clarify the discussion of the utilization of hybrid computers. No attempt is made, however, to delve into the details of implementing the various algorithms and techniques using specific hybrid computers. Rather, the accent is placed on generality and insight into the fundamental principles involved.

The text opens with an introductory chapter which briefly surveys the various ways in which hybrid computer concepts have been mechanized in the past and the various application areas which have benefited from the hybrid computer approach. The next six chapters are devoted to theory and mechanization. In Chapter 2 the hardware elements comprising the hybrid computer loop are described from an operational point of view. Particular emphasis is placed on analog-digital and digital-analog conversion and on the basic considerations that dictate a choice of one or another alternative approach. Chapters 3 and 4 provide the mathematical background for the solution of differential equations using digital and hybrid computer systems. Errors inherent in hybrid computers are discussed in Chapter 5. Chapter 6 describes complete hybrid computer systems and the technical, administrative, and economic considerations involved in their specifications and acquisition. In Chapter 7 the types of software packages required for large hybrid computers are discussed.

The second part of the book is devoted to the description of the application of hybrid computer systems. Each of the eight chapters comprising this part opens with a brief discussion of the mathematical formulation of typical engineering problems and follows with a description of the various hybrid computer techniques which have been useful in their solution.

Nearly a hundred problems provide for student practice in extending the material in the text. A number of the problems are designed to be used as laboratory exercises.

The authors are particularly indebted to Mr. Man T. Ung of Electronic Associates, Inc. and to Professor Robert M. Howe of the University of Michigan for their contributions to this book in Chapters 7 and 12, respectively. The helpful advice of Professor Granino A. Korn of the University of Arizona is also gratefully acknowledged. Finally, the authors would like to express their thanks to Simulation Councils, Inc., the publishers of *Simulation*, and to Spartan Books and Thompson Books, publishers of the AFIPS Conference Proceedings, for making available material that originally appeared in their publication.

GEORGE A. BEKEY

WALTER J. KARPLUS

May 1968

Contents

PART 1 THEORY AND MECHANIZATION

1 INTRODUCTION TO HYBRID COMPUTER TECHNIQUES   3

    1.1 The Trend to Hybridization   3
    1.2 The Spectrum of Hybrid Computing Techniques   6
    1.3 True Hybrid Systems   10
    1.4 Motivation for Hybridization   13
    1.5 Areas of Application of Hybrid Computer Techniques   16

2 HYBRID COMPUTER SYSTEM COMPONENTS   21

    2.1 Introductory Remarks   21
    2.2 Digital-Analog Conversion, General Principles   24
    2.3 Weighted-Resistor Digital-Analog Converters   26
    2.4 Ladder-Network Digital Analog Converters   29
    2.5 Analog-Digital Conversion-General Remarks   30
    2.6 Simultaneous Analog-Digital Conversion   35
    2.7 Closed-Loop Analog Digital Converters   36
    2.8 Analog-Digital Conversion by Subranging   39
    2.9 Time-Interval Analog-Digital Converter   41
    2.10 Multiplexing and Demultiplexing   42

3 NUMERICAL SOLUTION OF ORDINARY DIFFERENTIAL EQUATIONS — 46

- 3.1 General Remarks — 46
- 3.2 Considerations in Selecting an Integration Scheme — 49
- 3.3 A Useful Reformulation — 52
- 3.4 Taylor Series Methods — 54
- 3.5 The Runge-Kutta Methods — 58
- 3.6 Runge-Kutta Methods for Systems of Equations — 63
- 3.7 Noniterative Forward-Integration Methods — 67
- 3.8 Predictor-Corrector Methods — 70
- 3.9 Predictor-Modifier-Corrector Methods — 74
- 3.10 Boundary Value Problems — 83
- 3.11 Control of Errors — 85
- 3.12 Conclusions and Generalizations — 87

4 TRANSFORM ANALYSIS OF HYBRID COMPUTER SYSTEMS — 90

- 4.1 Introduction — 90
- 4.2 z-Transform Techniques — 91
- 4.3 z-Transforms of Numerical Integration Formulas — 94
- 4.4 Frequency Domain Characteristics of Integration Formulas — 96
- 4.5 z-Transform Analysis of the Numerical Integration of a First-Order Equation — 99
- 4.6 Solution of a Second-Order Equation — 106
- 4.7 Summary — 109

5 ERROR ANALYSIS OF HYBRID COMPUTER SYSTEMS — 113

- 5.1 Introductory Remarks — 113
- 5.2 Classification of Errors — 115

Subsystem Input-Output Errors — *117*
- 5.3 Time-Delay Compensation — 117
- 5.4 Sampler Errors — 124
- 5.5 Quantizing Errors — 125
- 5.6 Multiplexing and Distributing — 127
- 5.7 Digital-Analog Conversion and Data Reconstruction — 129

Solution Errors — *135*
- 5.8 The Concept of Sensitivity Functions — 135

5.9	Extension of Sensitivity Analysis to Hybrid Computer Systems	139
5.10	Compensation of Per-Step Errors	143
5.11	Sensitivity Analysis Applied to Sampling Rate	147
5.12	The Method of Corrected Inputs	149

6 COMPLETE HYBRID COMPUTER SYSTEMS   151

6.1	Introductory Remarks	151
6.2	Considerations in Specifying a Hybrid Computer System	151
6.3	Evolution of Hybrid Computer Systems	154
6.4	Differences in Philosophies of Operation	156
6.5	Differences in Backgrounds of the Staff	158
6.6	Operating-Cost Considerations	159
6.7	Administrative Considerations	160
6.8	Planning for Hybrid Computation	162
6.9	Typical Modern Hybrid Computer Systems	164

7 SOFTWARE FOR HYBRID COMPUTERS   177

7.1	Introduction	177
7.2	Software Requirements for Hybrid Computation	179
7.3	FORTRAN Compiler	182
7.4	HYTRAN Operation Interpreter (HOI)	183
7.5	HYTRAN Simulation Language (HSL)	186
7.6	Set-Up and Control Software	188
7.7	Checkout and De-Bugging Software	194
7.8	Utility Library	199
7.9	Conclusion	206

PART 2   APPLICATIONS

8 HYBRID COMPUTER SOLUTIONS OF FIELD PROBLEMS   211

8.1	General Remarks	211
Analog Methods		*213*
8.2	Continuous-Space-Continuous-Time	213
8.3	Discrete-Space-Continuous-Time	214
8.4	Continuous-Space-Discrete-Time	216
8.5	Discrete-Space-Discrete-Time	218

Digital Methods 220
8.6 Finite-Difference Techniques 220
8.7 Monte Carlo Methods 223

Hybrid Methods 228
8.8 Discrete-Space-Continuous-Time 228
8.9 Continuous-Space-Discrete-Time 229
8.10 Discrete-Space-Discrete-Time 234
8.11 Monte Carlo Method 239

9 PARAMETER OPTIMIZATION 244

9.1 The Nature of the Optimization Problem 244
9.2 Some Mathematical Considerations 247
9.3 A Geometrical Interpretation of the Parameter Optimization Problem 253
9.4 Optimization by Continuous Steepest Descent 256
9.5 Approximations to Continuous Steepest Descent for Dynamic Optimization 262
9.6 Dynamic Optimization by Discrete Gradient Methods 270
9.7 Iterative Optimization with Cyclical Parameter Adjustment 280
9.8 Optimization by One-Parameter Search Methods 282
9.9 Optimization by Random Search 284
9.10 A Detailed Example: A Satellite Acquisition Optimization 290
9.11 Dynamic Optimization in the Presence of Noise 297

10 OPTIMAL CONTROL PROBLEMS 301

10.1 Introduction 301
10.2 Mathematical Formulation 303
10.3 Computational Considerations 307
10.4 Linear Time-Optimal Control Problem 310
10.5 Solution by Means of Sensitivity Functions 316
10.6 Solution by Continuous Steepest Descent 318
10.7 Other Analog and Hybrid Techniques 322
10.8 Gradient Methods in Functional Optimization 322
10.9 Approximate Steepest Descent by Fast-Time Repetitive Computation 326
10.10 Summary and Evaluation 327

11	**RANDOM PROCESSES**	**331**
	11.1 Introduction and Basic Definitions	331
	11.2 Measurement of Probability Distributions	332
	11.3 Estimation of Averages	337
	11.4 Estimation of Correlation Functions	343
	11.5 Estimation of Spectral Density Functions	347
	11.6 Generation of Noise Signals in Hybrid Systems	350
	11.7 Linear System Studies	355
	11.8 Monte Carlo Techniques and Applications	360
12	**HYBRID COMPUTATION IN FLIGHT SIMULATION**	**363**
	12.1 Introduction	363
	12.2 Six-Degree-of-Freedom Airframe Equations	364
	12.3 Simulation of Space Vehicles	372
	12.4 Digital Function Generation	378
13	**MAN-MACHINE SYSTEMS**	**390**
	13.1 General Remarks	390
	13.2 Hybrid Simulation of Manned Aerospace Vehicles	391
	13.3 Simulation of the Human Operator	394
14	**BIOLOGICAL SYSTEMS**	**404**
	14.1 Computation and Data Processing in Biomedical Sciences	404
	14.2 The Nature of Biological Signals	406
	14.3 Hybrid Processing of Electrocardiogram (ECG) Data	407
	14.4 On-Line Hybrid Computer Applications	411
	14.5 A Hybrid Approach to the Study of Animal Locomotion	415
15	**SOLUTION OF INTEGRAL EQUATIONS**	**422**
	15.1 Introduction	422
	15.2 The Neumann Iteration Method	423
	15.3 The Fisher Iteration Method	424
	15.4 Illustrative Examples	426
	15.5 Discussion of Errors	430
	15.6 Extension to Other Types of Integral Equations	432
	15.7 Conclusion	432
	PROBLEMS	**435**
	INDEX	**455**

Part 1

Theory and Mechanization

1

Introduction to Hybrid Computer Techniques

1.1 THE TREND TO HYBRIDIZATION

The fields of analog and digital computation owe their present form and widespread popularity to electronic developments during and immediately after World War II. The late 1940's and early 1950's saw a period of intensive activity in both fields during which progressively more elegant and more widely applicable computers were designed, as computing techniques were extended to new areas of science and engineering. As a by-product of this rapid period of expansion, there appeared a rather unfortunate specialization of interest and skills on the part of computer engineers. So many new devices and techniques were being introduced that engineers tended to devote themselves exclusively either to digital or to analog computation. A number of factors contributed subsequently to the reinforcement of this type of specialization.

Once an engineer or scientist had become familiar with a specific computer, it was natural that he should prefer to attempt to apply this computer to all types of problems that came his way rather than invest additional time in the study of another type of machine. Many industrial and academic organizations acquired either an analog or a digital computer, and in this way cast their lot with one of the two major approaches to general-purpose computations. In time, there arose a wide-open competition between proponents of analog and digital computers, with each side having enthusiastic adherents and opponents. Digital computers gradually assumed a lion's share of the computer market. By the late 1950's the trend toward specialization began to reverse, and engineers began to realize that both analog and digital techniques had well-defined advantages as well as disadvantages, depending on the specific application. One result of this increased sophistication on the part of computer engineers is an emergent trend toward a blending of analog and digital techniques in computer systems.

The chief distinction between analog and digital computers lies in the manner in which the dependent variables are handled within the computer. In analog machines, the dependent variables (though not necessarily the independent variables) appear everywhere in continuous form and may be recorded with as many significant figures as the quality of the circuit components permits. In digital computers, all variables appear in discrete form, and the accuracy of the data which are manipulated and recorded as the output of the digital computation depends on the number of digits carried throughout the solution and is directly related to the capacity or size of the memory registers.

As a direct result of the difference in the handling of data within the machine, the basic organization or logic of analog and digital computers developed along radically different lines. Consequently, there is one series of characteristics associated with analog equipment and another series of characteristics associated with digital devices.

Over the years, the following attributes have come to be considered as characteristic of analog computer systems:

1. Dependent variables within the machine treated in continuous form.
2. Accuracy limited by the quality of the computer components, and rarely better than 0.01% of full-scale for electronic equipment.
3. Parallel operation, with all computational elements operating simultaneously.
4. High-speed or "real-time" operation, with computing speeds limited primarily by the bandwidth characteristics of the computing elements, and not by the complexity of the problem.
5. Ability to perform efficiently such operations as multiplication, addition, integration, and nonlinear function generations; on the other hand, very limited ability to make logical decisions, store numerical data, provide extended time delays, and handle nonnumerical information.
6. Programming techniques which consist largely of substituting analog computing elements for corresponding elements in a physical system under simulation, i.e., providing computer elements having transfer characteristics analogous to those of the original system under study.

1.1/THE TREND TO HYBRIDIZATION 5

7. Facility for including analog hardware from a system under study in the computer simulation.
8. Provisions to permit the engineer to experiment by adjusting coefficient settings on the computer, thereby gaining direct insight into system operation.

Similarly, the following attributes are commonly associated with digital computer systems:

1. Handling of dependent variables, and indeed all data within the computer, in quantized or discretized form.
2. Serial operation, involving the time-sharing of all operational and memory units, only one or a limited number of operations being carried out at one time.
3. Accuracy relatively independent of the quality of system components and determined primarily by the number of bits contained in memory registers and by the specific numerical technique selected for a specific problem.
4. Solution times relatively long and determined by the complexity (i.e., the number of arithmetic operations required for the solution) of a problem.
5. Ability to "trade off" solution time and accuracy, i.e., to reduce errors inherent in the computer solution by increasing the length of time required to obtain the solution on the computer.
6. Ability to perform a limited number of arithmetic operations including particularly addition and multiplication; more complex operations such as integration and differentiation must be performed by approximate (numerical) techniques.
7. Facility for memorizing numerical and nonnumerical data indefinitely.
8. Facility to perform logical operations and decisions utilizing numerical as well as nonnumerical data.
9. Facility for floating-point operation, thereby eliminating scale-factor problems.
10. Programming techniques, often bearing little direct relationship to the engineering problem under study, but facilitated by compilers and special interpretive routines.
11. Facility for automatically altering and controlling the topology of the data flow within the machine on the basis of calculations.

Hybrid computer techniques inevitably represent an effort to combine in one computer system some of the characteristics normally associated with analog systems and some of the characteristics associated with digital computer systems. In the broad sense of the term, a "hybrid computer system" can be purely analog or purely digital in its mechanization, with data in the machine being entirely continuous or entirely discrete in form. In a more narrow sense, hybridization involves the actual interconnecting of analog and digital portions within the system. The interconnecting type of hybrid devices is often termed "true hybrid" to distinguish it from those systems in which hybridization occurs at a conceptual rather than a mechanistic level. In this book, emphasis is placed on the true hybrid systems and on actual interconnections of analog and digital computers. For completeness, however, other more general forms of hybridization are also surveyed in this chapter.

1.2 THE SPECTRUM OF HYBRID COMPUTING TECHNIQUES

If the term "hybrid computation" is used in its broad general sense, a wide variety of computing techniques and computing systems developed in recent years can be considered as being hybrids. All these techniques can then be viewed as falling somewhere along the spectrum of combined analog-digital techniques shown in Figure 1.1. At one end of this spectrum are located the pure analog techniques possessing all the traits enumerated in Section 1.1 as characterizing analog computers and none of the traits characterizing digital computers. At the other end of the spectrum are the pure digital systems, manifesting all the characteristics listed for digital computers and none of those listed for analog computers. Ranged in between these two extremes are computer systems and techniques combining some of the characteristics of one category of computer devices with some of those of the other category. At the center are the true hybrid computers which are comprised of pure analog and digital computers interconnected by suitable linkage devices, such that both computers play an equally important role in the data-handling process. The following are the more important computational techniques which combine to comprise this spectrum of hybrid computations.

Pure Analog Computers. The "conventional" analog computers including the slow, one-shot variety[1,2]† as well as the high-speed[3] or repetitive computers. Prominent among the pure analog computer systems currently marketed in the United States are the basic consoles of Electronic Associates, Comcor/Astrodata, and Applied Dynamics. These manufacturers have made available optional auxiliary equipment to facilitate hybrid operation.

† Superscript numerals in parentheses are reference citations at the end of each chapter.

1.2/THE SPECTRUM OF HYBRID COMPUTING TECHNIQUES 7

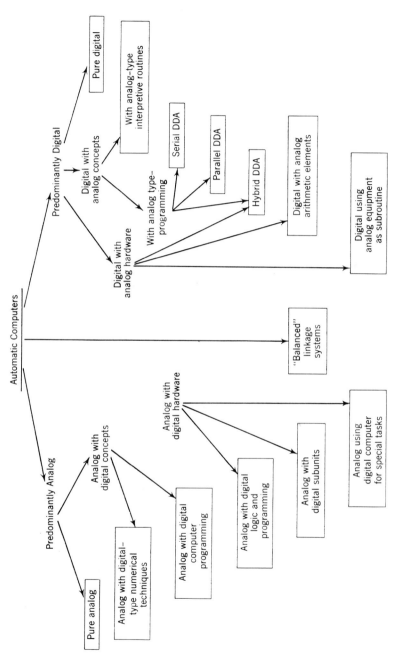

Figure 1.1 Spectrum of hybrid computer mechanizations.

Analog Computers Using Digital-Type Numerical Analysis Techniques. So-called iterative differential analyzers[4,5] have been developed to permit analog computers to operate in a repetitive mode and to memorize numerical data from one repetitive cycle for use in subsequent cycles. This permits the utilization on pure analog computers of techniques of numerical analysis ordinarily employed only when using digital computers. By the mid-1960's, all major manufacturers of analog computers made iterative operation available as an option.

Analog Computers Programmed with the Aid of Digital Computers. A number of digital computer programs have been developed to facilitate the programming of analog computers. For example, the APACHE program[6] developed at EURATOM, Ispra, Italy, is employed to obtain all scalefactors, potentiometer settings, and patchboard connections on a digital computer. However, no actual link between the digital and the analog computers exists. A number of other programs and devices have been developed to permit the digital computer to set automatically the initial condition and scalefactor potentiometers of an analog computer system, while the entire solution is carried out in analog fashion.

Analog Computers Using Digital Control and Logic. A number of devices have been developed to permit the utilization of digital logic circuits to control pure analog solutions. These devices can be employed to make automatic decisions in the course of an analog computer run and to modify certain parameters or initial conditions accordingly. The HYDAC system of Electronic Associates[7] represented the first elegant example of this type of hybridization. By the mid-1960's, all major manufacturers of hybrid computers had made logic equipment available with their systems. Gates, flip-flops, and counters are terminated at an auxiliary patchboard to permit interconnection with the analog units as required by a specific application.

Analog Computers Using Digital Subunits. Most large-scale analog facilities utilize certain digital instruments for specialized tasks. For example, digital voltmeters are frequently employed for precise measurements; digital function generators[8,9,10] and memory devices are also used on occasion to enhance the power of an analog computer system.

Analog Computers Using Digital Computers as Peripheral Equipment. In such systems[11] a small general-purpose digital computer or digital differential analyzer is employed, together with an extensive analog system, to perform specialized tasks which would be difficult or impossible if pure analog hardware were used alone. The digital computer can therefore be considered as serving as peripheral equipment for the analog system.

Balanced Hybrid Computer Systems. The most extensive and powerful of existing hybrid computer systems are comprised of general-purpose digital computers and general-purpose analog computers. Both major components of such hybrid systems could "stand alone" and solve a wide variety of important problems. By interconnecting them, using suitable linkage equipment, an even more powerful computer system is realized. Initially, most hybrid computer systems employed large digital computers such as the IBM 7094 or the Univac 1103 for the digital portion.[12] During the early 1960's a variety of smaller digital computers, such as the Scientific Data Systems 930, the Honeywell (Computer Control Corporation) DDP-24, and the Digital Equipment Corporation PDP-8, were introduced. These computers are considerably less costly and have input/output circuitry to facilitate operation on-line with an analog computer.

Digital Computers Using Analog Subroutines. An example of this approach is the UCLA DSDT system[13] for the solution of partial differential equations in which analog hardware is used only for certain matrix inversions required by the digital computer program. In this case, the analog computer acts as a peripheral element for the digital computer.

Digital Computers with Analog Arithmetic Elements. Here an attempt is made to increase the speed of pure digital computers by performing some of the calculations within the computer in parallel by using analog hardware. Such a system has been developed at the Massachusetts Institute of Technology.[14]

Digital Computers Designed to Permit Analog-Type Programming. Computers termed "digital differential analyzers"[15] have been developed by Northrop, Bendix, Litton, Raytheon, and at the University of Arizona. Three basic types of digital differential analyzers are in existence. These include the serial, the parallel, and the hybrid realizations. The first two of these are pure digital computers in which all logic and control circuitry not required for the solution of differential equations has been eliminated. In the parallel[16] category, each digital integrator is supplied with a separate memory unit so that all digital integrators can operate simultaneously. In hybrid digital differential analyzers[17,18] the digital integrators are supplemented by analog hardware to permit more rapid and economic integration.

Digital Computers with Analog-Oriented Compilers and Interpreters. A number of problem-oriented languages[19] have been developed to assist an analog specialist in programming a digital computer so as to simulate an analog computer. The first really successful program of this type was developed at Wright-Patterson Air Force Base and given the name MIDAS. A later, greatly improved version of this program carries the name MIMIC. In 1966, International Business Machines introduced a program called DSL-90 (later CSMP). All these programs are similar to the FORTRAN language but contain commands which permit a programmer to instruct the computer to "interconnect" elements such as integrators, summers, potentiometers, and multipliers. The entire solution is then carried out by pure digital methods.[20,21]

Pure Digital Computers. These include the well-known large digital computers made by such manufacturers as International Business Machines, Radio Corporation of America, Control Data, Sperry-Rand (UNIVAC), Burroughs, and General Electric. Often, small or medium-size computers can be used for engineering applications. These include the Scientific Data Systems' 9300, the smaller of the IBM computers such as the 360/40 and the 1800. In 1966, there appeared the first of a series of computers designed to operate in a time-shared mode. One example of such a machine is the Scientific Data Systems' Sigma 7. Such computers contained special hardware that permit their control by a large number of remotely stationed consoles. It is to be expected that computers of this type will play an increasingly important role in hybrid computation.

1.3 TRUE HYBRID SYSTEMS

The term *true hybrid* is applied to those combined computing systems containing analog as well as digital hardware in appreciable amounts. The ways in which analog and digital computer installations can be used together can be classified into two broad categories: (1) unilateral operation in which information flows across the interface between the analog and the digital sections in only one direction, and (2) bilateral operation in which the flow across this interface is in both directions. Both methods require conversion equipment at the interface. Figures 1.2a and 1.2b illustrate two types of unilateral systems. Note that only one type of converter, either analog-digital or digital-analog, is required at each interface. In such systems the analog or the digital computer can be regarded as playing the part of a complex and elegant input or output device. Figure 1.2c illustrates a bilateral hybrid system. Such system are characterized by a closed loop formed by the digital computer, the digital-analog conversion devices, the analog computer, and the analog-digital converters.

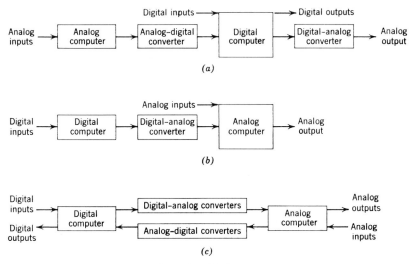

Figure 1.2 Unilateral and bilateral hybrid systems.

In addition to the major units shown in Figure 1.2, bilateral hybrid computer systems also include a number of other important devices. Multiplexers and demultiplexers are employed to permit the converters, which translate continuous d-c voltages into digital code and vice versa, to be time-shared. Hold devices are required to maintain the continuously varying analog signals at constant values for a time sufficiently long to permit conversion, and to maintain the output voltage of digital-analog converters at constant levels while conversion is taking place. Buffers are required to adjust voltage levels and pulse shapes in a manner compatible with the digital and analog computers in use. Finally, timing and control circuitry is required to synchronize the processes in the various units comprising the hybrid loop. Each of these units and subunits manifests input-output relationships which deviate from the ideal or specified behavior. In designing hybrid computer systems and in evaluating their performance, it is therefore necessary to determine the effect of the nonideal behavior of each subunit on the overall system dynamics.

An important major application of hybrid computer systems involves the interconnection of the analog-digital loop with major items of hardware. A number of aerospace companies and NASA centers have acquired hybrid facilities for the express purpose of providing external excitations for space vehicles so as to permit simulated flights under realistic conditions. This facilitates the training of astronauts as well as the designing of the capsules. Figure 1.3 shows the computer portion of such a system. Note that in addition to the conventional elements in the hybrid loop, as enumerated above, there is provision for outputting 128 digital and

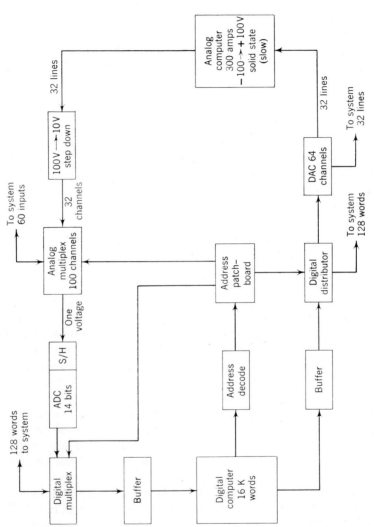

Figure 1.3 Hybrid computer system used in aerospace applications.

32 analog channels and inputting 128 digital and 60 analog channels from the space capsule.

Unilateral as well as bilateral systems have achieved considerable importance over the past ten years. Because of their widespread application in telemeter and communications systems, unilateral hybrids outnumber bilateral systems by a wide margin. Since most of the problems encountered in the design of unilateral systems are also encountered in the design of bilateral systems, while the reverse is generally not true, bilateral hybrid systems are emphasized in the present discussion.

1.4 MOTIVATION FOR HYBRIDIZATION

As indicated in Section 1.2, hybrid techniques have been introduced in a wide variety of computer systems for a large number of technical, psychological, and economic reasons. If attention is focused on those systems lying near the center of the spectrum of Figure 1.1, i.e., those systems containing appreciable analog and digital hardware, it is possible to identify certain general computational requirements which suggest the hybrid computer approach. In this connection, it should be noted that hybrid techniques are frequently employed to overcome certain shortcomings in present-day analog or digital computers. As these limitations are reduced or eliminated, it is possible that some applications for hybrid computing systems will disappear. For example, the increasing speed and decreasing cost of general-purpose digital computers, as well as the advent of time-shared on-line digital systems, may well narrow the range of applications for hybrid equipment. On the other hand, the introduction of integrated circuitry makes possible the realization of linkage equipment in increasingly simple, reliable, and economical form, thus making hybrid techniques increasingly attractive in areas in which pure analog or pure digital computers may be adequate. The following are, at present, the chief motivations for interconnecting digital and analog computers:

1. To combine the speed of an analog computer with the accuracy of a digital computer.
2. To permit the use of system hardware in a digital simulation.
3. To increase the flexibility of an analog simulation by using digital memory and control.
4. To increase the speed of a digital computation by utilizing analog subroutines.
5. To permit the processing of incoming data which are partially discrete and partially continuous.

In Figure 1.4, the motivations for the use of true hybrid computers

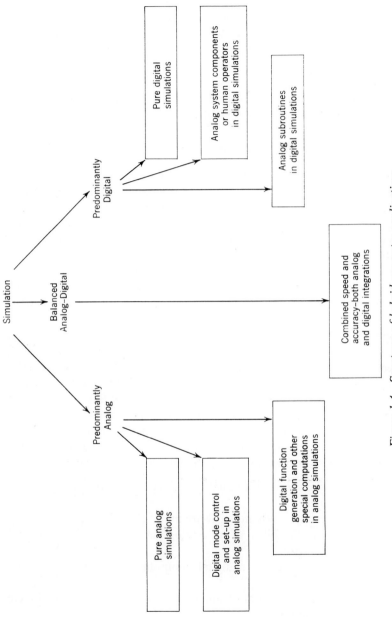

Figure 1.4 Spectrum of hybrid computer applications.

in the simulation of dynamic systems are arranged in a spectrum to illustrate the relative importance of the analog and digital computers in each case.

Some of the largest hybrid computer systems have been employed in ballistic-missile and space-vehicle research and development programs, in which guidance as well as control systems must be studied. In general, the accuracy demanded of the guidance computer exceeds the capabilities of available analog computers. At the same time, the dynamic characteristics of the on-board control systems and the vehicle itself are such that no digital computer can furnish response values sufficiently rapidly to permit real-time simulations.

The first major hybrid computer systems constructed in the late 1950's at the Ramo-Wooldridge Corporation (later called Space Technology Laboratories and TRW Systems) in Los Angeles and at Convair Astronautics in San Diego were motivated initially by a desire to include the autopilot of the guided missile in the full-fledged missile-systems simulation. Since this unit is an analog device, and since the trajectory calculation demanded digital accuracy, the hybrid computer approach seemed indicated. Since that time, the requirement to include in digital simulations analog elements comprising a part of the system being simulated has been one of the chief motivations for turning to hybrid computation. Not only analog hardware but also human operators make necessary the development of an analog-digital interface in such sophisticated computer systems. Once the necessity for an interface has been established, it becomes possible to examine each computational task to determine whether it can better be performed on the analog or the digital side of the hybrid computer system.

There exists a wide variety of tasks which are difficult to perform by using pure analog techniques even where the application does not demand digital computer accuracy. These operations include, among others, the control of the analog modes, multivariable function generation where table look-up techniques can be employed in a digital computer, variable time-delay realization, and the simulation of digital elements in the system under study. The latter requirement arises in the simulation of flight-control systems in which a digital computer forms an element of the system. It would be very difficult indeed to simulate such a computer by using only conventional analog techniques.

Frequently, hybrid computing techniques are employed to permit the utilization of analog computers or analog devices to perform a particular computation so as to speed-up an otherwise digital problem solution. For example, certain chemical processes such as fixed-bed catalytic reactors can be presented as a two-dimensional array of chemical reactions, including as many as several hundred cells. The original partial differential equation characterizing the chemical system is then replaced by a system or ordinary nonlinear differential equations characterizing each cell. Since moderate accuracy is adequate in the representation of each individual cell, an analog integration of the nonlinear differential equation is practical. From the point of view of computing time, the use of analog elements is desirable since analog solution time is independent of the degree of nonlinearity of the particular equation. It is therefore practical to formulate a hybrid computer program in which the digital computer performs all the control and decision functions, while an analog subroutine is employed to solve the simultaneous differential equations. In other applications of the hybrid concept, analog networks are employed to invert the matrices which arise in the digital treatment of partial differential equations

In many telemetry applications, data are transmitted to a computer system and must be processed immediately. Frequently, some of these data are in continuous form while others are in discrete or pulse-modulated form. Both types of data must be interpreted and processed simultaneously. This is greatly facilitated by hybrid computers including both analog and digital elements.

1.5 AREAS OF APPLICATION OF HYBRID COMPUTER TECHNIQUES

By far, the most widespread application for hybrid computer systems to date has been in the simulation of physical systems. A mathematical model of a system to be designed or to be studied is first formulated, and the equations constituting this model are programmed on a hybrid computer. The objective of such a simulation is usually to facilitate the study of the effect of parameter and topological modifications on the excitation/response relationship of the system. This objective can often be expressed as a parameter or function optimization problem. Another important application for hybrid systems has been in the analysis and reduction of data generated by experiments on actual engineering or biological systems. In this section, a brief description of some of the more important application areas is presented.

A. Sampled-Data System Simulation

The simulation of a digital computer or of digital components as elements of continuous systems usually requires a hybrid computer system. This application is of considerable importance because of the increasing utilization of digital computers, both in airborne and earth-bound control systems. Clearly, the digital computer has an effect on system performance quite apart from the equations which it solves. The sampling rate of the analog-to-digital converters and the rate at which the digital computer provides output information may be related in some way to a system resonant frequency. The quantization level of the digital computer may be related to particular levels of dead-zone or backlash present in the physical system itself.

B. Random Processes Simulation

A number of hybrid methods have been developed to investigate the response of physical systems to random excitations, random parameter variations, or random initial conditions. Such randomness may arise, for example, when we attempt to take into account the statistical distribution of manufacturing tolerances on the thrust of rocket vehicles, or when we desire to investigate the effects of thrust misalignment on a trajectory, or when a human pilot performs as an element of a control system and those components of his output which are apparently not related in a deterministic way to the input signals have a significant effect upon system variables. Hybrid computing techniques provide a convenient way of performing the evaluation of certain statistical parameters such as probability-density and distribution functions. They also comprise a convenient approach to the evaluation of system response to random disturbances by means of Monte Carlo methods and of correlation or spectral density functions.

C. Optimization

One major area of application of hybrid computing techniques is that of system optimization. It is characteristic of the optimization problems that the values of a set of parameters or functions must be determined in such a way that a particular performance function, or cost function, is maximized or minimized. For example, we may desire to determine a mathematical model which provides a best fit in an appropriately specified sense to a particular physical system, e.g., a biological system. The parameters of the mathematical model must be determined such that a function which defines the quality of the fit of the model is optimized. Typically, such an optimization problem consists of two portions: a simulation of the system itself and the solution of a system of equations which are used to determine the desired values for the parameters. The system simulation can, in many cases, be carried out conveniently on an analog computer while the logical decision functions, which are required for the computation of the optimum strategy of parameter adjustment, can be done conveniently with the aid of digital elements or digital computers.

D. Simulation of Distributed Parameter Systems

A large class of problems in engineering and applied physics is governed by partial differential equations. Typical of distributed parameter systems governed by such equations are heat-transfer problems, beam-vibration problems, electromagnetic-wave propagation problems, and nuclear-reactor problems. The difficulty in treating these problems by conventional computational techniques lies in the fact that they contain more than one independent variable. Electronic analog computers are generally limited to only one independent variable—time. All other independent variables of the field problem must then be approximated by finite-difference expressions. In pure digital simulations, all independent variables must be discretized. Particularly where the field parameters are nonlinear and where more than one space variable is involved, the equipment requirement in the case of analog computers and the computing-time requirements in the case of digital computers becomes uneconomically large. A considerable economic advantage can be obtained by forming a closed loop of analog and digital hardware such that the analog circuits serve as subroutines in a digital computation. In this way it is possible to obtain considerable speed with relatively small computers.

E. Studies in Guidance and Control of Missiles and Space Vehicles

Many large-scale combined analog-digital simulations have been undertaken in an attempt to study the complete trajectory of a long-range ballistic missile in such a way that the high accuracy required for the trajectory computation can be attained with the digital computer, while the high-frequency control system dynamics are handled on the analog computer. Such a division of labor between analog and digital computers characterizes the hybrid simulation of aerospace vehicle problems. The coordinate transformations required for the representation of the position of the vehicle in both inertial and earth-centered coordinates may be performed either on the analog or on the digital computer, depending on the requirements for speed and accuracy and on the availability of sufficient analog equipment.

F. Simulation of Man-Machine Systems

It is perhaps this area of application which has given the greatest impetus to hybrid simulation in recent years. While the simulation of piloted aircraft has traditionally been an area restricted to analog computers (at least insofar as control problems are concerned), as the aircraft

1.5/AREAS OF APPLICATION OF HYBRID COMPUTER TECHNIQUES

became increasingly complex the limitations of a purely analog simulation became readily apparent. It has been stated that the all-analog simulation, including pilot and cockpit of the X-15 aircraft, was probably the last of the pure analog simulations. Several hundred amplifiers were assigned to this simulation permanently, and much of the computing equipment was used near its limit of precision and performance. It therefore became apparent that the simulation of future missions such as the Apollo program required the development of new and more powerful simulators. These facilities are used for vehicle design as well as for the training of astronauts. A second area of man-machine interaction problems concerns the operation of manned detection and tracking systems such as those in early-warning networks. In this case, the human operator is presented with a large mass of coded information, generally by means of oscilloscope displays. He must make decisions based on this information and provide inputs to a large digital computer. Where, in addition to detection, the human operator is also required to perform closed-loop tracking or control functions, additional analog elements are required for a complete simulation.

G. Other Applications

Numerous other applications of the hybrid approach can be cited. The simulation of process control systems is logically suited to hybrid techniques since the steel mill or chemical reactor in question can be conveniently simulated on an analog computer while the digital computer performs the functions of the digital controller. The study of biological systems offers a wide variety of opportunities for hybrid computation since apparently both discrete and essentially continuous signal are present in a living organism. In some cases, operations on discrete entities, such as modulated pulse trains, are performed in the organism. In other cases, the actual dynamics of a chemical reaction occurring at a neuro-muscular junction or in the retina of the eye may be of interest. Hybrid computers offer the possibility of performing real-time auto-correlation of EEG recordings. The simulation of certain kinds of communication problems which involve modulation systems may be carried out conveniently by means of hybrid techniques. Problems of economics are characterized by difference-differential equations, since business decisions and economic policies result both in immediate effects as well as in a variety of delayed reactions. Furthermore, both discrete and continuous variables are present. It seems likely that problems in mathematical economics will, in the future, be studied by means of hybrid techniques.

References

1. Jackson, A. S., *Analog Computation*, McGraw-Hill, New York, 1960.
2. Fifer, S., *Analogue Computation: Theory, Techniques and Applications*, 4 vols., McGraw-Hill, New York, 1961.
3. Tomovic R., and W. J. Karplus, *High-Speed Analog Computers*, John Wiley, New York, 1962.
4. Gilliland, M. C., "The Iterative Differential Analyzer," *Instruments and Control Systems*, 675–680, April 1961.
5. Andrews, J. M., "Mathematical Application of the Dynamic Storage Analog Computer," *Proc. Western Joint Computer Conference*, **17**, 119–131, May 1960.
6. Green, C., H. D'Hoop, and A. Debroux, "APACHE—A Breakthrough in Analog Computing," *IRE Transactions on Electronic Computers*, **EC-11**, 699–706, Oct. 1962.
7. Electronic Associates, Inc., *Introduction to Hybrid Computations*, Electronic Associates, Inc., Long Branch, N.J., 1963.
8. Schmidt, H., "Linear-Segment Function Generator," *IRE Transactions on Electronic Computers*, **EC-11**, 780–788, Dec. 1962.
9. Hurney, P. A., Jr., "Combined Analog and Digital Technique for the Solution of Differential Equations," *Proc. Western Joint Computer Conference*, San Francisco, pp. 64–66, Feb. 1956.
10. Chapelle, W. J., "Hybrid Techniques for Analog Function Generation," *Proc. Spring Joint Computer Conference*, Detroit, May 1963.
11. *Proceedings of the Combined Analog-Digital Computer Systems Symposium*, Simulation Councils, Inc., Philadelphia, Dec. 1960; 18 technical papers.
12. King, C. M., and R. Gelman, "Experience with Hybrid Computations," *Proc. AFIPS Fall Joint Computer Conference*, Philadelphia, **22**, 36–43, 1962.
13. Karplus, W. J., "A Hybrid Computer Technique for Treating Nonlinear Partial Equations," *IEEE Transactions on Electronic Computers*, **EC-13**, 597–605, Oct. 1964.
14. Connelly, M. E., "Real-Time Analog-Digital Computation," *IRE Transactions on Electronic Computers*, **EC-11**, 31–41, Feb. 1962.
15. Palevsky, M., "The Digital Differential Analyzer," *Computer Handbook* (Huskey and Korn, Eds.), McGraw-Hill, New York, 1962.
16. Bradley, R. E., and J. F. Genna, "Design of a One-Megacycle Iteration Rate DDA," *Proc. Spring Joint Computer Conference*, San Francisco, pp. 353–364, 1962.
17. Urban, W. D., W. R. Hahn, Jr., and H. K. Skramstad, "A Combined Analog-Digital Differential Analyzer (CADDA)," *Proc. Combined Analog-Digital Computers Symposium*, Simulation Councils, Inc., Philadelphia, Dec. 1960.
18. Wait, J. V., "A Hybrid Analog-Digital Differential Analyzer System," *Proc. Fall Joint Computer Conference*, Las Vegas, Nov. 1963.
19. Linebarger, R. N., and R. E. Brennan, "A Survey of Digital Simulation," *Simulation*, **3**, 22–36, Dec. 1964.
20. Clancy, J. J., and M. S. Fineberg, "Digital Simulation Languages: A Critique and a Guide," *Proc. AFIPS Fall Joint Computer Conference*, **27**, Part I, 23–36, Nov. 1965.
21. Tiechroew, D., and J. F. Lubin, "Discussion of Computer Simulation Techniques and a Comparison of Languages," *Simulation*, **9**, 181–190, Oct. 1967.

2
Hybrid Computer System Components

2.1 INTRODUCTORY REMARKS

Modern general-purpose hybrid computer systems can be considered to have three major sections: the analog computer, the digital computer, and the linkage system. At the present time, virtually all hybrid-computer systems employ so-called "stand-alone" analog and digital computers which can function independently as well as in the hybrid mode. The detailed structural characteristics of the linkage system employed to couple the analog and digital computers depend, to a large extent, on the nature of these computers as well as on the specific application for which the system is designed. Most large-scale hybrid computer systems can, however, be represented as shown in Figure 2.1. In addition to the analog and digital computers, the following major components can be identified:

Multiplexer. A device for converting data from parallel form to serial form. The multiplexer accepts at its input a variety of simultaneously occurring analog voltages and emits over its single output line a sequence of pulses, each having an amplitude proportional to one of the input voltages. Usually the continuous voltages appearing at a number of points within the analog computer system must be processed by the hybrid computer loop. Since analog-digital converters are relatively expensive

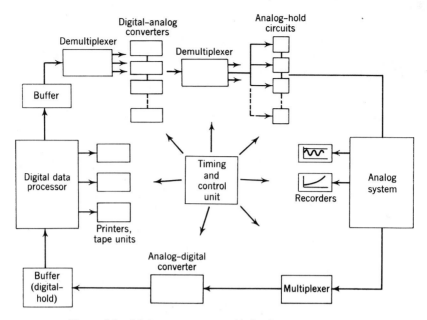

Figure 2.1   Major components of hybrid computer systems.

devices, it is usually desirable to employ time-sharing techniques. By means of a multiplexer, the analog outputs to be processed are sequentially connected to the analog-digital converter input. To reduce the time between successive samples, it is occasionally feasible to use two or more analog-digital converters, each handling a fraction of the total analog outputs. In that case, each analog-digital converter may have an associated multiplexer.

Analog-Digital Converter. A device which translates a pulse having an amplitude proportional to an analog voltage into the binary-coded equivalent of this amplitude. The digital output of the analog-digital converter appears in parallel form as "ones" and "zeros" stored in an array of register cells—one for each bit of the digital word. Once the conversion is complete, the number must be "read out" of the converter to make room for the next conversion. Usually the converter emits a pulse as soon as the conversion is complete in order to initiate this read-out.

Digital-Hold. A device which stores a digital number for a limited time. Digital computers can usually accept data only at specific instants of time, determined by the clock generator synchronizing data flow within the computer. It is therefore necessary to make provisions for retaining or holding the analog-digital converter output until the precise instant that the digital computer is ready for the information. At the same time it is usually necessary to adapt the format of the data to that employed within the digital computer. This may involve the changing of voltage levels, negative-number representation, and sometimes even the changing from a parallel to a serial form. Modern digital computers are usually equipped with input and output buffers for this purpose.

Digital-Analog Converter. A device which accepts input data in binary form and generates an analog voltage proportional to the binary number. Frequently, a number of digital-analog converters work in parallel, each providing an analog voltage for a different point in the analog computer system.

Demultiplexer (Distributor). A device for distributing digital data, sequentially generated in the digital computer, to a number of digital registers. This unit has a single input register which accepts a sequence of digital data and applies them in turn to an array of output registers, each controlling a separate digital-analog converter. Demultiplexers, therefore, permit the time-sharing of the single digital output channel. Occasionally, analog demultiplexers are also provided in a hybrid computer system. The output of a single digital-analog converter can then be applied sequentially to a number of parallel analog channels.

Analog-Hold (Sample-Hold). A device which stores a constant in analog form for a limited period of time. In order to convert analog data in pulse or sampled form into continuous voltages, it is necessary to provide such memory units. Also, analog-holds are often used at the inputs of the multiplexer or at the inputs of the analog-digital converter to obviate errors resulting from uncertainties as to sampling time and because of slewing. Frequently, the sample-hold units also act as buffers to change the voltage level of analog signals.

Timing and Control. A device which synchronizes and controls the operation of the hybrid computer loop. Frequently, the timing and control function is vested in the digital computer. The digital computer then emits control pulses which activate the multiplexer, the analog-digital converter, the demultiplexer, and the sample-hold units. In other systems, a separate clock and control unit is provided for this purpose.

This chapter is devoted to a discussion of those digital-analog and analog-digital converters and those multiplexers which are most widely used in large hybrid computing systems. In this connection, it should be recognized that analog and digital data co-exist in many engineering systems other than hybrid computer systems, and that many devices have been developed specifically for special-purpose applications. For example, in many electromechanical control systems, analog information is represented by shaft rotations or angular positions. A wide variety of electromechanical analog-digital and digital-analog converters has been developed for such systems. Because of the inertia inherent in mechanical devices, the bandwidths of such converters are far too limited for hybrid computer applications. Again, in many communication applications, particularly in the video area, information must be transmitted with extremely great rapidity, but a relatively low precision, say 1 part in 100, is acceptable. Such devices would introduce excessive errors into hybrid loops where 12-, 13-, or 14-bit accuracies are required. The following discussion is therefore limited to all-electronic converters possessing speed and accuracy characteristics compatible with the operation of high-accuracy hybrid computers.

2.2 DIGITAL-ANALOG CONVERSION, GENERAL PRINCIPLES

A digital-analog converter accepts a number in digital form, either directly from the digital computer or from the digital distributor, and generates at its output a d-c voltage directly proportional to this number. Most commercial hybrid computer systems employ a dynamic range of either ± 100 volts or ± 10 volts in their analog sections. For the sake of simplicity, the following discussion is directed to digital-analog converters which employ a digital number format as shown in Figure 2.2. The first binary digit of the digital word is termed the "sign-bit"; if this bit is "zero," the number is considered to be positive; a negative number is indicated by making the sign-bit "one." The radix point is assumed to be to the extreme left of the mantissa, so that the absolute value of all numbers represented must fall in the range of zero to unity. The handling of numbers outside this range requires amplitude scaling in the digital computer program. Thus a "one" in the first or most significant bit (the second bit from the left in Figure 2.2) represents 2^{-1} or $1/2$; the next bit represents 2^{-2} or $1/4$; the third bit represents 2^{-3} or $1/8$; etc. While

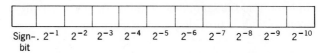

Figure 2.2   Ten-bit register with radix point at extreme left.

2.2/DIGITAL-ANALOG CONVERSION, GENERAL PRINCIPLES

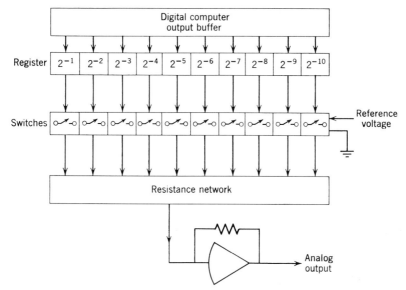

Figure 2.3   Digital-analog converter: general organization.

all hybrid computer systems employ some form of a binary number system, there exists a variety of detailed modifications of the format shown in Figure 2.2. Some computers, for example, represent negative numbers in complement form, while others employ a binary-coded decimal representation. The adaptation of the principles described below to such number systems is entirely straightforward and presents no special problems.

The basic principle of converting a positive digital number into an analog voltage is illustrated in Figure 2.3. The number, here assumed to consist of 10 binary digits with the radix point at the extreme left, is placed in the output buffer of the digital computer or the output register of the digital distributor. When conversion is desired, this output register is connected through 10 parallel lines to the digital register of the digital-analog converter. The flip-flops comprising the latter register automatically act to set an array of 10 switches in such a way that each "one" causes the reference voltage (usually $+10$ or $+100$ volts) to be applied to a specific point in a resistance network. Since this resistance network forms the input impedance to an operational amplifier, the output voltage of this amplifier is determined by the specific setting of the switches, hence by the digital input. The arrangement of the switches and the topology of the resistance network is such that as the digital number to be converted is increased from zero to unity, in steps of one least-significant bit, the analog output changes in steps of 0.01 volt from 0 volt to 10 volts. If the analog dynamic range is from 100 volts, the step size is 0.1 volt. The detailed functioning of the resistance network as well as the treatment of negative numbers is described in Sections 2.3 and 2.4.

Occasionally in hybrid computer systems, it is practical to combine the operation of digital-analog conversion with that of multiplication. In that case, an analog variable is applied to the switches shown in Figure 2.3, in place of the analog reference voltage. The analog output of the unit is then proportional to the product of this analog variable and the digital computer output.

In specifying a digital-analog converter, the chief considerations are resolution, accuracy, and speed. The resolution is determined by the number of binary digits which are converted. Hybrid computer systems employ a minimum of 10 binary digits in addition to the sign-bit. Frequently, 12, 13, or even 14 binary digits are employed. The greater the number of bits, the greater is the number of register cells, switches, and resistors that must be supplied and, hence, the greater the cost.

The accuracy of digital-analog converters is determined primarily by the quality of the resistors which comprise the resistance network as well as that of the feedback resistor and operational amplifier generating the analog output. The quality of the switches must, of course, also be sufficiently high so that noninfinite open-resistance and nonzero closed-resistance introduce negligible errors.

Two separate measures of conversion speed must be employed. The first involves the setting speed of the input register and the switches. By careful design and the use of very high-quality solid-state components, this setting time can be kept below 1 microsecond, and most commercial digital-analog converters have setting times below 2 or 3 microseconds. The second measure of speed involves the settling time of the resistance network and the operational amplifier. The operational amplifier is usually similar to that employed in an analog computer system and has a performance strongly determined by the dynamic range and the loading requirements.

2.3 WEIGHTED-RESISTOR DIGITAL-ANALOG CONVERTERS

The most widely used method of converting a number in binary form into a proportional analog voltage is to use a network of weighted resistors in combination with switches activated by the binary digits. A typical weighted-resistor digital-analog converter is illustrated in Figure 2.4. Amplifier 1, connected as a summer, has n input resistors, where n is the

2.3/WEIGHTED-RESISTOR DIGITAL-ANALOG CONVERTERS 27

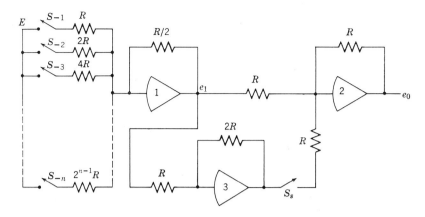

Figure 2.4 Weighted-resistor digital-analog converter.

number of binary digits employed to represent the number D to be converted. These input resistors vary in magnitude over the range $2^0 R$, $2^1 R, \ldots, 2^{n-1} R$. The input end of each of these resistors is connected to the pole of a single-pole, single-throw switch. The position of each of the switches is determined by the binary number to be converted, each binary digit controlling one switch. When the binary digit is "one," the switchpole is connected to the E supply, while a binary digit of "zero" causes the switch to be open. The most-significant binary digit, immediately to the right of the radix point, controls switch S_{-1} connected to input resistor R; the next most-significant binary digit controls switch S_{-2} connected to resistor $2R$, etc. Each binary number therefore produces a different combination of switch settings, and hence a different voltage at the output of amplifier 1. In particular, if all switches except switch S_{-1} are in the "zero" position, the voltage at the output of amplifier 1 becomes $-0.5E$. If switches S_{-1} and S_{-2} are in the "one" position, while all other switches are in the "zero" position, the voltage at the output of amplifier 1 is $-0.75E$. In general, the voltage e_1 at the output of amplifier 1 is given by

$$e_1 = -DE$$

where D is the absolute value (in the range 0 to 1.0) of the binary number to be converted.

The sign-bit of the binary number controls switch S_s. If the sign bit is "zero," identifying a positive number, switch S_s is in the open position. Under these conditions amplifier 2 acts as a simple sign-changer, producing an output voltage e_0 which is the negative of e_1. If the sign-bit is "one," the output of amplifier 3, which is equal to $-2e_1$, is added to e_1 before inverting. The output e_0 then becomes $+e_1$, which is equal to $-DE$.

In some digital-analog converters, the weighted resistors are placed in the feedback rather than the input path of an operational amplifier. Such a circuit is shown in Figure 2.5. The switches S, controlled by the binary digits, act to short out a specific resistor if the corresponding binary digit is "zero," and to connect it into the circuit if the binary digit is "one." The output voltage of amplifier 1 then is given again by Equation 2.1.

The big difficulty in realizing satisfactory digital-analog converter performance by using weighted-resistor networks is due to the range of resistance values required. Consider the circuit shown in Figure 2.4. Since the feedback resistor, $R/2$ in magnitude, acts as a load on amplifier 1, it is rarely practical to give this resistor a value of less than 50 K-ohms in a computer having a dynamic range of 100 volts. For a 10-bit converter, the resistance values in the weighted-resistor network must then range from 100 K-ohms to 51.2 M-ohms. If a 14-bit resolution is required, this range becomes 100 K-ohms to 819.2 M-ohms. Near the upper end of the range, the magnitude of the resistances approaches attainable insulation and switch resistances. This difficulty can be obviated in part by making the switches in the input network double-throw rather than single-throw switches. If the binary digit is a "one," the switchpole of the corresponding switch is connected to the reference supply, E; if the digit is a "zero," the switchpole is connected to ground. Under these conditions, an application of Thévenin's theorem results in the equivalent circuit shown in Figure 2.6a, where R_p is the parallel resistance of all the resistors making up the input network (this is slightly greater than $R/2$). The series voltage-source is equal to the product of the reference supply, E, and the digital

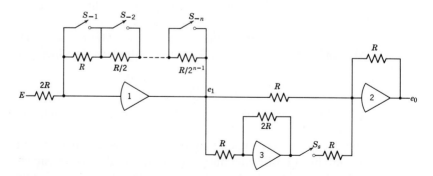

Figure 2.5 Alternative realization of weighted-resistor digital-analog converter.

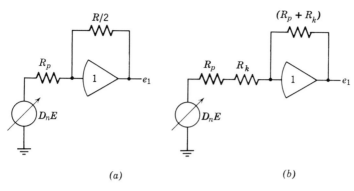

Figure 2.6 Thévénin's equivalent circuits for weighted-resistor digital-analog converter.

number, D, to be converted. It then becomes possible to modify R in the input network and at the same time to connect a resistor R_k between the weighted-resistor network and the summing point of amplifier 1, as shown in Figure 2.6b. For a 10-bit converter, R can then be made 1 K-ohm, so that the input impedances vary from $1K$ to $512K$, by placing a resistor slightly smaller than 49.5 K-ohms in series with the network; a resistance of 50 K-ohms can then be used as the feedback impedance of amplifier 1. Now all resistance values are in a comfortable range, but at the price of more elaborate and expensive switches.

2.4 LADDER-NETWORK DIGITAL-ANALOG CONVERTERS

Another widely used approach to digital-analog conversion involves the use of a ladder network of resistors at the input of an operational amplifier. A typical network of this type is shown in Figure 2.7. An analysis of this network reveals again that the voltage e_1 is directly proportional to D, the digital number in the range 0 to 1.0 to be converted, as well as to the reference voltage E. Note that in this case, all resistors have magnitudes of R or $R/2$ so that the need for a wide range of resistance values, as in the case of the weighted-resistor network converter, is obviated. On the other hand, all switches must be of the double-throw type.

Schmid[4] has described a novel digital-analog converter. In his system, precision current-dividers are used in place of the precision resistors which are required in conventional converters. This approach is particularly attractive if integrated-circuit techniques are to be employed, since the realization of precision resistors is then particularly difficult.

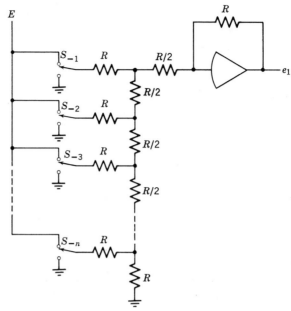

Figure 2.7   Ladder network digital-analog converter.

2.5   ANALOG-DIGITAL CONVERSION—GENERAL REMARKS

The conversion of an analog voltage into a binary code is a much more difficult operation than is digital-analog conversion. In fact, for many years the analog-digital converter constituted the "weakest link" in the hybrid computer loop. Although analog-digital converters are now available with accuracies of the same order as those of the best analog computers, a reasonably rapid converter of this type costs from five to ten times as much as a digital-analog converter with approximately the same precision. An analog-digital converter must accept a continuous transient voltage at its input and furnish in its output register, at equally spaced instants of time, a binary number corresponding to the magnitude of the voltage input. Usually, analog-digital converters operate under external control and receive control pulses either from the digital computer or from a separate timer. Accordingly, most analog-digital converters have a "start conversion" channel, which is activated by an externally applied pulse or logic level, and emit a "conversion complete" signal when the correct value has been placed in the converter output register. Analog-digital converters also often come equipped with an array of lightbulbs to provide a visual indication of the digital output.

2.5/ANALOG DIGITAL CONVERSION—GENERAL REMARKS

As described in more detail in succeeding sections, analog-digital converters contain analog as well as digital circuitry. Most types contain at least one digital-analog converter, associated flip-flop registers, and digital-logic circuitry as well as at least one comparator. In many respects, it is the comparator that is the heart of the analog-digital converter and determines its performance. In essence, a comparator is a high-gain summing circuit for adding two input voltages. This is achieved by employing a very large feedback resistance if an operational amplifier is employed. The two inputs must have opposite polarity; if the positive input is slightly larger than the negative input, the output will approach the negative reference level; if the negative input becomes larger than the positive input, the output will immediately become a large positive voltage. Comparators, therefore, are devices which are very sensitive to the relative magnitudes of two inputs. Of key importance in determining the quality of comparator performance is the switching time and the stability of the switching point. Commercially available comparators have switching times of less than 0.1 microsecond, although for large voltage ranges and conservative construction, switching times of 1 or more microseconds are not uncommon. The precise point at which switching from a large negative to a large positive output occurs is affected to some extent by offset and drift in the operational amplifier, which in turn is affected by changes in ambient temperature and by power-supply fluctuations. Monolithic integrated comparators are now becoming commercially available and promise considerable improvements in converter performance.

In addition to errors introduced by component inaccuracies and drift, analog-digital converters suffer from so-called "aperture errors." Every analog-digital converter requires a certain amount of time to make a conversion. Depending on the specific design of the converter, this time may be constant from conversion to conversion or it may undergo rather wide variations. In any event, if the analog signal to be converted undergoes appreciable variations during this aperture time, there will be uncertainty as to the precise significance of the digital output.

To minimize errors due to aperture uncertainty, a number of manufacturers provide special sample-hold circuits. As an example of such an approach, consider the Texas Instruments converter shown in simplified form in Figure 2.8a. The sample-hold unit which precedes the analog-digital converter contains four amplifiers, four solid-state switches, and a capacitor. These are tied into the loop which contains a successive approximation analog-digital converter of the type described in Section 2.7. The comparator and a digital-analog converter (not shown in the figure but included in the block labelled "analog-digital converter") are

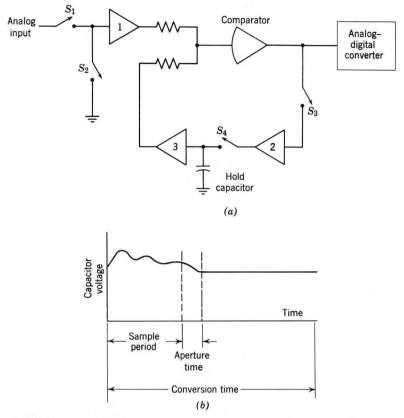

Figure 2.8 (a) *Sample-hold circuit for analog-digital converter.* (b) *Significance of "aperture time" in analog-digital conversion.*

actually a part of this converter. At the start of conversion, the digital-analog converter is set to zero and the switches are set in the following positions; S_1 closed, S_2 open, S_3 closed, and S_4 closed. The input is therefore connected through amplifier 1 to the comparator. The output of this comparator is applied through amplifier 2 to the hold capacitor. The capacitor voltage forms the other input to the comparator so that the comparator will act to charge the hold capacitor and cause it to track the analog input voltage. Three microseconds after the "start conversion" signal, the position of the four switches is reversed in the following order: S_4, S_3, S_1, and finally S_2. This switching sequence prevents transients from affecting the hold-capacitor voltage, and the time required for switching represents most of the aperture time. When all switches have been reversed, the input has been effectively disconnected, and the analog-digital converter operates on the constant voltage which has been imposed on the hold capacitor. Figure 2.8b shows the full conversion cycle. In the converter under discussion, the sample period is 3 microseconds, the aperture time is 100 nanoseconds, and the total conversion period is 18 microseconds. An additional refinement of this circuit permits the compensation of errors arising from offsets within the analog-digital converter.

In Table 2.1, the specifications of a number of commercially available analog-digital converters are presented. This table is limited to those units which have word-lengths, accuracies, and conversion rates compatible with high-quality hybrid computer operation. Many of the manufacturers listed have available a number of models and a number of options. In general, the models listed in Table 2.1 are those with the highest resolution (longest word-length) and highest conversion rate available from each manufacturer. Some of the manufacturers provide sample-hold units as standard equipment in their analog-digital conversion systems. In these cases, the characteristics of the overall system are listed, together with the aperture time of the sample-hold circuit.

2.6 SIMULTANEOUS ANALOG-DIGITAL CONVERSION

The most rapid and most direct approach to rapid analog-digital conversion would be to use an array of comparators, all operating

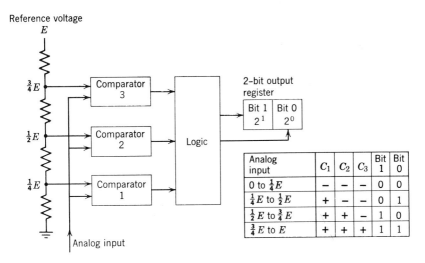

Figure 2.9   Multicomparator-type analog-digital converter.

Table 2.1

Manufacturer Model Number	Adage Inc. VT 13-AB	Brown Engineering Co.	Control Data 214	Control Equipment Corp. 3042	Digital Equipment Corp. 142	Dynamics Instrumentation Co. 6539	Dynamic System Electronics 824300	Electronic Development Corp. VBC 12
Standard input voltage range	±5v ±10v ±100v	±5v	±5v ±10v	±1v ±10v ±100v	±10v	optional	±1v ±10v ±100v	±1v ±10v ±100v
Number of bits (including sign)	14	10	14	14	10	15	11	12
Approximate conversion time (microseconds)	5	40	19	30	6	32	60	50
Approximate maximum conversion rate (conversions per second)	200,000 (125,000 with SH)	25,000	60,000 (40,000 with SH)	31,000	166,000	31,200	16,000	20,000
Output logic "one" levels "zero"	+12v 0v	+15v 0v	0v +6v	−6v 0v	−3v 0v	+1.0 to 28v 0v	−6v 0v	+3v −3v
Quoted accuracy ±½ LSB	0.01% (0.015% with SH)	0.05%	0.015% (0.025% with SH)	0.005%	0.15%	0.012%	0.05%	0.025%
Specifications for conversion with sample-hold (aperture time)	Yes (20 n sec)	No	Yes (100 n sec)	No	No	Yes (100 n sec)	No	No

simultaneously. Figure 2.9 illustrates such a simultaneous converter for a 2-bit digital output. Note that three comparators are required to determine into which of four possible levels the analog input falls. A logic network is employed to translate the comparator output levels (either positive or negative) into pulses which act to set the output register. Since all comparators work in parallel, the conversion speed of such a converter can be made less than 1 microsecond. To provide a digital output of n bits, 2^{n-1} comparators are required. In most hybrid systems, it is the practice to choose a word-length such that the quantization errors (one-half

2.6/SIMULTANEOUS ANALOG-DIGITAL CONVERSION

Table 2.1 (Contd.)

Electronic Engineering Co. EECO 76 DA	Lancer Electronic Corp. AD-20R	Navigation Computer Corp. 2201	Pastoriza Electronics ADC 10$_{\text{IC}}$	Preston Scientific 8500-VHS	Radiation Inc. 5516	Redcor Corp. 663	Scientific Data Systems AD-35	System Engineering Labs. 521	Texas Instruments 846
±5v	±10v	±10v	0 to −10v	±5v ±10v	±5v	±10v ±100v	±10v	±4v ±10v	±10v
14	14	10	10	15	12	15	15	15	13
25	6	125	10	4	20	24	5.2	20	17.5
39,500	150,000	8000	100,000	200,000	50,000	36,000	192,000	50,000	57,000
0v −1v	+5v 0v	−6.8v −0.2v	+2v 0v	+4v 0v	−10v 0v	−6v 0v	+8v 0v	−6v 0v	+7v 0v
0.01%	0.01%	0.5%	0.05%	0.01%	0.025%	0.01%	0.01%	0.015%	0.025%
Yes (100 n sec)	Yes (50 n sec)	No	No	No	Yes (100 n sec)	Yes (100 n sec)	No	No	Yes (100 n sec)

the least significant bit) are of the same order of magnitude as those introduced by analog-component tolerances. For this reason, word-lengths of 12 to 14 bits are employed. To produce a 14-bit digital output by the simultaneous conversion technique would require over 16,000 comparators. It is clear, therefore, that the simultaneous conversion approach is economically impractical in hybrid computer applications. The conversion methods described in the succeeding sections obviate this difficulty by using one or a limited number of converters repeatedly in each conversion, sacrificing conversion speed for greatly reduced cost.

2.7 CLOSED-LOOP ANALOG-DIGITAL CONVERTERS

The design of the analog-digital converter most widely used in present-day hybrid computing systems is based on the recognition that digital-analog conversion is a much more simple process than analog-digital conversion. Accordingly, a digital-analog converter is connected into a closed loop as shown in Figure 2.10. The digital output of the converter is stored in a digital register, and changes in the analog input voltage produce corresponding changes in the contents of this register. The digital number contained in the register serves as the input for a digital-analog converter. This converter then produces an analog output voltage proportional to the number contained in the digital register. This analog voltage constitutes one input, input B, of a comparator; the other input of the comparator, input A, is the analog voltage to be converted. The comparator output is therefore an error signal, proportional to the difference between the two inputs.

The comparator output is processed by a simple logic network, usually consisting of a number of gate circuits, such that a sequence of pulses is produced which causes the contents of the digital output register to be modified so as to reduce the error signal. Usually, the analog-digital converter is under control of an external clock, so that one modification in the digital register is effected for each clock pulse. This change in the digital output produces a change in the digital-analog converter, which in turn modifies the output of the comparator circuit. The error signal is thereby reduced progressively until it falls within a specified tolerance. The conversion is then complete, and the logic circuit emits a control pulse to convey this information to the rest of the hybrid loop. The

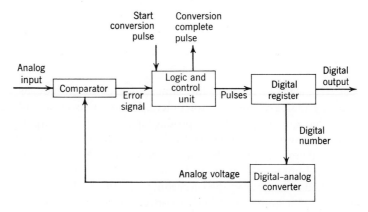

Figure 2.10 Closed loop-type analog-digital converter.

2.7/CLOSED-LOOP ANALOG-DIGITAL CONVERTERS

number within the digital register is then read out, perhaps directly into the digital computer, and a new conversion can begin. The rate at which conversion can be accomplished, i.e., the number of analog samples which can be handled per second, is determined by the manner in which the contents of the digital register are modified in response to an error signal.

There are two major alternatives in the implementation of the closed-loop conversion method. These relate to the manner in which the pulses emitted by the logic circuit are utilized. In the so-called "successive approximation" method, the most significant bit in the register is modified first, while the second approach, the so-called incremental method, the least significant bit in the register is modified first. In the successive-approximation converter, the start pulse, initiating the conversion, causes the contents of the digital register to be reset to zero. The most significant bit is then changed from "zero" to "one." If this change in the contents of the register causes the input B of the comparison circuit to be larger than the input at A, the most-significant bit is reset to zero; if not, the most-significant bit remains at "one." In either case, the second most-significant bit is then changed from "zero" to "one" and the analog voltages A and B are again compared. This procedure is repeated for all of the bits in the digital register. It is evident that the length of time required for a complete conversion is proportional to the word-length. Since the digital register is reset to zero before the start of each conversion, the conversion speed is unaffected by the quantity which was stored in the digital register at the completion of the preceding conversion.

In an incremental converter, the digital register is not reset to zero at the start of a conversion, and the least-significant bit is changed in response to an error signal. Where the analog input voltage undergoes relatively small changes from one conversion to the next, an incremental computer is able to produce an output much more rapidly, since only a very few of the digital register bits have to be modified. Where large changes in analog input voltages occur, however, as in the case of multiplexed inputs, the incremental converter must laboriously step from one value to another, one least-significant bit at a time. The length of time required for a given conversion is therefore proportional to the change in the analog input from one conversion to the next.

The difference in operation of the successive-approximation and incremental methods of analog-digital conversion is illustrated in Figure 2.11 for a 6-bit analog-digital converter and a voltage range such that one least-significant bit corresponds to 0.1 volt. While the analog signal is varying slowly, the incremental converter can keep up with the input and provides many more outputs than does the successive-approximation converter. On the other hand, where the analog signal undergoes abrupt

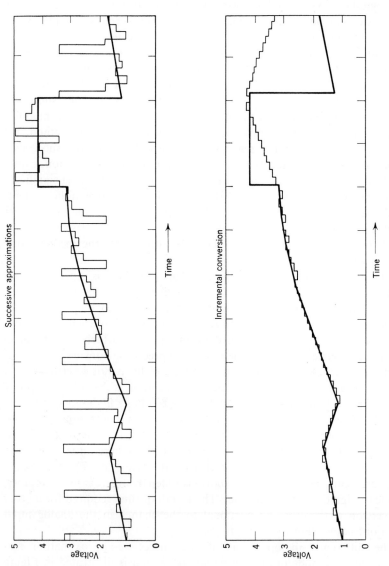

Figure 2.11   Alternative modes of operation of closed loop analog–digital converter.

changes, the incremental converter must laboriously approach the new value in steps of one least-significant bit, at a great expenditure of time. For the successive-approximation converter, abrupt changes in input voltage present no special problem, and a read-out is available after every six clock pulses regardless of the nature of the input. Since most hybrid systems are designed so as to time-share the relatively expensive analog-digital converter by multiplexing, the successive-approximation converter is used almost universally in this application. Typically, such a converter can perform 20,000 to 50,000 conversions per second.

2.8 ANALOG-DIGITAL CONVERSION BY SUBRANGING

The closed-loop converters described in the preceding section use only one comparator and are therefore much more economical than the simultaneous converter described in Section 2.6. The time for an n-bit conversion, however, is necessarily at least n-cycle times (or the time for n-clock pulses). To increase the conversion speed, a new class of analog-digital converters was introduced in 1966. These represent a compromise between the closed-loop and the simultaneous-conversion approaches. A simplified block diagram of a subranging analog-digital converter is shown in Figure 2.12.

During the first conversion cycle, digital-analog converter 1 is set to the maximum value while digital-analog converter 2 is set to zero. The full voltage range, E, therefore appears across the resistance network and comparators $C_1, C_2, C_3, \ldots, C_7$ receive inputs of $\frac{1}{8}E, \frac{2}{8}E, \frac{3}{8}E, \ldots, \frac{7}{8}E$, respectively. The other input of each comparator is the analog voltage to be converted. The voltage levels assumed by the seven comparators are processed by the logic network and translated into the digital output. To this point, the system acts exactly like a simultaneous analog-digital converter with a 3-bit digital register, and in fact the result of the first conversion cycle determines the three most significant bits of the digital output. In the next conversion step, however, the entire resistance network is employed to subdivide a small portion of the full voltage range. Consider, for example, that the analog input voltage is $0.525E$. In that case, the analog input is greater than the reference input for converters C_1, C_2, C_3, and C_4, while it is less than the reference input for comparators C_5, C_6, and C_7. The last three comparators will therefore have negative output levels while the first four comparators will have positive output levels. At this juncture, the converter has established that the analog input lies between $0.5000E$ and $0.6250E$.

In the next step of the conversion cycle, digital-analog converter 1 is set to $\frac{5}{8}E$, and digital-analog converter 2 is set to $\frac{1}{2}E$. The reference

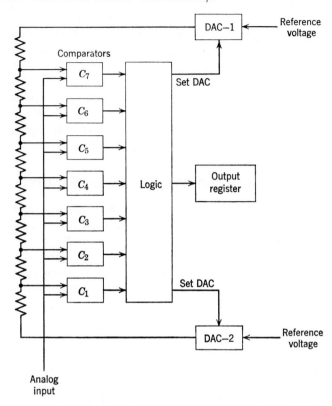

Figure 2.12 Subranging analog-digital converter.

inputs of the comparators C_1, C_2, C_3, ..., C_7 now become approximately 0.5156E, 0.5312E, 0.5468E, ..., 0.609E, respectively, so that the voltage range that formerly existed between the reference voltages of C_4 and of C_5 is now divided into eight equal steps. The analog input has remained at 0.525E, so that the second conversion step determines that the unknown falls between .5156E and 0.5312E, effectively establishing the fourth, fifth, and sixth most significant bits of the digital output. The two values, 0.5156E and 0.5312E, are now applied to digital-analog converter 2 and digital-analog converter 1, respectively, to start the third conversion step. In this way, each successive conversion step determines an additional three digits of the digital output. If seven comparators are used, a subranging converter is approximately three times as rapid as a closed-loop successive-approximation converter.

It is possible, of course, to use a greater or smaller number of comparators. In general, the total resolution of conversion is $[1/(M+1)]^S$, where M is the number of comparators and S is the number of conversion steps. Thus the system shown in Figure 2.12 could provide resolution of 1 part in 4096 (corresponding to 12 bits) in four steps and a resolution of 1 part in 32,768 (or 15 bits) in five steps.

2.9 TIME-INTERVAL ANALOG-DIGITAL CONVERTER

The last method of analog-digital conversion to be described in this chapter is too slow to be useful for closed-loop hybrid systems. Time-interval analog-digital converters are, however, very widely used in the digital voltmeters which are furnished as peripheral equipment in most hybrid systems and, therefore, merit some discussion. This approach to the conversion of an analog voltage into digital code includes two steps. The analog input is first converted into a time interval such that the length of time between a "start" pulse and a "stop" pulse is proportional to the magnitude of the analog signal. This time interval is also converted into a digital number by applying to a digital register a number of counting pulses proportional to the length of the time interval. Figure 2.13 illustrates one manner of realizing such a conversion scheme. A comparator is employed to compare the analog input, A, with the output, B, of a staircase generator. The latter unit generates an output voltage which is zero at time t_1 when the "start" pulse is applied, and increases in small increments each time a pulse is applied to its input. Input B of the comparator is therefore a staircase function as shown. When input A and input B are equal to each other, the comparator changes its output level.

The "start" pulse also causes a flip-flop to assume an output level so as to open a gate circuit. The flip-flop maintains this level until the

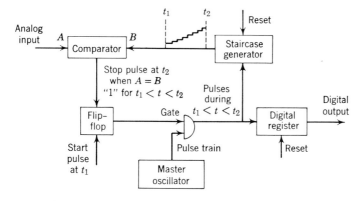

Figure 2.13 Time-interval analog-digital converter (Susskind[1]).

comparator output level changes, causing a stop signal to be applied to the flip-flop. During the interval $t_1 \leq t_0 \leq t_2$, the "and" gate permits pulses from a constant-frequency master oscillator to pass through. At t_2, when $A = B$, the gate is shut and no further pulses are transmitted. The pulses passing through the gate are accumulated in a digital register, which is reset to zero before the "start" pulse is applied, such that each arriving pulse causes the number stored in the digital register to increase by one least-significant bit. The pulses passing through the gate also constitute the input to the staircase generator such that each arriving pulse causes the staircase generator output to increase by one step. At any time t, the number in the digital register and the voltage level at the output of the staircase generator are both proportional to the time that has elapsed since the "start" pulse at t_1. The time interval between the "start" and the "stop" pulses $(t_2 - t_1)$ is therefore directly proportional to the analog input A, as is the number contained in the digital register. The number in the digital register is therefore proportional to the analog input A. The conversion rate of such a circuit is determined by the magnitude of the analog input A and the frequency of the master oscillator and is usually limited to approximately 50 conversions per second.

2.10 MULTIPLEXING AND DEMULTIPLEXING

The functions of the multiplexer and the demultiplexer or distributor units are exact inverses of each other. The multiplexer converts information from parallel form into serial form, while the demultiplexer converts data from serial into parallel form. The same general techniques therefore serve for both operations. The detailed nature of these units depends, however, to a very large extent on whether the information to be multiplexed or distributed is in analog or in digital form. If the data are in digital form, as many parallel channels as there are bits in the digital word are usually required in hybrid linkage systems, since the distribution of serial digital data would be too time consuming; in an analog multiplexer scheme, on the other hand, only one multiplexing circuit need be supplied for each data channel. In a digital multiplexer, each channel must have a quality sufficient only to assure that a "zero" not be mistaken for a "one" in the process of multiplexing; in an analog multiplexer, care must be taken that zero-offsets and noise do not introduce errors exceeding the specified analog tolerance, usually 0.01 % of full-scale.

A multiplexer can be visualized as having two major portions: the switching unit and the addressing unit. The switching unit is discussed first, followed by a description of the more important addressing schemes. Inherently, a multiplexer is nothing more than an array of switches which act to connect each data channel to a common transmission line. The number of switches which must be provided is therefore equal to the number of channels which are to be multiplexed. These switches can be either electromechanical or solid-state in nature, though for reasons of speed, solid-state multiplexing is now used in virtually all high-quality hybrid systems. The development of high-quality multiplexers was greatly aided in the 1960's by the introduction of the reed relay and the field-effect transistor for electromechanical and electronic switching, respectively.

The reed relay is an electromechanical device with a geometry radically different from that of conventional relays. The relay contacts and the switching pole are housed in a sealed glass cylinder filled with an inert gas to minimize contact corrosion and dirtying. Switching is produced by means of an external coil, concentric with the glass housing, the entire relay pole acting as part of the magnetic circuit. If the relay is to be polarized, a small permanent magnet is placed outside the glass housing, immediately adjacent to the relay contacts. This magnet is sufficiently strong to hold the relay pole in either position, but not strong enough to impede switching. Compared to conventional relays, reed relays offer the advantages of small size, very rapid switching action, very reliable operation, and considerable economy. Reed relays can operate at speeds of 100 switchings per second, and occasionally as many as 1000 such relays are housed together in a compact container.

The field-effect transistor is a monolithic, three-terminal, solid-state device, not unlike a transistor amplifier. The field-effect transistor is distinguished, however, in that it is possible by the application of a small control signal to modify the impedance in the source-drain circuit over an extremely wide range. With one polarity of control signal, the source-drain impedance can be made less than 10 ohms, while a reversal of the control signal causes the source-drain impedance to become well above 10^9 ohms. The controlled signal in the source-drain circuit may have either polarity, and switching can be accomplished in under 1 microsecond. In recent years, high-quality field-effect transistors have become progressively less expensive, and therefore promise to play a bigger and bigger role in high-quality, high-speed multiplexing applications.

The purpose of the addressing unit of a multiplexer is to permit the closure of selected multiplexer switches in a specified sequence. If each multiplexer channel is given an identifying number or address, the function of the addressing unit can be considered to be to translate this identifying number into a control signal acting to close the corresponding switch. Three general methods are available for the performance of this function. These include sequential scanning, the selection pyramid, and the selection matrix, as presented in an excellent discussion by Susskind.[1]

In sequential scanning devices, switches, each connecting one of the parallel inputs (or outputs) to the serial output (or input), are closed in turn. A familiar and widely used example of such a device is the stepping switch, which has found wide utilization in telephone applications. Such switches are essentially single- or multi-pole, n-throw switches. A control pulse acts through a relay to step the switching pole from one contact to the next. A number of such poles may be stepped in parallel and as many as 80 contact points may be scanned sequentially. In specifying stepping switches, it is possible to demand "make-before-break" or "break-before-make" operation depending on the specific application. Stepping switches have the advantage relative to other multiplexing devices of being relatively economical and of providing a very sharp distinction between "open" and "closed" resistance. In addition to their limited speed of operation, their biggest disadvantage is that random access is not possible, so that the parallel inputs must always be scanned in the same sequence. A similar scanning action can be obtained by arranging a number of reed relays in star fashion upon a horizontal surface. These relays have no associated coils. Instead, a permanent magnet, mounted on a moving bar similar to the second-hand of a clock, is made to pass over each relay in turn, thereby closing it for a brief period of time. As many as 50 relay closures per second can be effected in this manner.

The application of selection pyramids usually demands that the "address" identifying the parallel input (or output) be available in digital form. This information is usually furnished by the digital computer, or by a counter, and successive addresses may come in any desired order. A selection pyramid for connecting one input to one of eight outputs, labeled 0, 1, 2, ... , 7, is shown in Figure 2.14. The address of the parallel channel to be selected is placed in a 3-bit address register, with the least-significant bit on the right. Each of the register cells is a flip-flop with two stable states—"zero" and "one." These flip-flops operate in such a way that when a given flip-flop is in the "one" position, an energizing signal is applied to all lines connected to the "one" terminal of the flip-flop. Similarly, if the flip-flop is in the "zero" position, all lines connected to the "zero" terminal are energized. Thus when flip-flop 0 in Figure 2.14 is in the "zero" position, gates a, c, e, and g receive an excitation. If the other inputs of these gates are excited simultaneously, these gates will generate output signals. Otherwise, no outputs are generated. Similarly, if flip-flop 0 is in the "one" position, gates b, d, f, and h receive input signals. Flip-flops 1 and 2 act in a similar manner to control one input of gates i, j, k, l, m, and n. An analysis of the circuits will show that any 3-bit address placed in flip-flops 0, 1, and 2 causes one and only one closed path to exist between the input and the output terminals of the pyramid.

2.10/MULTIPLEXING AND DEMULTIPLEXING

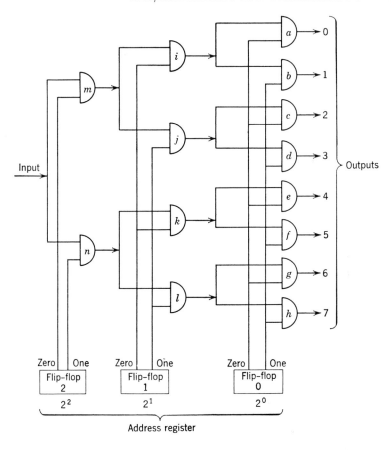

Figure 2.14 Pyramid-type multiplexer.

Consider, for example, that the address register contains the number 5, the binary representation of which is 101. The most significant binary digit then acts to permit gate n to transmit an input signal, the second binary digit permits an input signal to pass through gates i and k, while the least significant bit opens gates d, e, f, and h. The only continuous path from input to output is then through gates n, k, and f. The parallel line 5 is therefore selected as desired. Pyramid circuits of the type shown in Figure 2.14 can be expanded to handle any number of parallel lines with a concomitant increase in the number of logic circuits required.

In matrix selection circuits, the separation of the parallel input channels is effected in a different manner. Each parallel line is connected to one terminal of a semiconductor diode. If the other terminal of this diode is placed at ground potential, the corresponding parallel channel is effectively

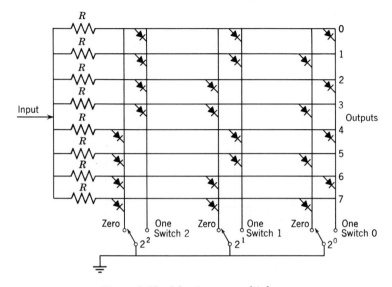

Figure 2.15 Matrix-type multiplexer.

shorted to ground and is thereby disconnected from the serial input (or output) line. Figure 2.15 illustrates how a diode matrix is employed to connect one input line to any one of eight parallel output lines. Again the address identifying the parallel line is contained in a 3-bit register, with the most-significant bit on the left. The register switches are designed such that if a "zero" is stored in a given bit location, all points connected to the "zero" output line are placed at ground potential. Similarly, if the register cell contains a "one," all points connected to the "one" output line are at ground potential. This can be accomplished by utilizing mechanical switches as shown in the figure or by means of flip-flops.

With reference to Figure 2.15, note that each output line of each of the three switches is connected through diodes to four of the parallel lines. Therefore, each switch position of each of the three switches connects four of the input lines to ground. The input signal can appear only at an output line which is not grounded, and for each combination of switch positions of the three switches, there is only one such output line. Consider again the selection of output line 5, represented by 101 in binary notation. Switch 2 is then in the "one" position, grounding output lines 0, 1, 2, and 3; switch 1 is in the "zero" position, grounding lines 2, 3, 6, and 7; switch 0 is in the "one" position, grounding lines 0, 2, 4, and 6. Of the eight output lines, only line 5 remains ungrounded and therefore has an output voltage equal to the input voltage applied to the matrix. The purpose of the resistors R is to prevent each diode from shorting out the entire input signal when its other terminal is connected to ground. If desired, each of the output lines can be employed to activate a separate switch. Selection of a specific output line then causes one relay or solid-state switch to close, thereby making a direct connection between two points in the hybrid system.

A number of elegant multiplexer units have appeared on the market. The more economical of these employ diode-relay matrices and are termed "cross-bars." Cross-bar units are available to handle as many as 1200 input lines. These devices include a two-dimensional array of six-pole electromechanical relays, and can scan up to 50 points per second. By contrast, available solid-state multiplexing units can scan as many as 1,000,000 points per second at accuracies of .01%. However the per channel cost of such all-electronic multiplexers is much greater than that of the relay scanners.

References

1. Susskind, A. K., *Notes on Analog-Digital Conversion Techniques*, John Wiley, New York, 1958.
2. Digital Equipment Corp., *The Digital Logic Handbook*, Digital Equipment Corp., Maynard, Mass., 1966.
3. Korn, G. A., and T. M. Korn, *Electronic Analog and Hybrid Computers*, McGraw-Hill, New York, 1964.
4. Schmid, H., "Current Dividers Convert Digital Signals into Analog Voltages," *Electronics*, pp. 142–148, Nov. 14, 1966.

3

Numerical Solution of Ordinary Differential Equations

3.1 GENERAL REMARKS

As indicated in Section 1.4, digital computers are used in a hybrid computer loop for a variety of reasons. Frequently, the task is only to control the modes of the analog computer, to set potentiometers, and to perform other peripheral tasks. Often they are employed for the generation of nonlinear functions, to realize time-delays, and to perform logic and decision operations. In many hybrid simulations, however, all or some of the differential equations characterizing system dynamics are solved on a digital computer, the results of this integration being applied in an on-line fashion to the analog computer or to external systems. Whenever a digital computer is used in a hybrid loop, considerable care and finesse must be employed in programming to assure that the digital computer results are ready at precisely those instants of time when the analog computer requires them. Since digital computers operate serially, performing only one arithmetic operation at a time, the computing time required for a specified calculation is determined by the number of arithmetic steps, hence the complexity of the computation. Usually, the programmer is forced to make an engineering compromise between accuracy and speed. The task of the programmer in selecting suitable algorithms and in implementing these in a program is particularly difficult where the digital computer is employed to treat systems of simultaneous differential equations. Here, a wide variety of algorithms is available, each

3.2/CONSIDERATIONS IN SELECTING AN INTEGRATION SCHEME 49

possessing certain advantages and disadvantages, and a judicious choice from among these possibilities is difficult indeed. This chapter is devoted to a discussion of the most important considerations entering into the choice of a differential equation integration routine and to a survey of the most widely used solution methods. The accent in this presentation is on engineering applications rather than on detailed numerical analysis.

The differential equations usually solved in the digital portion of the hybrid computer system take the general form:

$$a_{1,n}\frac{d^n y_1}{dx^n} + a_{1,n-1}\frac{d^{n-1} y_1}{dx^{n-1}} + \cdots + a_{1,1}\frac{dy_1}{dx} = a_{1,0}F_1(x, y_1, y_2, \ldots, y_m)$$

$$\vdots \qquad\qquad\qquad\qquad\qquad\qquad\qquad\qquad \vdots$$

$$a_{m,n}\frac{d^n y_m}{dx^n} + a_{m,n-1}\frac{d^{n-1} y_m}{dx^{n-1}} + \cdots + a_{m,1}\frac{dy_m}{dx} = a_{m,0}F_m(x, y_1, y_2, \ldots, y_m)$$

(3.1)

In Equation 3.1 the parameters $a_{m,n}$ may be constants or they may be functions of the dependent and independent variables or their derivatives. To permit their solutions, equations of this type must be accompanied by additional information in the form of boundary or initial conditions. If all this information is provided for a single value of the independent variable, i.e., if y, y', y'', etc. are specified for $x = 0$, the problem is termed an *initial* value problem; if, on the other hand, y or its derivatives are specified for two values of x, the problem is termed a *boundary* value problem.

3.2 CONSIDERATIONS IN SELECTING AN INTEGRATION SCHEME

All practical methods for the solution of initial value problems of the type of Equation 3.1 involve stepwise integration. Starting with specified values of the dependent variable and its derivatives at an initial value of x, designated as x_0, an effort is made to determine y and some of its derivatives at the next step in the x domain, that is, at $x = x_0 + h$ where h is the step size. Once y and its derivatives at $x = x_0 + h$ are available, the same or similar methods can be employed to find y and its derivatives at $x = x_0 + 2h$. This stepwise solution is continued until a desired range of x has been covered. In essence, numerical methods for the solution of initial value problems therefore involve the prediction of successive values of y and its derivatives. For this purpose a wide variety of techniques is available. From the point of view of the engineer, each of these has certain advantages and disadvantages, and the choice of a particular method therefore involves a judicious compromise. Of particular importance in the discussion and selection of numerical methods are considerations of computing time, accuracy, and ease of programming.

The time required to obtain the solution of a problem on a digital computer is generally a function of the number of arithmetical or logical operations which must be performed. In estimating the computing time, it is necessary to consider the number of steps in the domain of the independent variable for which solutions are to be obtained, as well as the number of operations required to obtain the desired solution at each step. Although digital computers perform individual arithmetic operations with great rapidity, it is quite possible in the case of relatively complex nonlinear equations, and in the case of large systems of such equations, that hours or even days of running time may be required even on the fastest computers.

It is a characteristic of engineering problems, as distinct from scientific or purely mathematical problems, that a certain error in the solution is tolerable. A well-formulated engineering problem includes the specification of the solution error which is acceptable to the engineer. This tolerance depends to a large extent on the practical use to which the solution is to be put as well as on the confidence the engineer has in the mathematical model. Thus, if certain parameters in a mathematical model can only be specified with an accuracy of 1%, it makes little sense to demand computer accuracies of 0.1%. Since there are nearly always inherent errors in numerical solutions, a comprehensive consideration of these errors is of particular importance in engineering applications. In most problems, the two main sources of errors in the numerical treatment of differential equations are those errors resulting from the discretization, the representation in quantized form, of the independent and dependent variables. These errors are termed *truncation* errors and *round-off* errors, respectively. Truncation errors incurred at any point in the computation are generally a function of the step size h as well as of the dynamics of the solution—the magnitude of the higher derivatives of y at the point in the x domain under consideration. A valuable feature of certain numerical methods is that they facilitate the control of the truncation error at each step in the course of the calculation in a manner that will assure that the truncation error will not exceed a specified limit. In most cases, the round-off errors made at a specific step of the calculation are negligible. In the case of many numerical methods it is possible, however, that round-off errors made at one step of a computation sequence grow in subsequent steps until the cumulative round-off error becomes sufficiently large to make the solution worthless. This phenomenon must be carefully considered in evaluating any numerical analysis procedure. Evidently, the fewer the number of calculations performed in the course of a solution, the smaller the chance for such an accumulation of round-off errors; hence, from this point of view it is desirable to employ as large a step size h as possible. The requirements imposed upon step size by the existence of truncation and round-off errors are therefore contradictory, and the selection of an optimum h involves careful consideration and compromise.

Almost invariably in the treatment of differential equations, the effort expended in programming and coding the problem prior to its application to the digital computer requires a time several orders of magnitudes larger than the actual computer running time. While the programming time is not nearly as "expensive" as computer running time, ease in programming and coding nevertheless becomes an important consideration in the selection of a numerical method. To minimize the programming effort, most computing facilities encourage engineers to formulate their problems in such a manner that their program can then become a part of the facility's library of subroutines and can be used to solve similar problems in the future. It is therefore expedient for an engineer with a problem to be solved to inquire whether some program in the facility's library of routines can be adapted to his problem. Frequently, he will prefer to employ such an already prepared and "checked-out" program even if it does not constitute the optimum numerical method from his point of view. If, on the other hand, there is a likelihood of the recurrence of similar problems in the future, an engineer may well decide to undertake a complete programming and coding effort.

The numerical methods for the solution of ordinary differential equations described in this chapter fall into two broad categories:

1. Self-starting methods.
2. Non-self-starting methods.

A self-starting method involves the use of integration formulas for the determination of the dependent variable y_{i+1}, corresponding to successive points in the domain of the independent variable x, in which only data corresponding to the value of y and its derivatives at x_i are required. The function is thus expressed entirely in terms of data obtained at the step immediately preceding the current calculation. Employing the specified initial condition, it is therefore possible, in a self-starting method, to calculate directly the dependent variable and its derivatives at the first step in the x domain. The results of this calculation then become the initial condition for obtaining the dependent variable and its derivatives at the succeeding step in x. The Taylor series method described in Section

3.4 and the Runge-Kutta methods described in Sections 3.5 and 3.6 are self-starting methods.

Non-self-starting methods, on the other hand, require for the calculation of y_{i+1} data corresponding to the value of y and its derivatives at x_i, x_{i-1}, x_{i-2}, x_{i-3}, etc. That is, it is necessary to have available data corresponding to the solution of three or four preceding steps. Since this information does not exist at the start of a calculation, it is necessary to precede the application of a non-self-starting method by a self-starting method for the determination of the first several solution points. The forward integration method, the predictor-corrector method, and the predictor-modifier-corrector method described respectively in Sections 3.7, 3.8, and 3.9 fall into this category of numerical techniques. The disadvantage inherent in their non-self-starting feature is counterbalanced by the fact that their application generally requires considerably less computing time.

To facilitate a general discussion of numerical methods, it is customary to emphasize the treatment of sets of simultaneous first-order differential equations. In Section 3.3 it is demonstrated how a set of n-th order differential equations of the type of Equation 3.1 can be transformed readily into a system of first-order differential equations. The discussion in Sections 3.4 through 3.9 is then limited to the treatment of such equations.

In Section 3.10, a special situation is considered—that involving differential equations in which the dependent variable is specified at two points in the domain of the independent variable.

3.3 A USEFUL REFORMULATION

The general ordinary differential equation, Equation 3.1, may, in practice, assume a wide variety of forms; n may vary over a wide range, and the analytic expressions for the parameters $a_{n,m}$ and the driving functions F_m may likewise vary widely. To facilitate discussion and permit the development of methods with a sufficient generality, it has become general practice to reformulate the set of ordinary differential equations to be solved as a set of first-order differential equations. Although from the point of view of the engineer such a reformulation is not absolutely necessary and may entail occasional disadvantages, this traditional approach is nevertheless followed in the succeeding discussion since it permits the reader to make more effective use of the published literature. The procedure employed in reformulating the original equations involves the following:

Given the n-th order differential equation

$$a_n \frac{d^n y}{dx^n} + a_{n-1} \frac{d^{n-1} y}{dx^{n-1}} + \cdots + a_1 \frac{dy}{dx} + a_0 y = f(x, y) \qquad (3.2)$$

a set of n first-order equations is defined such that

$$\frac{dy_{(1)}}{dx} = f_1(x, y_{(1)}, y_{(2)}, \ldots, y_{(n)})$$

$$\vdots \qquad \vdots \qquad (3.3)$$

$$\frac{dy_{(n)}}{dx} = f_n(x, y_{(1)}, y_{(2)}, \ldots, y_{(n)})$$

where $y_{(1)} \equiv y$, and where

$$\frac{dy_{(1)}}{dx} = y_{(2)} \qquad \frac{dy_{(2)}}{dx} = y_{(3)} \qquad \cdots \qquad \frac{dy_{(n-1)}}{dx} = y_{(n)}$$

$$\frac{dy_{(n)}}{dx} = \frac{1}{a_n} \{ F(x, y_{(1)}, y_{(2)}, \ldots, y_{(n)})$$
$$- (a_{n-1} y_{(n)} + a_{n-2} y_{(n-1)} + \cdots + a_0 y_{(1)}) \} \quad (3.4)$$

Provided certain very general limitations on the coefficients a_n are met it is always possible to effect the above transformation.

Consider, for example, the equation

$$\log y \frac{d^3 y}{dx^3} + 3 \frac{d^2 y}{dx^2} + 2x \left(\frac{dy}{dx} \right)^2 + e^y = \left[\log \left(2x + \frac{dy}{dx} \right) \right] \qquad (3.5)$$

the three first-order differential equations become

$$\frac{dy_{(1)}}{dx} = y_{(2)} \qquad \frac{dy_{(2)}}{dx} = y_{(3)}$$

$$\frac{dy_{(3)}}{dx} = \left(\frac{1}{\log y_{(1)}} \right) [\log (2x + y_{(2)}) - 3 y_{(3)} - 2x (y_{(2)})^2 - e^{y_{(1)}}] \qquad (3.6)$$

The problem of solving sets of ordinary differential equations of the type of Equation 3.1 is therefore reduced to the solution of sets of simultaneous equations of the type

$$\frac{dy}{dx} = y' = f(x, y) \qquad (3.7)$$

where the term on the right-hand side will generally be nonlinear and quite complex. As a result of the basic nature of digital computations a solution will be obtained in the form

$$y_i = y(x_i) \quad i = 0, 1, 2, \ldots \quad (3.8)$$

where x_i is the value of the independent variable x at the i-th step in the computation; x_i is separated from the value of x of the preceding step, x_{i-1}, by the step size $h = x_i - x_{i-1}$.

3.4 TAYLOR SERIES METHODS

Of fundamental importance in the numerical treatment of differential equations is the Taylor series expansion about a point. If, in treating a differential equation of the type of Equation 3.7, y_0 is known to be a solution of the equation at x_0, then the magnitude of the dependent variable y_1 at a neighboring point x_1 can be expressed as

$$y_1 = y_0 + h y_0^{(1)} + \frac{1}{2!} h^2 y_0^{(2)} + \frac{1}{3!} h^3 y_0^{(3)} + \frac{1}{4!} h^4 y_0^{(4)} + \cdots \quad (3.9)$$

where $h = x_1 - x_0$ and $y_0^{(n)} \equiv d^n y / dx^n$ evaluated at $x = x_0$.

Although Equation 3.9 is an infinite series, it generally converges sufficiently rapidly so that a limited number of terms suffices to express y_1 to within a specified tolerance. That is, the terms on the right side of Equation 3.9 become successively smaller and smaller until they become negligible.

In principle, Taylor series expansions represent a method for solving ordinary differential equations. For any specified x_0, y_0, and h, it remains only to find a sufficient number of higher derivatives of y and to insert these values in Equation 3.9. This process can then be repeated successively to find $y_2, y_3, \ldots$, etc.

Consider for example the applications of the Taylor series method to the equation governing an RLC circuit with $L = C = 1$, and $R = 2$, subject to a sinusoidal voltage excitation $e = \sin \omega t$. The equation governing such a circuit is

$$\frac{d^2 y}{dt^2} + 2 \frac{dy}{dt} + y = \sin \omega t \quad (3.10)$$

Assume that the initial conditions are

$$\frac{dy}{dt} = 2, \quad y = 3 \quad \text{at } t = 0$$

3.4/TAYLOR SERIES METHODS   55

Equation 3.10 is reduced to two first-order equations by introducing the auxiliary variable z such that

$$\frac{dy}{dt} = z \qquad \frac{dz}{dt} = \sin \omega t - 2z - y \qquad (3.11)$$

The Taylor series expression for y and z at point t_1 is

$$y_1 = y_0 + \Delta t \frac{dy}{dt} + \frac{\Delta t^2}{2!} \frac{d^2 y}{dt^2} + \frac{\Delta t^3}{3!} \frac{d^3 y}{dt^3} + \cdots$$

$$z_1 = z_0 + \Delta t \frac{dz}{dt} + \frac{\Delta t^2}{2!} \frac{d^2 z}{dt^2} + \frac{\Delta t^3}{3!} \frac{d^3 z}{dt^3} + \cdots \qquad (3.12)$$

where $\Delta t = t_1 - t_0$. The higher derivatives required in Equations 3.12 are found by successive differentiations of Equations 3.11 to be

$$y^{(2)} = z^{(1)} \qquad z^{(2)} = \omega \cos \omega t - 2z^{(1)} - y^{(1)}$$

$$y^{(3)} = z^{(2)} \qquad z^{(3)} = -\omega^2 \sin \omega t - 2z^{(2)} - y^{(2)} \qquad (3.13)$$

$$z^{(4)} = -\omega^3 \cos \omega t - 2z^{(3)} - y^{(3)}$$

$$\vdots \qquad \vdots$$

Employing the specified initial conditions and substituting in Equation 3.13, the numerical coefficients to be employed in Equation 3.12 are seen to be for $\omega = 1$:

$$\begin{aligned} y_0 &= 3 & z_0 &= 2 \\ y_0^{(1)} &= 2 & z_0^{(1)} &= -7 \\ y_0^{(2)} &= -7 & z_0^{(2)} &= 13 \\ y_0^{(3)} &= 13 & z_0^{(3)} &= -19 \end{aligned} \qquad (3.14)$$

If the spacing in the time domain, Δt, is selected to be 10^{-2} second, then substitution in Equation 3.12 yields

$$y_1 = 3 + (10^{-2} \times 2) - \left(\frac{10^{-4}}{2} \times 7\right) + \left(\frac{10^{-6}}{6} \times 13\right) + \cdots$$

$$z_1 = 2 - (10^{-2} \times 7) + \left(\frac{10^{-4}}{2} \times 13\right) - \left(\frac{10^{-6}}{6} \times 19\right) + \cdots \qquad (3.15)$$

It can be seen that as a result of the relatively small Δt the higher-order terms of Equation 3.15 rapidly become negligible. Had the frequency ω

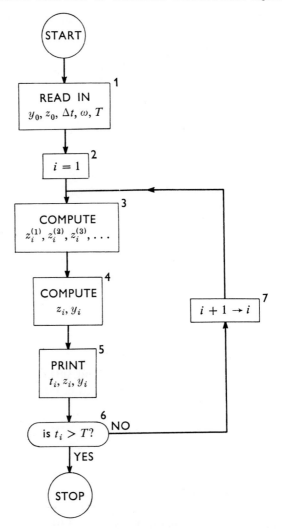

Figure 3.1

been specified to be 100 instead of 1, the higher-order terms in Equation 3.14 would have much larger magnitudes and the convergence of Equation 3.15 would be much less rapid. Under these conditions a smaller Δt might be indicated.

Figure 3.1 illustrates a computer flow chart for the solution of Equation 3.10 by the above method. The program has the generality of permitting the use of arbitrary initial conditions, Δt and ω. The following is the significance of each step in the computation:

BOX 1. Initial conditions including y_0 and z_0 for $t = t_0$ as well as Δt, ω, and the limit of the integration interval $t = T$ are read into the machine.
BOX 2. The index i is set to unity to indicate that y_1 and z_1 are to be determined.
BOX 3. The derivatives of z are computed by using formulas of the type of Equation 3.13. This step involves a number of substeps including the taking of sines, cosines, and exponents. To simplify the block diagram, these substeps have not been shown separately.
BOX 4. y_i and z_i are computed as specified by Equation 3.12.
BOX 5. The results for the i-th step of the calculation, that is, t_i, y_i, and z_i are printed out.
BOX 6. A decision is made whether to continue the computation by comparing t_i with T. If t_i is sufficiently large the computation comes to an end.
BOX 7. If the answer to the question posed in Box 6 is negative, the index i is increased by unity and the steps in Boxes 3, 4, 5, and 6 are repeated, using t_{i+1} instead of t_i.

Although self-starting and relatively straightforward, the above method has one major disadvantage which precludes its application to the solution of most engineering problems. This difficulty lies in the fact that the derivatives $z^{(2)}$, $z^{(3)}$, $z^{(4)}$, etc., must be calculated at each step of the integration procedure. In the above simple example, it was possible to obtain analytical expressions for these derivatives, so that their evaluation involved merely a substitution in an algebraic expression. In most applications this type of differentiation is very difficult or even impossible. For example, the excitation of a system is frequently an arbitrary function provided only in graphical or tabular form. Under these conditions the analytical differentiations as in Equation 3.13 must be replaced by numerical differentiation methods.

Whereas, in analytical methods, differentiation is a relatively simple operation, while integration entails considerable difficulties and is frequently impossible, precisely the reverse condition applies when numerical methods are involved. In both digital and analog computations the integration of a variable is a relatively straightforward process whereas the operation of differentiation is fraught with the possibilities of serious errors. Using a digital computer, differentiation implies the calculation of the rate of change of a variable. Derivatives therefore are relatively small numbers obtained by subtracting two relatively large numbers; a small error, such as a round-off error, in these large numbers is reflected as a relatively large percentage error in the derivative. Integration on the other hand implies addition and averaging and can therefore be considered as a smoothing operation. The direct application of the Taylor series method is therefore not practical. The Euler method presented below is an exception.

Nonetheless Taylor series expansions are basic to virtually all numerical integration methods because they provide a measure of the errors inherent in the calculation. The approximations involved in the various methods to be discussed can generally be regarded as the truncation of the Taylor series expansion after a certain number of terms. The magnitude of the first several terms thus ignored therefore provides a measure of the *truncation error*. The decision to use a more complex and hence more accurate approximation, or the use of a smaller step size, can be based on an estimate of this truncation error.

For example, Euler's method is based on ignoring all terms in the Taylor series containing second or higher powers of h. That is,

$$y_1 = y_0 + h \frac{dy_0}{dt} \tag{3.16}$$

Evidently the predicted value for y_1 will be in error ϵ by

$$\epsilon = \frac{h^2}{2!} y_0^{(2)} + \frac{h^3}{3!} y_0^{(3)} + \cdots \tag{3.17}$$

A similar error will be introduced at each step in the calculation. Note, however, that whereas in Equation 3.16 the only error involved is that represented by Equation 3.17 this will not be true in the calculation of y_2 or y_3, etc. While in Equation 3.16 the term y_0 is known accurately, the corresponding terms in subsequent calculation become more and more inaccurate. The truncation errors therefore tend to accumulate in the course of the calculation and eventually reach a point exceeding the tolerable error limit. Nonetheless, because of its simplicity, the Euler method is widely used in applications in which this error growth can be closely monitored.

3.5 THE RUNGE-KUTTA METHODS

Over the years a wide variety of numerical methods have been developed for the purpose of retaining the advantageous self-starting feature of the Taylor series method described above while obviating its major disadvantage—that of requiring the calculation of higher derivatives. The most important and widely used of these methods are known as the Runge-Kutta methods, although in addition to Runge and Kutta, investigators such as Heun, Nystrom, and many others have made important contributions. In essence, the basic approach involved in these methods is to treat an equation of the type

$$y' = f(x, y) \quad \text{given} \quad y = y_0 \quad \text{at} \quad x = x_0 \tag{3.18}$$

by calculating dy/dx not only at y_0 but also at three or more closely adjacent points in the x and y domains. The solution for the first step of the calculation, i.e., y_1, is obtained by means of a weighted sum of these expressions for dy/dx. Evidently, a chief disadvantage of the method is the fact that the functional relationship expressed by the right side of Equation 3.18 must be evaluated for four or more values of x and y at each step of the calculation. For complex $f(x, y)$ these calculations can be most time-consuming and constitute a major stumbling block. Nonetheless, as already stated, among the various self-starting methods available, the Runge-Kutta methods constitute the most powerful and widely used approach.

In all these methods the expression for the dependent variable at the $(i + 1)$ step of the computation takes the form

$$y_{i+1} = y_i + \mu_0 k_0 + \mu_1 k_1 + \mu_2 k_2 + \cdots + \mu_p k_p \tag{3.19}$$

where k_i are obtained by substitution in Equation 3.18 and μ_i are weighting factors to be determined such that $\mu_0 + \mu_1 + \cdots + \mu_p = 1$. The values k_i are defined by the following set of equations

$$\begin{aligned} k_0 &= hf(x_i, y_i) \\ k_1 &= hf(x_i + \alpha_1 h, y_i + \beta_{1,0} k_0) \\ k_2 &= hf(x_i + \alpha_2 h, y_i + \beta_{2,0} k_0 + \beta_{2,1} k_1) \\ k_3 &= hf(x_i + \alpha_3 h, y_i + \beta_{3,0} k_0 + \beta_{3,1} k_1 + \beta_{3,2} k_2) \end{aligned} \tag{3.20}$$

$$\cdot$$
$$\cdot$$
$$\cdot$$

$$k_p = hf(x_i + \alpha_p h, y_i + \beta_{p,0} k_0 + \beta_{p,1} k_1 + \cdots + \beta_{p,p-1} k_{p-1})$$

where h is the step size $(x_{i+1} - x_i)$ and α_i and β_{ij} are to be determined. Each of the terms $k_0 \cdots k_p$ in Equation 3.19 is proportional to $f(x, y)$ in the range x_i to x_{i+1}. In essence, therefore, the Runge-Kutta methods involve the calculation of the first derivative of y p times for each step in the x domain rather than the evaluation of higher derivatives as is required in the Taylor series methods. Note that at the start of the calculation x_i and y_i are known so that k_0 can be determined directly. The result of this calculation is then employed to calculate k_1, permitting the calculation of k_2, etc. The problem now becomes to select values of μ, α, and β so that Equation 3.19 becomes identical to the first several terms of the Taylor series Equation 3.9. This task, though straightforward, involving the expansion in Taylor series of functions of two variables and a considerable amount of algebraic manipulation, is most laborious for the general case. It develops, moreover, that the problem of selecting these coefficients has no unique solution—that is, some of the coefficients may be chosen arbitrarily. The derivation of the pertinent formulas will not be detailed here but can be found in the references.[1,2,3] Rather, several of the more basic Runge-Kutta methods will be described, followed by a more general program for the application of the method to sets of simultaneous differential equations.

The Runge-Kutta methods are termed third-order, fourth-order, fifth-order, etc., methods, depending on the number of terms in the Taylor series Equation 3.9 which are matched by Equation 3.19. Thus a third-order method matches all terms up to and including that proportional to h^3; a fourth-order method matches all terms up to and including h^4, etc. Evidently, the higher the order of the method, the smaller the error made at each step but, on the other hand, the more complex and lengthy the calculation required to attain successive values y_i. In practice, Runge-Kutta methods higher than the fifth order are rarely employed.

One example of a third-order Runge-Kutta formula is given by the expression

$$y_{i+1} = y_i + \tfrac{1}{6}k_0 + \tfrac{2}{3}k_1 + \tfrac{1}{6}k_2 \tag{3.21}$$

where

$$\begin{aligned} k_0 &= hf(x_i, y_i) \\ k_1 &= hf(x_i + \tfrac{1}{2}h, y_i + \tfrac{1}{2}k_0) \\ k_2 &= hf(x_i + h, y_i + 2k_1 - k_0) \end{aligned} \tag{3.22}$$

An alternative third-order formula obtained by a different selection of some of the arbitrary constants among μ, α, and β is given by

$$y_{i+1} = y_i + \tfrac{1}{4}k_0 + 0k_1 + \tfrac{3}{4}k_2 \tag{3.23}$$

where

$$\begin{aligned} k_0 &= hf(x_i, y_i) \\ k_1 &= hf(x_i + \tfrac{1}{3}h, y_i + \tfrac{1}{3}k_0) \\ k_2 &= hf(x_i + \tfrac{2}{3}h, y_i + \tfrac{2}{3}k_1) \end{aligned} \tag{3.24}$$

Although Equations 3.21 and 3.23 have approximately the same accuracy with error terms proportional to h^4 and higher powers of step size, one or the other may have certain computational advantages in specific cases.

A very familiar and widely used fourth-order Runge-Kutta formula is of the form
$$y_{i+1} = y_i + \tfrac{1}{6}k_0 + \tfrac{1}{3}k_1 + \tfrac{1}{3}k_2 + \tfrac{1}{6}k_3 \tag{3.25}$$
where
$$\begin{aligned} k_0 &= hf(x_i, y_i) \\ k_1 &= hf(x_i + \tfrac{1}{2}h, y_i + \tfrac{1}{2}k_0) \\ k_2 &= hf(x_i + \tfrac{1}{2}h, y_i + \tfrac{1}{2}k_1) \\ k_3 &= hf(x_i + h, y_i + k_2) \end{aligned} \tag{3.26}$$

In the special case where f does not contain the dependent variable y, this method is equivalent to the application of the familiar Simpson rule.

An example of a fifth-order Runge-Kutta formula is given by
$$y_{i+1} = y_i + \tfrac{1}{192}[23k_0 + 0k_1 + 125k_2 + 0k_3 - 81k_4 + 125k_5] \tag{3.27}$$
where
$$\begin{aligned} k_0 &= hf(x_i, y_i) \\ k_1 &= hf(x_i + \tfrac{1}{3}h, y_i + \tfrac{1}{3}k_0) \\ k_2 &= hf(x_i + \tfrac{2}{5}h, y_i + \tfrac{6}{25}k_1 + \tfrac{4}{25}k_0) \\ k_3 &= hf(x_i + h, y_i + \tfrac{15}{4}k_2 - \tfrac{12}{4}k_1 + \tfrac{1}{4}k_0) \\ k_4 &= hf(x_i + \tfrac{2}{3}h, y_i + \tfrac{8}{81}k_3 + \tfrac{50}{81}k_2 + \tfrac{90}{81}k_1 + \tfrac{6}{81}k_0) \\ k_5 &= hf(x_i + \tfrac{4}{5}h, y_i + \tfrac{8}{75}k_3 + \tfrac{10}{75}k_2 + \tfrac{36}{75}k_1 + \tfrac{6}{75}k_0) \end{aligned} \tag{3.28}$$

It becomes clear that the point of diminishing returns is soon reached, beyond which a desired increase in accuracy can more readily be attained by decreasing h than by increasing the order of the method.

From the point of view of constructing a digital computer program, the basic Runge-Kutta methods present no special problems. The chief difficulty lies in the calculation of the k's at each step. To conserve memory space, a usual practice is to designate the obtaining of $f(x, y)$ given x and y as a special subroutine which is employed in calculating each of the k's. A computer flow chart for a fourth-order Runge-Kutta method is shown in Figure 3.2. The calculation involves the following steps:

BOX 1. The specified initial conditions x_0 and y_0, the step size h, and the final value of x for which a solution is sought, X_L, are fed into the machine.

BOX 2. The index i is set to zero.

BOXES 3–6. The terms k_0, k_1, k_2, and k_3 are calculated by direct substitution of previously calculated values of x, y, and k.

BOX 7. y_{i+1} is calculated by substituting the values of k obtained in Boxes 3 through 6 in Equation 3.25.

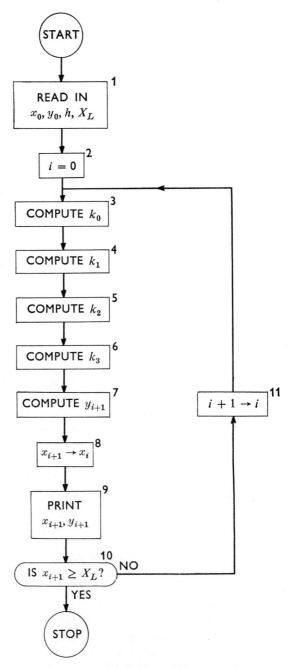

Figure 3.2

BOX 8. The value of x is increased by h.
BOX 9. The results of this step of the calculation are printed out.
BOX 10. A decision is made whether the calculation is complete.
BOX 11. If more data are to be obtained the index i is advanced by unity and Boxes 3 through 10 are repeated as long as necessary.

One advantageous feature of the above methods is that the net interval h can be changed without difficulty after any step in the calculation. Thus the programmer can reduce the net size for those values of the independent variable for which abrupt changes in the dependent variable and its derivatives are expected. In an engineering problem this is occasionally a practical procedure. On the other hand, it is a distinct weakness of the Runge-Kutta methods that no simple estimates of the truncation error can be made. In the non-self-starting methods described in Sections 3.8 and 3.9, estimates of the truncation error can be obtained directly from calculated data. This makes it feasible to adjust h in the course of the calculation in order to keep this error within specified limits; this is impossible in the case of the Runge-Kutta methods.

3.6 RUNGE-KUTTA METHODS FOR SYSTEMS OF EQUATIONS

The above method is readily generalized to apply to a set of m simultaneous first-order differential equations. For example, to apply the fourth-order formula discussed above to two simultaneous equations

$$\frac{dy}{dx} = f(x, y, z)$$
$$\frac{dz}{dx} = g(x, y, z) \qquad (3.29)$$

successive values of the dependent variables y and z are calculated according to the formulas:

$$y_{i+1} = y_i + \tfrac{1}{6}(k_0 + 2k_1 + 2k_2 + k_3)$$
$$z_{i+1} = z_i + \tfrac{1}{6}(l_0 + 2l_1 + 2l_2 + l_3) \qquad (3.30)$$

where

$k_0 = hf(x_i, y_i, z_i)$ $\qquad l_0 = hg(x_i, y_i, z_i)$
$k_1 = hf(x_i + \tfrac{1}{2}h, y_i + \tfrac{1}{2}k_0, z_i + \tfrac{1}{2}l_0)$ $\qquad l_1 = hg(x_i + \tfrac{1}{2}h, y_i + \tfrac{1}{2}k_0, z_i + \tfrac{1}{2}l_0)$
$k_2 = hf(x_i + \tfrac{1}{2}h, y_i + \tfrac{1}{2}k_1, z_i + \tfrac{1}{2}l_1)$ $\qquad l_2 = hg(x_i + \tfrac{1}{2}h, y_i + \tfrac{1}{2}k_1, z_i + \tfrac{1}{2}l_1)$
$k_3 = hf(x_i + h, y_i + k_2, z_i + l_2)$ $\qquad l_3 = hg(x_i + h, y_i + k_2, z_i + l_2)$

$$(3.31)$$

For large sets of equations, the large number of numerical values which must be memorized and combined at each step in the calculation may well exceed available core memory capacity. Accordingly, special methods have been developed to optimize the application of Runge-Kutta methods for the utilization of high-speed digital computers. Such programs are characterized by requiring a minimum number of storage registers, and minimize the danger of accumulating round-off errors. One such procedure was developed by Gill[6] and is described by Romanelli.[5]

The basic Runge-Kutta formula employed by Gill is

$$y_{i+1} = y_i + \tfrac{1}{6}k_1 + \tfrac{1}{3}(1 - \sqrt{\tfrac{1}{2}})k_2 + \tfrac{1}{3}(1 + \sqrt{\tfrac{1}{2}})k_3 + \tfrac{1}{6}k_4 \quad (3.32)$$

where

$$\begin{aligned}
k_1 &= hf(x_i, y_i) \\
k_2 &= hf(x_i + \tfrac{1}{2}h, y_i + \tfrac{1}{2}k_1) \\
k_3 &= hf[x_i + \tfrac{1}{2}h, y_i + (-\tfrac{1}{2} + \sqrt{\tfrac{1}{2}})k_1 + (1 - \sqrt{\tfrac{1}{2}})k_2] \\
k_4 &= hf[x_i + h, y_i + (-\sqrt{\tfrac{1}{2}})k_2 + (1 + \sqrt{\tfrac{1}{2}})k_3]
\end{aligned} \quad (3.33)$$

To increase the efficiency of the computational process, subsidiary quantities l_1, l_2, l_3, and l_4 as well as q_0, q_1, q_2, q_3, and q_4 are introduced as follows:

$$\begin{aligned}
l_1 &= y_i + \tfrac{1}{2}(k_1 - 2q_0) & q_1 &= q_0 + 3[\tfrac{1}{2}(k_1 - 2q_0)] - \tfrac{1}{2}k_1 \\
l_2 &= l_1 + (1 - \sqrt{\tfrac{1}{2}})(k_2 - q_1) & q_2 &= q_1 + 3[(1 - \sqrt{\tfrac{1}{2}})(k_2 - q_1)] \\
& & & \quad - (1 - \sqrt{\tfrac{1}{2}})k_2 \\
l_3 &= l_2 + (1 + \sqrt{\tfrac{1}{2}})(k_3 - q_2) & q_3 &= q_2 + 3[(1 + \sqrt{\tfrac{1}{2}})(k_3 - q_2)] \\
& & & \quad - (1 + \sqrt{\tfrac{1}{2}})k_3 \\
l_4 &= l_3 + \tfrac{1}{6}(k_4 - 2q_3) & q_4 &= q_3 + 3[(\tfrac{1}{6})(k_4 - 2q_3)] - \tfrac{1}{2}k_4
\end{aligned}$$
$$(3.34)$$

An examination of the equation for l_4 in the above set of equations reveals that if this expression is developed by substituting the defined quantities for l_3 and q_3, l_4 becomes identical to y_{i+1} as defined by Equation 3.32. Therefore, l_4 represents the solution for the $i + 1$ step of the calculation. In this procedure, the quantities q_0 and q_4 are introduced solely to compensate round-off errors accumulated in each step of the calculation. Initially, q_0 is given a value of 0. If no round-off errors are committed in the first step of the calculation, q_4 will turn out to be 0. If this is not the case, then q_4 is used in place of q_0 in the next step.

As an example of the application of the above method, consider the system of M simultaneous first-order differential equations defined by

$$y'_{(m)} = f_m(y_{(0)}, y_{(1)}, y_{(2)}, \ldots, y_{(m)}) \quad \text{where} \quad m = 0, 1, 2, \ldots, M$$
$$(3.35)$$

3.6/RUNGE-KUTTA METHODS FOR SYSTEMS OF EQUATIONS

In this case, an additional equation

$$y'_{(0)} = f_0 = 1 \tag{3.36}$$

is included in the set of Equation 3.35 to simplify the calculation process. To describe the computational procedure, it is necessary to employ a triple-subscript notation. Accordingly, let m identify the equation among the set of equations of Equation 3.35; let i specify the step in the computation, i.e., the number of steps along the domain of the independent variable which have been covered; and let j represent the subscripts 1, 2, 3, and 4 in Equation 3.34.

The calculation therefore proceeds in the following general manner:

STEP 1. $i = 0$, indicating that a solution is to be obtained for $x = x_0 + h$.
STEP 2. k_1 is calculated for each of the m simultaneous equations. This involves only direct substitution in the specified Equations 3.35.
STEP 3. l_1 and q_1 are calculated in accordance with Equation 3.34 by substituting previously calculated values. As indicated above, q_0 is made 0 for this first step, but during any step i it takes the value obtained for q_4 during the $i - 1$ step.
STEP 4. Steps 2 and 3 are now repeated for $j = 2, 3,$ and 4.
STEP 5. The values l_4 obtained for each of the m equations are printed out and constitute the solution.

Figure 3.3 constitutes a partial flow chart for the application of the Gill method. In this figure:

BOX 1. Specified numerical data including the initial conditions for all the equations, the step size h, and the domain of the independent variable x are applied.
BOXES 2–4. Initial settings for the i, j, and m indices are introduced.
BOX 5. The k's are calculated by substitution in the specified differential equations.
BOXES 6–7. The calculation in Box 5 is repeated for each of the M simultaneous equations.
BOX 8. The index m is reset to 0.
BOX 9. l and q for each of the n simultaneous equations are computed. This involves successive applications of Equations 3.34.
BOXES 10–11. This process is repeated for all the simultaneous equations.
BOXES 12–13. The calculations represented by Boxes 4 to 11 are repeated for $j = 2, j = 3, j = 4$.
BOX 14. The stored value of the independent variable x is increased from x_i to x_{i+1}.

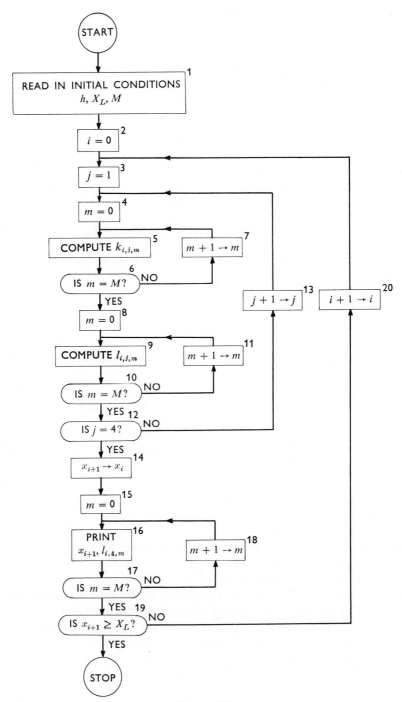

Figure 3.3

BOX 15. The m index is reset to zero.

BOX 16. The solution for this the i-th step in the calculation is printed out. The read-out consists of x_i and the solution y_i for each of the equations (note that this value of y corresponds to the $l_{i,4}$, obtained in Box 9).

BOXES 17–18. This process is repeated until data corresponding to all the simultaneous equations have been read out.

BOXES 19–20. The process included by Boxes 2 through 18 is repeated until the entire specified interval of x has been traversed. The computer then stops.

An estimate of the computer requirements for the application of the Gill method to a system of M simultaneous equations can be obtained from the following considerations. $(M + 1)$ words of memory spaces are required for each l_{mj}, k_{mj}, and q_{mj}. The program itself requires approximately 50 words so that

$$3(M + 1) + 50 + B \text{ words}$$

of memory space are required where B represents the number of words required for the calculation of the k's in Box 5. B will evidently depend on the specific nature of the equations and to what extent they are similar to each other.

3.7 NONITERATIVE FORWARD-INTEGRATION METHODS

An important class of numerical methods for the solution of differential equations and systems of differential equations involves the application of formulas of the type

$$y_{i+1} = a_i y_i + a_{i-1} y_{i-1} + \cdots + a_{i-p} y_{i-p}$$
$$+ h(b_{i+1} y'_{i+1} + b_i y'_i + \cdots + b_{i-p} y'_{i-p}) \quad (3.37)$$

Provided the coefficient b_{i+1} is equal to zero, all the terms on the right-hand side of Equation 3.37 are known at a given step in the calculation. The solution can then be obtained by a linear combination of the solutions for the p preceding steps. This method can therefore not be used at the beginning of the solution procedure when only the initial conditions are available. Instead, a method such as the Runge-Kutta method described in the preceding sections must be employed for the first p steps.

Since the forward-integration methods require fewer calculations at each step along the independent variable, they have the advantage of greatly reduced computing time for a specified accuracy. The accuracy of methods employing Equation 3.37 is considerably increased by including the term $y_{i+1}^{(1)}$; that is, when $b \neq 0$. Since, for any given step in the calculation, $y_{i+1}^{(1)}$ is not known, one of a variety of prediction-correction or iterative procedures must be employed These are considered in Sections 3.8 and 3.9.

Forward-integration equations are derived from well-known interpolation formulas. The so-called Adams-Bashforth predictors are formulas expressing y_{i+1} in terms of y_{i-p} and values for y' corresponding to y_i', $y_{i-1}', \ldots, y_{i+1-p}'$. It is customary in this formulation to include an explicit expression for an approximate value of the truncation error. Thus for p equal to 1, 3, and 5,

(a) $y_{i+1} = y_{i-1} + 2hy_i' + \dfrac{h^3}{3} y^{(3)}(\epsilon)$

(b) $y_{i+1} = y_{i-3} + \tfrac{4}{3}h(2y_i' - y_{i-1}' + 2y_{i-2}') + \tfrac{14}{45}h^5 y^{(5)}(\epsilon)$ (3.38)

(c) $y_{i+1} = y_{i-5} + \tfrac{3}{10}h(11y_i' - 14y_{i-1}'$
$\quad + 26y_{i-2}' - 14y_{i-3}' + 11y_{i-4}') + \tfrac{41}{140}h^7 y^{(7)}(\epsilon)$

where ϵ lies in the range x_{i-p} and x_{i+1} The errors in Equation 3.38a, b, and c are proportional respectively to the third, fifth, and seventh derivative of the dependent variable. Equations 3.38 suffice for most forward integration purposes. Only very rarely are the values for p greater than 5 employed. Formulas for which $p = 2$, $p = 4$, and $p = 6$ are never used for, as Hildebrand[2] demonstrates, they have no greater accuracy than formulas for which $p = 1$, $p = 3$, and $p = 5$, respectively.

The forward-integration methods described above have the advantage of simplicity, ease of programming, and relatively short computing time. At each step in the calculation, only one value of $y' = f(x, y)$ need be obtained. At the same time the method is free from the necessity of iterating as in the methods described in subsequent sections. On the other hand, the accuracy of the forward-integration methods is relatively low and the chances of error accumulation relatively high.

The computer flow chart corresponding to the application of Equation 3.38b to the solution of $y' = f(x, y)$ is shown in Figure 3.4. The operation of this program may be summarized as follows:

BOX 1. Specified initial conditions, step size h, and the limit of the x domain X_L are read into the machine.

BOX 2. A self-starting method such as for example the Runge-Kutta methods described in the preceding section is employed to obtain the necessary set of starting values. These include in this case three values of y and y' in addition to the specified initial conditions.

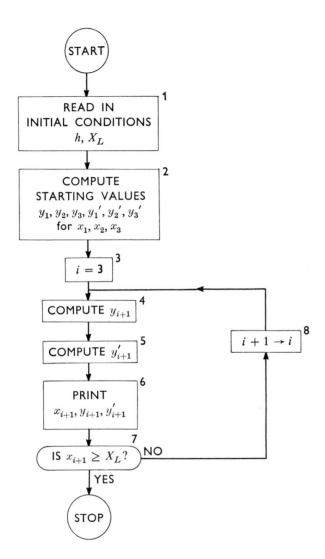

Figure 3.4

BOX 3. The index i identifying the step along the x domain is set at 3.
BOX 4. y_{i+1} is now calculated employing Equation 3.38b.
BOX 5. y'_{i+1} is now calculated by insertion of y_{i+1} and x_{i+1} in the equation being solved. (Equation 3.7.)
BOX 6. The solution for the i-th step in the calculation is printed out.
BOX 7. x_{i+1} is examined to determine whether any more calculations are necessary. If $x = X_L$ the calculation is stopped; if not,
BOX 8. The index i is increased by unity.

The above method is readily extended to systems of simultaneous first-order equations.

3.8 PREDICTOR-CORRECTOR METHODS

Superior accuracy can be obtained in methods analogous to those in the preceding section if the coefficient b_{i+1} in Equation 3.37 is not made equal to 0; that is, if a term proportional to y'_{i+1} is retained in calculating y_{i+1}. For example, when $p = 1, 3$, and 5 respectively, it can be shown that

(a) $y_{i+1} = y_{i-1} + \dfrac{h}{3}(y'_{i+1} + 4y'_i + y'_{i-1}) - \tfrac{1}{90}h^5 y^{(5)}(\epsilon)$

(b) $y_{i+1} = y_{i-3} + \tfrac{2}{45}h(7y'_{i+1} + 32y'_i + 12y'_{i-1}$
$\qquad + 32y'_{i-2} + 7y'_{i-3}) - \tfrac{8}{945}h^7 y^{(7)}(\epsilon)$ (3.39)

(c) $y_{i+1} = y_{i-5} + \tfrac{3}{10}h(y'_{i+1} + 5y'_i + y'_{i-1} + 6y'_{i-2} + y'_{i-3}$
$\qquad + 5y'_{i-4} + y'_{i-5}) - \tfrac{1}{140}h^7 y^{(7)}(\epsilon) - \tfrac{1}{1400}h^9 y^{(9)}(\epsilon)$

Equation 3.39a is sometimes termed the Adams-Moulton formula. A comparison of Equations 3.39a and 3.38b reveals that although these two equations are of the same order, involving a like number of terms and both having error terms proportional to the fifth derivative of y, Equation 3.39a is considerably more accurate; in particular, the error term in Equation 3.38b is 28 times as large as that in Equation 3.39a. A similar comparison of Equations 3.39b and 3.38c, equations of like order, indicates that the error term, proportional to the seventh derivative of y, is approximately 35 times as large in Equation 3.38c. It is therefore apparent that the inclusion of a term proportional to y'_{i+1} effects an appreciable increase in accuracy.

The chief difficulty in employing integration formulas of the type of Equations 3.39 is that the term proportional to y'_{i+1} is not known at the time that y_{i+1} is to be calculated. It is therefore necessary to employ iterative, trial-and-error techniques. More specifically it is necessary to employ a forward-integration formula to *predict* a value for y_{i+1}, and subsequently to *correct* this value by using the more accurate formulas of the type of Equations 3.39. For this reason, this approach is termed a predictor-corrector method and takes the following general steps:

STEP 1. A self-starting method, such as the Runge-Kutta method, is employed to obtain the first n starting values required to apply a forward-integration formula.
STEP 2. Using a formula of the type of Equations 3.38, a first prediction of y_{i+1} is made.
STEP 3. The value for y_{i+1} calculated in Step 2 is inserted in the specified first-order differential equation $y' = f(x, y)$ to obtain a value for y'_{i+1}.
STEP 4. Using a formula of the type of Equations 3.39, a new value for y_{i+1} is calculated.
STEP 5. Steps 3 and 4 are repeated until successive values of y_{i+1} are sufficiently similar. Provided the step size h is sufficiently small, each sequence of Steps 3 and 4 can be expected to reduce the difference between successive values of y_{i+1}.

A number of practical predictor-corrector methods are available. The difference between these methods lies primarily in which combination of predictor and corrector formula is employed. A very widely used method identified with Milne[4] employs predictor and corrector equations of the same order. Thus Equation 3.38b might be used as the predictor and Equation 3.39a as the corrector; or Equation 3.38c might be used as the predictor and Equation 3.39b as the corrector. This approach has the important advantage that the truncation error inherent in the method can be estimated very readily. This, in turn, makes it feasible to adjust h, the step size, in the course of the calculation to control truncation errors in any specified fashion. Provided the predictor and corrector equations are both of the same order, the difference between the predicted and corrected value of y_{i+1} can be employed to obtain an estimate of the truncation error. Consider, for example, a predictor-corrector formula consisting of Equations 3.38b and 3.39a. Then y_{i+1} as predicted by Equation 3.38b will have a truncation error T_p:

$$T_p = \tfrac{28}{90} h^5 y^{(5)}(\epsilon_p) \tag{3.40}$$

where ϵ_p is in the range between x_{i-3} and x_{i+1}. Similarly, the corrected value of y_{i+1} as obtained from Equation 3.39a will have a truncation error T'_p:

$$T_c = -\tfrac{1}{90}h^5 y^{(5)}(\epsilon_c) \qquad (3.41)$$

where ϵ_c is in the range x_{i-1} to x_{i+1}. Assuming that the fifth derivative of y remains relatively constant over the range x_{i-3} to x_{i-1}, the difference between the corrected value $(y_{i+1})_c$ and the predicted value $(y_{i+1})_p$ is approximately

$$(y_{i+1})_c - (y_{i+1})_p = \tfrac{1}{90}h^5 y^{(5)}\epsilon_c + \tfrac{28}{90}h^5 y^{(5)}\epsilon_p = \tfrac{29}{90}h^5 y^{(5)}\epsilon_p \qquad (3.42)$$

The difference between the predicted and corrected value for y at any step in the calculation therefore constitutes a direct measure of the truncation error. This difference can be examined at each step in the calculation, and if it is too large, the calculation can be repeated with a reduced h.

A program for the solution of $y' = f(x, y)$ using a predictor-corrector method is illustrated in Figure 3.5. This program does not include provision for varying the step size h, since that point is considered in more detail in Section 3.9. The following is the significance of the various steps included in the flow chart:

BOX 1. The specified initial conditions, the step size h, the domain of the independent variable X_L, and a constant α are read into the computer. The α constitutes a specification of the point at which the iteration at each step will be considered to have converged sufficiently. This value should evidently be greater than the round-off errors to be expected.

BOX 2. Starting values corresponding to y and y' for the first three increments of x are computed employing a self-starting formula such as, for example, a Runge-Kutta method.

BOX 3. The index i is set to 3 indicating that y_4 is to be calculated.

BOX 4. The predicted value for y_{i+1} is computed employing a predictor formula such as Equation 3.38b.

BOX 5. Then y'_{i+1} is computed by inserting $(y_{i+1})_p$ in $y' = f(x, y)$.

BOX 6. A corrected value for y_{i+1} is computed employing a corrector formula such as Equation 3.39a.

BOX 7. The difference between the computed and corrected value is computed. Since this value is proportional to the truncation error, it could be employed to control the step size h.

BOX 8. The difference computed in Box 7 is compared with α.

BOX 9. If the difference computed in Box 7 is too large, the iterative process must be continued and the corrected value of y calculated in Box 9 is applied to Box 5 in place of the previously predicted value for y_{i+1}.

3.8/PREDICTOR-CORRECTOR METHODS 73

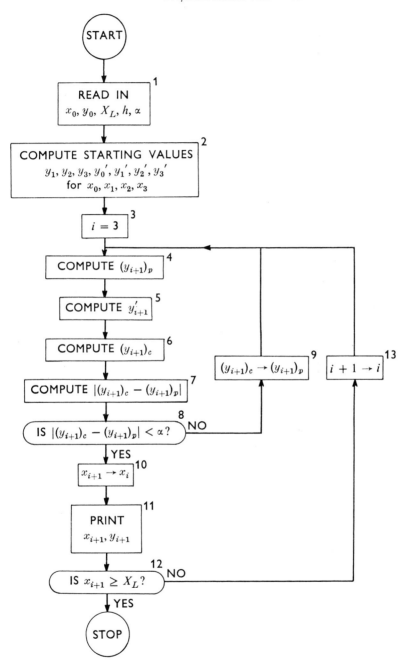

Figure 3.5

BOX 10. If the answer to the question in Box 8 is affirmative the value in the x register is advanced one step.
BOX 11. The results of the calculation, that is x_{i+1} and y_{i+1}, are printed out.
BOX 12. Then x_{i+1} is examined to determine whether the specified range of x has been covered. If the answer is affirmative, the computation comes to an end.
BOX 13. If the answer is negative the index i is advanced by unity and the operations in Boxes 4 to 12 are repeated.

An important disadvantage of the predictor-corrector methods employing Equations 3.38 and 3.39 is that they tend to be computationally unstable. Computational instability, a topic discussed in more detail in Section 3.11, implies that any error made at the i-th step of the calculation is transmitted with an increased magnitude in succeeding steps. Thus a small truncation or round-off error in step i may have an intolerably large effect at step $i + 100$. Milne's method is therefore particularly useful where only a relatively small number of steps in the x domain needs be taken.

To obviate this accumulation and propagation of errors, it is frequently advantageous to replace the corrector formulas of Equations 3.39 by corrector formulas which lead to computationally stable solutions. In these formulas, accuracy is sacrificed to a certain extent for the sake of stability. One such stable method due to Hamming[7] uses for the corrector

$$y_{i+1} = \tfrac{9}{8}y_i - \tfrac{1}{8}y_{i-2} + \tfrac{3}{8}h(y'_{i+1} + 2y'_i - y'_{i-1}) - \tfrac{1}{40}h^5 y^{(5)}(\epsilon) \quad (3.43)$$

where ϵ is in the range x_{i-2} to x_{i+1}. This formula is derived by setting $p = 2$ in Equation 3.37 and by employing a Taylor series to determine the coefficients a_i, a_{i-1}, a_{i-2}, b_{i+1}, b_i, and b_{i-1}. If it is desired to match only the terms up to and including h^4 in the Taylor series, one of these six coefficients can be chosen arbitrarily. By setting $a_{i-1} = 0$, the resulting predictor-corrector method becomes stable for virtually all practical values of h. On the other hand, the truncation error is more than twice as large as the error in the corresponding Equation 3.39a. For equivalent per-step accuracy, it is therefore necessary to employ a somewhat smaller h when using Equation 3.43.

3.9 PREDICTOR-MODIFIER-CORRECTOR METHODS

One of the difficulties in applying the predictor-corrector methods to complex problems lies in the necessity for successive iterations. As a part of each iterative substep, it is necessary to evaluate $y' = f(x, y)$. Where $f(x, y)$ is a complex or lengthy expression and where relatively large systems of simultaneous equations must be solved, the computing time required for this method may very well become excessive. In fact, one of the principal advantages of the forward-integration and predictor-corrector methods over the Runge-Kutta methods was stated to be that they obviate the necessity for numerous calculations of $f(x, y)$ at each step in the x domain. If numerous iterations are required in a predictor-corrector method, this advantage is lost. Accordingly, compromise methods have been developed which effect a considerable improvement of accuracy over the noniterative forward-integration methods without themselves requiring iterations. As described by Ralston[5] these methods involve only one application of the corrector formula. The predicted value employed in the corrector formula is first modified, however, by taking into account the truncation error committed in the preceding step of the calculation.

For example, in place of Milne's method using Equations 3.38b and 3.39a, the following predictor-modifier-corrector set is employed:

Predictor: $\quad p_{i+1} = y_{i-3} + \frac{4}{3}h(2y_i' - y_{i-1}' + 2y_{i-2}')$ (a)

Modifier: $\quad m_{i+1} = p_{i+1} - \frac{28}{29}(p_i - c_i)$ (b)

$\quad\quad\quad\quad m_{i+1}' = f(x_{i+1}, m_{i+1})$ (c) $\quad\quad$ (3.44)

Corrector: $\quad c_{i+1} = y_{i-1} + \dfrac{h}{3}(m_{i+1}' + 4y_i' + y_{i-1}')$ (d)

Final value: $\quad y_{i+1} = c_{i+1} + \frac{1}{29}(p_{i+1} - c_{i+1})$ (e)

In these equations p_{i+1}, m_{i+1}, and c_{i+1} represent, respectively, the predicted, modified, and corrected values for y_{i+1}. These are the successive preliminary estimates employed in calculating the final value. Equations 3.44a and 3.44d correspond to the standard Milne predictor-corrector equations. Equation 3.44d is derived from Equations 3.40, 3.41, and 3.42. Thus,

$$p_i - c_i \approx -\tfrac{29}{28}T_p \quad\quad (3.45)$$

so that the term $\frac{28}{29}(p_i - c_i)$ is a measure of the truncation error at step i. The assumption is now made that this truncation error remains approximately constant from step i to step $i + 1$, so that Equation 3.45 can be employed to minimize or counteract the truncation error incurred in applying Equation 3.44a. Similarly, the corrected value c_{i+1} is modified by the term $\frac{1}{29}(p_{i+1} - c_{i+1})$ to counteract the truncation error incurred in the corrector formula. Equations 3.44 constitute the entire sequence of calculation for each step in the x domain—hence, no iterations are required.

If we desire to assure computational stability at some expense of accuracy for a given h, Equation 3.43 may be employed in place of Equation 3.39a. The predictor-modifier-corrector equations then become:

Predictor: $\quad p_{i+1} = y_{i-3} + \frac{4}{3}h(2y_i' - y_{i-1}' + 2y_{i-2}')$ \hfill (a)

Modifier: $\quad m_{i+1} = p_{i+1} - \frac{112}{121}(p_i - c_i)$ \hfill (b)

$\quad m_{i+1}' = f(x_{i+1}, m_{i+1})$ \hfill (c) \quad (3.46)

Corrector: $\quad c_{i+1} = \frac{9}{8}y_i - \frac{1}{8}y_{i-2} + \frac{3}{8}h(m_{i+1}' + 2y_i' - y_{i-1}')$ \hfill (d)

Final value: $\quad y_{i+1} = c_{i+1} + \frac{9}{121}(p_{i+1} - c_{i+1})$ \hfill (e)

A computer flow chart for the application of the predictor-modifier-corrector method to a first-order differential equation is shown in Figure 3.6. In this diagram:

BOX 1. The initial conditions y_0 and x_0, the limit of the x domain X_L, and the step size h are read into the machine.

BOX 2. Using a self-starting method such as the Runge-Kutta method, values for y and y' corresponding to the first three steps x_1, x_2, and x_3 are calculated.

BOX 3. Using a formula such as 3.46a, a predicted value for y for the third step is calculated.

BOX 4. Using a formula such as formula 3.39b, a corrected value for step y_3 is calculated (p_3 and c_3 are necessary to permit the formation of m_4 in the subsequent step).

BOX 5. The index i is set to 3, indicating that values corresponding to the fourth increment in x are to be calculated.

BOX 6. The value of x stored in the x register is increased by one increment. Thus in the first step of the calculation it will change from x_3 to x_4.

BOX 7. The predicted value for y_{i+1} is calculated by using Equation 3.46a.

BOX 8. A modified value for y_{i+1} is calculated by using Equation 3.46b. This calculation involves the use of data obtained in Boxes 3, 4, and 7.

BOX 9. The modified value of y_{i+1}' is calculated by using Equation 3.46c. This involves inserting the value of x obtained in Box 6 and the value of m_{i+1} obtained in Box 8 into the specified differential equation.

BOX 10. The corrected value for y_{i+1} is obtained by using Equation 3.46d.

BOX 11. The final value for y_{i+1} is obtained by using Equation 3.46e.

BOX 12. The final value for y_{i+1}' is obtained by substituting x from Box 6 and y from Box 11 into the original differential equation.

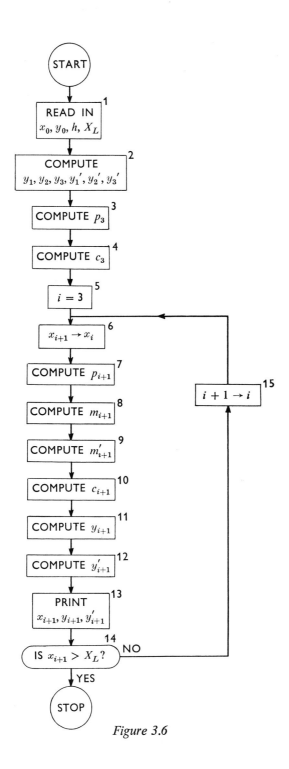

Figure 3.6

BOX 13. The solution for the i-th step is printed out.

BOX 14. x_{i+1} is examined to determine whether the specified range of x has been covered. If the answer to the question posed is affirmative the calculation comes to an end.

BOX 15. If the answer to the question in Box 14 is negative, the index i is increased by 1 and the calculations in Boxes 6 through 14 are repeated.

The application of the above method to systems of simultaneous first-order differential equations is straightforward. In that case, a set of ordinates is calculated in place of each of the ordinates y_{i+1}, y'_{i+1}, p_{i+1}, etc., of Figure 3.6. The calculations indicated in Boxes 2, 3, 4, 7, 8, 9, 10, 11, 12, and 13 are carried out for all of the simultaneous equations before proceeding to the next box.

The fact that the predictor-corrector and predictor-modifier-corrector methods are capable of providing an estimate of the truncation error at each step of the calculation is a very important feature, because this permits the variations of h in the course of a computer run to keep the truncation error at each step within prespecified upper and lower limits.

The upper limit on the truncation error must be chosen in such a fashion that if this per-step error is multiplied by the total number of steps taken in the x domain, the resulting accumulated error will be within a specified tolerance. The lower limit on the truncation error must likewise be chosen with care. Too low a limit leads to too small values for h and, hence, to an excessive number of steps required to cover a specified range of x. This, in turn, is reflected in excessively long computing times and entails the possibility of error growth if the computing method has a tendency toward computational instability. In practice the minimum truncation error is generally made somewhat larger than the per-step round-off errors resulting from the limit on the number of significant figures employed in storing calculated values in the memory registers of the computer. It is therefore expedient to increase h if the truncation error becomes smaller than a specified value, and to decrease h if the per-step truncation error grows too large. To simplify the necessary manipulations, the general practice is to increase or decrease h by a factor of 2.

In order to permit the doubling of the step size h when required, it is necessary to provide a somewhat increased memory capacity in the computer. Since predictor-corrector formulas are not self-starting but employ the results of several preceding calculations, it is necessary to have available the solutions over enough preceding steps to permit the change of h. For example, if the predictor formula 3.38b is to be employed, it is ordinarily necessary to have available at the $(i + 1)$ step only y_{i-3}, y_i',

y'_{i-1}, and y'_{i-2}. If, however, it is desired at the $(i + 1)$ step to change the step size from h to $2h$, the predictor formula required to predict y_{i+2} would be

$$(y_{i+2})_p = y_{i-6} + \frac{8h}{3}(2y'_i - y'_{i-2} + 2y'_{i-4}) \tag{3.47}$$

It therefore becomes necessary to store y'_{i-4} and y_{i-6} in addition to the quantities which are ordinarily stored. Thus the six past values of y must be stored at each step in x instead of only 4.

To decrease the step size by a factor of 2, we must determine values for y and y' midway between previously calculated values. For example, in order to halve h in Equation 3.38b, that is, to predict a value for y at $i + \frac{1}{2}$,

$$(y_{i+\frac{1}{2}})_p = y_{i-\frac{3}{2}} + \frac{2h}{3}(2y'_i - y'_{i-\frac{1}{2}} + 2y'_{i-1}) \tag{3.48}$$

It is therefore necessary to calculate $y_{i-\frac{3}{2}}$ and $y'_{i-\frac{1}{2}}$. To obtain these values, recourse is made to interpolation formulas having error terms no larger than those inherent in the predictor-corrector procedure. Keitel[8] provides equations which make it unnecessary to store additional values to permit doubling h. Also, $y_{i+\frac{1}{2}}$ can be predicted for a predictor-corrector method, using Equations 3.38b and 3.39a, using only the solution for the preceding four steps. His expressions for the predicted values are

$$y_{i-\frac{1}{2}} = \frac{1}{256}(80y_i + 135y_{i-1} + 40y_{i-2} + y_{i-3})$$

$$+ \frac{h}{256}(-15y'_i + 90y'_{i-1} + 15y'_{i-2}) \tag{3.49}$$

and

$$y_{i-\frac{3}{2}} = \frac{1}{256}(12y_i + 135y_{i-1} + 108y_{i-2} + y_{i-3})$$

$$+ \frac{h}{256}(-3y'_i - 54y'_{i-1} + 27y'_{i-2}) \tag{3.50}$$

A flow chart program illustrating the application of the predictor-modifier-corrector method to the first-order differential equation

$$y' = f(x, y) \qquad y = y_0 \quad \text{for} \quad x = x_0 \tag{3.51}$$

is shown in Figure 3.7. This program makes provision for doubling h if the truncation error drops below one specified constant α_1 and the halving of h if the truncation error exceeds another specified constant α_2. The

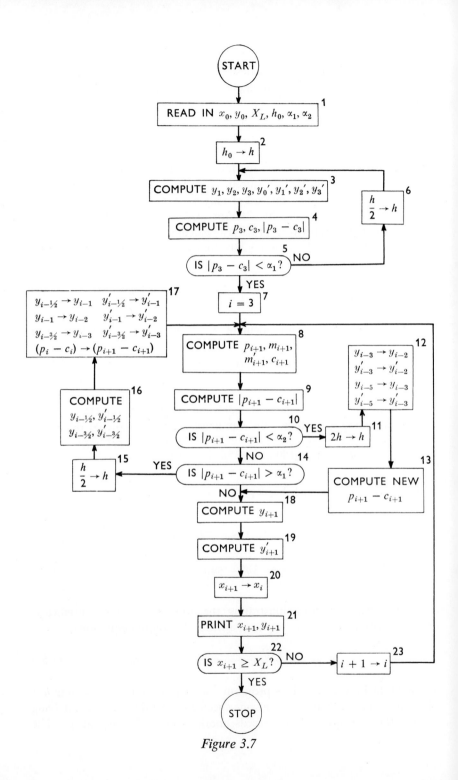

Figure 3.7

3.9/PREDICTOR-MODIFIER-CORRECTOR METHODS 81

following are the general steps involved in this program:

BOX 1. The specified initial conditions, y_0 and x_0, the range X_L of the variable x, the constants α_1 and α_2, as well as an initial value of h, are read into the machine.

BOX 2. The storage register containing h is given the initial guess for the step size h_0.

BOX 3. Starting values consisting of y and y' for x_1, x_2, and x_3 are calculated by using a self-starting method such as, for example, the Runge-Kutta method.

BOX 4. Employing a pair of predictor-corrector equations such as, for example, Equations 3.38b and 3.39a values for p_3, c_3, and $|p_3 - c_3|$ are calculated.

BOX 5. The value $|p_3 - c_3|$ is compared with α_1 to determine whether the starting method has an excessive truncation error.

BOX 6. If the truncation error calculated in Box 5 is too large, h is replaced by $h/2$ reducing the step size and resulting in a closer spacing between successive values of x. The computation of starting values in Boxes 3 and 4 is then repeated until the answer to the question in Box 5 is affirmative.

BOX 7. The index register i is set to 3, indicating in this case that the $i_1 = $ 4-th step along the x coordinate is to be computed.

BOX 8. Employing a suitable predictor-modifier-corrector formula, such as for example Equations 3.44, values for p_{i+1}, m_{i+1}, m'_{i+1}, and c_{i+1} are computed.

BOX 9. The truncation error $|p_{i+1} - c_{i+1}|$ is computed.

BOX 10. This truncation error is compared with the specified constant α_2 to determine whether h has been made too small.

BOX 11. If the answer to the question in Box 10 is affirmative, the step size is increased by giving h a value twice its former value.

BOX 12. In accordance with Equation 3.47, it is necessary to assign a new significance to the solutions for the preceding steps. Thus, in place of the values identified by the subscript $i - 2$, the values formerly identified by the subscript $i - 3$ must be used; similarly, in place of the values formerly identified by the subscript $i - 3$, the values which were formerly identified by the subscript $i - 5$ are employed.

BOX 13. Using the newly defined values of Box 12, a new value for $p_{i+1} - c_{i+1}$ is computed. This value is then employed to calculate y_{i+1} in Box 18.

BOX 14. If the answer to the question posed in Box 10 is negative, indicating that the truncation error is not too small, the truncation error is now compared with the specified constant α_1 to determine whether it is too large.

BOX 15. If the truncation error is too large, the step size is reduced by employing $h/2$ in place of h.

BOX 16. It is now necessary to interpolate between previously calculated values to obtain y and y' for $i - \frac{1}{2}$ and $i - \frac{3}{2}$. This can be accomplished by using for example Equations 3.49 and 3.50.

BOX 17. To permit prediction and correction employing the newly reduced interval, it is necessary to change the significance of a number of numerical quantities. Thus the quantities formerly identified with the subscripts $i - 1$, $i - 2$, $i - 3$, are now represented by quantities formerly identified by the subscripts $i - \frac{1}{2}$, $i - 1$, and $i - \frac{3}{2}$ respectively. These new values are introduced into Box 8 and the computations in Boxes 8, 9, 10, and 14 are repeated.

BOX 18. If the answer to the question posed in Box 14 is negative, the final value for y_{i+1} is calculated using for example Equation 3.44e. As indicated above, data calculated in Box 13 are used for this, provided the answer to the question in Box 10 was affirmative.

BOX 19. The value of y'_{i+1} is computed by substitution in Equation 3.51.

BOX 20. The value of x at the $i + 1$ step, corresponding to the value at the i step plus the most recently used h, is placed in the register containing x.

BOX 21. The solution in the form of x_{i+1} and y_{i+1} is printed out.

BOX 22. x_{i+1} is examined to determine whether the specified computing range has been completely covered. If the answer is affirmative, the calculation comes to an end.

BOX 23. If the answer to the question posed in Box 22 is negative, the number in the i register is increased by unity and the operations in Boxes 8 through 22 are repeated.

The method just described is readily extended to systems of simultaneous first-order differential equations. In that case, the criterion for specifying upper and lower bounds on the truncation error must be given additional thought, since the truncation error will, in general, be different for each of the simultaneous equations. One approach recommended by Ralston[5] is to employ a weighted sum of the truncation errors in all the simultaneous equations. In this way the programmer has the facility of weighting more important equations more heavily to assure that they will be treated in an optimum manner possibly at the expense of other less important simultaneous equations. Ralston also supplements a program of the type shown in Figure 3.7 by making provisions to detect gross errors in the application of the predictor-modifier-corrector formulas. For this purpose, a third constant α_3 is specified. This constant is considerably larger than α_1. If the answer to the question posed in Box 14 is affirmative, the quantity $(p_{i+1} - c_{i+1})$ is compared with this constant α_3; if the truncation error is larger than α_3, it is assumed that a computing error has been made, and the calculations involved in Boxes 8, 9, 10, and 14 are repeated. If the truncation error, proportional to $|p_{i+1} - c_{i+1}|$, is larger than α_1 but smaller than α_3, the operations contained in Boxes 15, 16, and 17 of Figure 3.7 are performed.

3.10 BOUNDARY VALUE PROBLEMS

All differential equations treated so far in this chapter fall into the category of initial value problems. That is, all auxiliary conditions involve specification of y, y', y'', etc., at a single point of the independent variable at $x = x_0$. Occasionally in engineering work, problems arise which are characterized by two-point boundary conditions. For example, in a second-order differential equation, y may be specified at $x = a$ and $x = b$, with no specification of y' at $x = a$. Such problems present peculiar difficulties in the application of digital computers. This difficulty arises because no effective use can be made of the boundary condition at point b. Among the numerical methods that have been applied in the past to the treatment of two-point boundary conditions, the following three are of greatest importance in the application of digital computers:

1. Replacement of the differential equation by a set of simultaneous finite difference equations.
2. Successive trial-and-error procedures.
3. Solution of the problem as two initial value problems.

The finite difference method involves the subdividing of the entire range of the independent variable x. Attention is then limited to the resulting discretely spaced series of points $x_0, x_1, x_2, \ldots, x_n$, and a separate algebraic equation is formulated for each point in the x domain. In each of these equations, the variable y is expressed as a linear combination of values for the variable y at adjacent points in the x domain. Therefore, a set of simultaneous equations in n unknowns results. These are solved by matrix inversion or other appropriate methods. Wachspress in Ralston[5] describes a program for applying the finite difference technique to second-order, linear, constant-coefficient, ordinary differential equations with very general types of two-point boundary conditions.

The trial-and-error method involves the rather obvious approach of transforming the boundary value problem into an initial value problem by guessing the missing initial conditions. Thus, if the dependent variable y is specified for $x = 0$ and $x = X$, a judicious guess is made as to the value of dy/dx at $x = 0$. The equation is then solved by any one of the methods described in this chapter, and the solution y obtained at $x = X$ is compared with that specified as the boundary condition. On the basis of this comparison a new estimate of dy/dx at $x = 0$ is made and the equation is again solved as an initial value problem. This process is continued until the difference between the value for y at $x = X$ and the specified boundary condition is sufficiently small. This method has the evident disadvantage that numerous time-consuming solutions must be made. Furthermore, in the case of many nonlinear problems, convergence to the correct solution is by no means easy to obtain. In fact, it is frequently difficult to decide, even after a relatively large number of trial solutions, whether the unspecified initial condition should be increased or decreased. Evidently, the better the initial guess as to the unspecified initial condition, the greater is the chance of obtaining a satisfactory solution without excessive computing effort. It is therefore expedient to apply to a maximum any insight or outside knowledge of the physical system described by the differential equation to make this first guess as "reasonable" as possible.

The third method for treating boundary value problems is applicable only to linear ordinary differential equations. The coefficients however, may, be functions of the independent variable. In accordance with the development presented in more detail by Hildebrand[2] and Milne,[4] consider a second-order equation of the form

$$y'' + P(x)y' + Q(x)y = F(x) \tag{3.52}$$

subject to the boundary conditions

$$y = A \quad \text{at} \quad x = a \qquad y = B \quad \text{at} \quad x = b \tag{3.53}$$

Define two subsidiary second-order equations with independent variables $u(x)$ and $v(x)$

$$\begin{aligned} u'' + P(x)u' + Q(x)u = F(x) \\ v'' + P(x)v' + Q(x)v = F(x) \end{aligned} \tag{3.54}$$

with initial conditions according to

$$\left. \begin{aligned} u &= A \\ u' &= C \\ v &= A \\ v' &= D \end{aligned} \right\} \quad \text{for} \quad x = a \tag{3.55}$$

C and D may be chosen arbitrarily but may not be equal to each other, nor may they lead to a trivial solution; for example, if $F(x) = 0$ and $A = 0$, C or D cannot also be equal to zero. Equations 3.54, subject to boundary conditions 3.55 are now solved separately by using any of the methods described in the preceding sections, and $u(x)$ and $v(x)$ are obtained over the range $a < x < b$. These solutions provide the values of $u(b)$ and $v(b)$—the values of u and v at $x = b$. The application of the superposition theorem now permits the expression of the solution $y(x)$ as a linear combination of $u(x)$ and $v(x)$ according to

$$y(x) = \alpha u(x) + \beta v(x) \qquad (3.56)$$

The coefficients α and β are determined by substituting in Equation 3.56 first the values for y, u, and v at $x = a$ and then at $x = b$. There result then two simultaneous equations

$$A = \alpha A + \beta A$$
$$B = \alpha u(b) + \beta v(b) \qquad (3.57)$$

from which

$$\alpha = \frac{B - v(b)}{u(b) - v(b)} \qquad \beta = \frac{B - u(b)}{v(b) - u(b)} \qquad (3.58)$$

α and β are therefore entirely determined by the specified end condition B and the end conditions obtained by solving Equations 3.54. The application of this method involves the following steps:

STEP 1. Define subsidiary equations of the type of Equation 3.54 subject to boundary conditions of the type of Equation 3.55.
STEP 2. Solve these equations on the digital computer.
STEP 3. Employ the values obtained in these two solutions at $x = b$ to determine α and β.
STEP 4. Substitute these two solutions in Equation 3.56 to obtain the final solution.

This method involves no iterations and only two solutions of the specified differential equation. It therefore has important advantages over the trial-and-error methods and is always preferred in the case of linear differential equations. Collatz[1] demonstrates how this method may be applied to linear differential equations of higher order.

3.11 CONTROL OF ERRORS

The digital computer solution of ordinary differential equations, such that results of a specified accuracy are obtained in a minimum of computing time, demands careful consideration and control of errors inherent in the numerical process. Only in this manner can we select judiciously the step size h and the complexity of the integration formula, that is, how many terms of the Taylor series expansion are to be matched. In treating computing errors, two types of effects must be considered: per-step errors and accumulated or inherited errors.

To characterize the error introduced at a specific step in the calculation, it is necessary to take into account round-off errors and truncation errors. Round-off errors arise because the size of the registers on digital computers is limited. It thus becomes necessary to round off to the lowest available integer all numbers arising in the calculation. As a rule, modern digital computers employ word lengths such that the round-off error at a specific step is negligible. The only concern then need be whether the per-step round-off errors accumulate and grow as the calculation proceeds over many steps. Truncation errors result from the fact that practical integration formulas constitute approximations to the desired computational process. In general, this approximation can be interpreted as constituting the truncation or ignoring higher-order terms in a Taylor series expansion. These errors are therefore proportional to a power of h and higher derivatives of the dependent variable.

The accumulation of round-off, truncation, and possibly other random machine errors as the calculation proceeds is a distinct possibility in the application of the numerical methods described in this chapter. Although the per-step, round-off, and truncation errors may be small or negligible, conditions may arise under which the error terms grow after several hundred steps along the domain of the independent variable until the errors completely mask the solution, making it worthless. The term *computationally stable* is employed to describe a numerical process in which such accumulation does not take place. The term *relative computational stability* is applied to those procedures in which errors accumulate but at a rate which is less rapid than the rate of increase of the dependent variable being determined. For example, in the study of a dynamic system with negative damping, the dependent variable grows exponentially with time; under these conditions error terms can also be expected to increase, but under conditions of relative computational stability this growth will be less rapid than the rate of increase of the dependent variable so that the percentage error will not become excessive.

The quantitative description of truncation errors and, to a lesser extent, round-off errors, as well as the description of error accumulation has been the subject of intensive study by mathematicians. From an engineer's point of view, these investigations have not yet come to a fully satisfying conclusion. There exist in fact no simple or direct methods for predicting the accuracy of solutions nor for specifying optimum step sizes or computing procedures. Moreover, most of the available procedures of error analysis applicable to equations of the type of Equation 3.3 involve prior knowledge of the $\partial f(x, y)/\partial y$. Particularly where $f(x, y)$ contains nonanalytic terms, this quantity may be impossible to establish in advance. For this reason only some general guides and "rules of thumb" applicable to the specific methods recommended in this chapter are given below. The reader is referred to texts in numerical analysis such as those by Collatz[1] and Hamming[11], as well as to the discussion in Chapter 4, for additional information.

3.12 CONCLUSIONS AND GENERALIZATIONS

The digital computer is unrivalled and unexcelled in its power and generality in treating all types of mathematical and scientific problems. Nevertheless, its application to the solution of systems of ordinary differential equations is neither entirely natural nor direct. The difficulty lies in the fact that digital computers are capable only of treating discretized variables, whereas the dependent and independent variables arising in engineering analysis are generally in continuous form. Thus the utilization of digital computers involves inevitably the application of approximations. These in turn lead to errors which may frequently outweigh the advantage gained by the great accuracy of the digital computer in performing arithmetic operations. Of particular significance in this connection is the accumulation of truncation and round-off errors, as discussed briefly in the preceding section. In turning to the digital computer method and in deciding to employ a specific numerical approach, it is important to recognize that these two sources of errors make themselves felt only when all other difficulties in digital programming have been overcome.

Frequently, combinations of errors compensate and mask each other to such an extent that a "reasonable" though entirely incorrect solution is obtained. The application of digital computers to engineering problems therefore invariably involves a considerable amount of testing, reconciliation of computer solutions with experimental data, and patient review of each step in the programming and coding operation. To minimize this latter effort, most computer users find it expedient to maximize as much as possible the utilization of tried and tested subroutines and existing programs. Often it is advisable to use such an approach even where considerations based purely on numerical analysis indicate that another method is preferable. Thus a Runge-Kutta method may be employed because a tested subroutine of this method is available, even though considerable computing time can be saved in applying the Milne method. These considerations must be taken into account in utilizing the general guidelines presented below, in which it is assumed that the engineer and programmer are working in a "vacuum."

In selecting a specific numerical method for the solution of an ordinary differential equation or a set of simultaneous ordinary differential equations of the type of Equation 3.3, the following questions must be asked:

1. How complex is the function $f(x, y_1, y_2, \ldots, y_n)$? How many calculations are required to evaluate this function, given $x, y_1, y_2, \ldots, y_n$?
2. How many simultaneous equations are there?
3. How long is the integration range? How many steps $x_0, x_1, x_2, \ldots, x_N$ need be taken?

In all the computer programs described, the evaluation of $f(x, y_1, y_2, \ldots, y_n)$ at each step in the x domain accounts for from 85 to 95% of the entire computing time. Where this function is relatively complex, or where a large number of simultaneous equations must be solved, the necessity of limiting the overall computing time to a practical value often outweighs all other considerations and leads to a choice of a method in which these time-consuming operations are minimized. In the fourth-order Runge-Kutta method, the function f must be calculated four times at each step in x. In the predictor-corrector methods, this calculation must be repeated until a sufficient convergence has been reached. Only in the predictor-modifier-corrector methods is there assurance that the function f need only be calculated twice for each step in the x domain. These methods will therefore be preferred despite the fact that their accuracy is somewhat lower for a specific h than predictor-corrector or Runge-Kutta methods of the same order and despite the fact that they are not self-starting and therefore require a more extensive programming effort.

If N, the total number of steps to be taken in the x domain, is relatively small, a predictor-modifier-corrector method using Equations 3.44 will be preferred since it gives relatively higher accuracy than a method based on Equations 3.46. On the other hand, the former method is computationally unstable. If N is relatively large, the more stable method of Equations 3.46 has to be employed. Thus if it is desired in analyzing an underdamped dynamical system to obtain only the first few cycles of the system response, Equations 3.44 are used; if, on the other hand, a large number of cycles are to be plotted, Equations 3.46 are used.

In cases where the function f is not highly complex and where there are not a very large number of simultaneous equations to be treated, the Runge-Kutta methods are preferred. Since the predictor-corrector and predictor-modifier-corrector methods require the use of a formula of the Runge-Kutta type to obtain starting values, the solution of the entire problem by a Runge-Kutta method requires no greater programming effort than the programming of the first step of the non-self-starting methods. Particularly where relatively little programming and coding experience is available and where computer time is not excessive, the Runge-Kutta method, particularly that of the fourth order described by Equations 3.25 and 3.26, is preferred.

References

1. Collatz, L., *Numerische Behandlung von Differentialgleichungen*, Springer Verlag, Berlin, 1951.
2. Hildebrand, B., *Introduction to Numerical Analysis*, McGraw-Hill, New York, 1956.
3. Kunz, K. S., *Numerical Analysis*, McGraw-Hill, New York, 1957.
4. Milne, W. E., *Numerical Solution of Differential Equations*, John Wiley, New York, 1953.
5. Ralston, A., and H. S. Wilf, *Mathematical Methods for Digital Computers*, John Wiley, New York, 1960.
6. Gill, S., "A Process for the Step by Step Integration of Differential Equations in an Automatic Digital Computing Machine," *Proc. Cambridge Phil. Soc.*, **47**, 96–108, 1951.
7. Hamming, R. W., "Stable Predictor-Corrector Methods for Ordinary Differential Equations," *J. Assoc. for Computing Machinery*, **6**, 37–47, 1959.
8. Keitel, G. H., "An Extension of Milne's Three Point Method," *J. Assoc. for Computing Machinery*, **3**, 212–222, 1956.
9. Kuo, S. S., *Numerical Methods and Computers*, Addison-Wesley, Reading, Mass., 1965.
10. Carr, J. W., III, "Error Bounds for the Runge-Kutta Single Step Integration Process," *J. Assoc. for Computing Machinery*, **5**, 39–45, Jan. 1958.
11. Hamming, R. W., *Numerical Methods for Scientists and Engineers*, McGraw-Hill, New York, 1962.

4

Transform Analysis of Hybrid Computer Systems

4.1 INTRODUCTION

The study of linear continuous dynamical systems is greatly facilitated by the use of frequency domain techniques involving the Laplace and Fourier transforms. Similarly, discrete systems can be studied by means of a special form of the Laplace transform known as the z-transform. The use of such techniques not only provides insight into system behavior in the frequency domain, but makes available a number of procedures developed for the study of feedback-control systems in such areas as stability, compensation, and design to meet desired performance specifications. The purpose of this chapter is to illustrate the applicability of frequency domain methods to the study of hybrid computer systems. By viewing the solution of differential equations as a feedback process, transform methods provide an alternative point of view to the numerical methods described in the preceding chapter. The z-transform methods make it possible to study the stability and truncation errors associated with the solution of particular differential equations by given numerical integration formulas.

The first portion of this chapter provides a brief introduction to the z-transform and its relation to the Laplace transform. No attempt is made to provide a complete review of the theory of sampled data systems, especially since certain other aspects associated with sampling and data reconstruction are more fully discussed in Chapter 5 in the general context of error analysis. Section 4.2 concerns the application of z-transforms to the study of numerical integration techniques. This is followed by a brief discussion of the theory needed to treat mixed continuous and discrete data systems. Throughout the chapter the theory is illustrated with examples.

Some material in this chapter has been discussed in earlier publications by Howe[1] and Gilbert,[2] but was developed independently in courses on hybrid computation at USC and UCLA.*

4.2 z-TRANSFORM TECHNIQUES

The application of transform techniques to digital signals can be approached from two points of view, either from an examination of sampled signals or from a consideration of number sequences. We begin by looking at the analog-to-digital converter as a mathematical device which transforms a continuous function $x(t)$ into a sequence of numbers $\{x(n)\}$, as shown in Figure 4.1. The index n refers to equally spaced points sampled from the continuous function, that is,

$$\{x(nT)\} = \{x(0), x(T), x(2T), x(3T), \ldots\} \tag{4.1}$$

where T represents the time interval between samples. Since T is a constant, it may be omitted and the entire sequence of numbers represented by $\{x(n)\}$. The z transform of the sequence of numbers $\{x(nT)\}$ is defined as

$$\mathscr{Z}\{x(nT)\} = X(z) = \sum_{n=0}^{\infty} x(nT)z^{-n} \tag{4.2}$$

where z is a complex variable. The questions of convergence of this summation are treated in the literature.[3-5]

To illustrate the significance of the above definition, consider a number of simple examples.

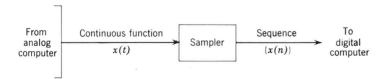

Figure 4.1 Transformation of continuous function to number sequence.

---

* Some of the examples in this chapter were prepared by Dr. M. J. Merritt of the Aerospace Engineering Department, University of Southern California.

EXAMPLE 4.1. Let the continuous function $x(t)$ at the input to the analog-to-digital converter be simply a unit step, that is,

$$x(t) = u_1(t) \tag{4.3}$$

then the output sequence is simply

$$\{x(nT)\} = \{1, 1, 1, \ldots\} \tag{4.4}$$

Substituting this sequence into the definition 4.2, we obtain

$$X(z) = \sum_{n=0}^{\infty} z^{-n} \tag{4.5}$$

which is simply the geometric series. Under appropriate conditions, this series can be summed and converges to

$$X(z) = \frac{1}{1 - z^{-1}} = \frac{z}{z - 1} \tag{4.6}$$

EXAMPLE 4.2. Consider now the samples obtained from an exponential

$$x(t) = e^{-at} \tag{4.7}$$

then

$$\{x(nT)\} = \{e^{-anT}\} \tag{4.8}$$

Substituting this sequence into the defining Equation 4.2, we obtain

$$\mathscr{L}\{e^{-anT}\} = \sum_{n=0}^{\infty} e^{-anT} z^{-n} = \sum_{n=0}^{\infty} (e^{-aT} z^{-1})^n \tag{4.9}$$

which again is a geometric series, so that

$$\mathscr{L}\{e^{-anT}\} = \frac{1}{1 - e^{-aT} z^{-1}} = \frac{z}{z - e^{-aT}} \tag{4.10}$$

By similar procedures it is possible to derive z-transform expressions corresponding to a large variety of number sequences. Several z transforms are shown in Table 4.1. Item number 5 in this table is known as the "delay property" of z transforms. It can be seen that this property is analogous to the effect of integrations on the Laplace transforms of continuous functions. In the case of continuous functions, each integration corresponds to a multiplication by s^{-1} in the s domain. In the case of sequences, each shift or time delay by one sampling interval corresponds to a multiplication by z^{-1} in the z domain. This property can be used to transform difference equations into algebraic expressions.

Table 4.1 Table of z Transforms

Transform Number	$x(nT)$	$X(z)$
1	1	$\dfrac{1}{1-z^{-1}}$
2	e^{-anT}	$\dfrac{1}{1-e^{-aT}z^{-1}}$
3	nT	$\dfrac{Tz^{-1}}{(1-z^{-1})^2}$
4	a^{nT}	$\dfrac{1}{1-a^{T}z^{-1}}$
5	$x[(n-k)T]$	$z^{-k}X(z)$
6	$x[(n+1)T]$	$zX(z) - zx(0)$

A. Solution of Difference Equations by z-Transforms

In order to apply z-transform techniques to the evaluation of numerical solutions of differential equations, it is necessary to examine the solution of the corresponding difference equations (since these are in fact the equations solved by digital computers). Consider the difference equation

$$y(n) - y(n-1) = x(n-1) \qquad (4.11)$$

To solve this difference equation by means of z-transforms, we apply the property given in entry number 5 of Table 4.1 and write

$$Y(z) - z^{-1}Y(z) = z^{-1}X(z) \qquad (4.12)$$

If we now let $X(z)$ be the z transform of a unit step input (Table 4.1, Item 1):

$$Y(z) = \frac{z^{-1}}{1-z^{-1}} X(z) = \frac{z^{-1}}{(1-z^{-1})^2} \qquad (4.13)$$

The sequence corresponding to the solution of Equation 4.13 can be obtained in two ways. An easy way is to examine Table 4.1 or other larger tables of z transforms and attempt to find an entry corresponding to a given z-transform expression. A partial fraction expansion of a given expression will lead to simple forms which can always be found in the tables. In the case of the present example, it can be seen to correspond to entry number 3 in the table with $T = 1$. Therefore, the solution to the difference Equation 4.11 is

$$y(n) = n \qquad (4.14)$$

An alternative method of obtaining a number sequence corresponding to a given z transform is based on simply dividing the numerator by the denominator, using long division. In the case of the example discussed in the preceding paragraph, such a division results in the sequence

$$Y(z) = z^{-1} + 2z^{-2} + 3z^{-3} + 4z^{-4} + \cdots \qquad (4.15)$$

Each term of this sequence can be inverted by using entry number 5 of Table 4.1 (the delay property), thus resulting in the time domain sequence

$$\{y(n)\} = \{0, 1, 2, 3, 4, \ldots\} \qquad (4.16)$$

which is seen to correspond to Equation 4.14.

4.3 z-TRANSFORMS OF NUMERICAL INTEGRATION FORMULAS

Since numerical integration formulas are difference equation approximations used in the solution of differential equations, it is reasonable to expect that they could be represented by z-transform techniques. Consider first the simple Taylor series formulas discussed in Section 3.4. Let the differential equation to be solved have the form

$$y' = \frac{dy}{dx} = f(x, y); \qquad y(x_0) = y_0 \qquad (4.17)$$

Then the simplest Taylor series formula is the one obtained by using only the first term of the series, namely Euler's method (Equation 3.16) which can be written as

$$y_{n+1} = y_n + Tf(x_n, y_n) \qquad (4.18)$$

where T is the integration step size, denoted by h in Chapter 3 and the subscript n denotes the number of the step along the x domain. In Chapter 3 the first step of this equation was given by Equation 3.16 as

$$y_1 = y_0 + hy_0'$$

In Equation 4.18 the right-hand side of the differential equation, $f(x, y)$, is used to represent the derivative y' in order to avoid confusion in z-transform operations. If we denote $f(x_n, y_n)$ by f_n, then the z-transform expression corresponding to Equation 4.18 can be written, with the aid of the advance property given in Item 6 of Table 4.1, as

$$zY(z) - zy_0 = Y(z) + TF(z) \qquad (4.19)$$

4.3 / z-TRANSFORMS OF NUMERICAL INTEGRATION FORMULAS

Solving for $Y(z)$ we obtain

$$Y(z) = \frac{T}{z-1} F(z) + \frac{zy_0}{z-1} \quad (4.20)$$

The coefficient of $F(z)$ in this equation can be viewed as the "transfer function" of the Euler integrator. The second term of Equation 4.20 indicates the effect of initial conditions. Note that the first term on the right-hand side of Equation 4.20 represents the forced solution, while the second term represents the homogeneous solution of the difference Equation 4.18. The solution of the original differential Equation 4.17 by continuous and discrete methods may be compared with the aid of the block diagrams of Figure 4.2. In Section 4.6 we shall return to the closed-loop representations of Figure 4.2 in examining the stability of the difference equation approximations.

As indicated above, Euler or rectangular integration is based on taking only the first term of the Taylor series. If two terms of the Taylor series are used, we obtain the so-called trapezoidal integration formula, namely,

$$y_{n+1} = y_n + Tf_n + \frac{T^2}{2} f_n' \quad (4.21)$$

where the derivative f_n' is approximated by

$$f_n' \cong \frac{f_n - f_{n-1}}{T} \quad (4.22)$$

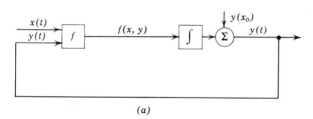

(a)

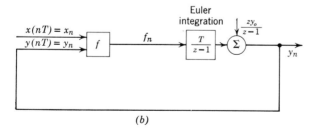

(b)

Figure 4.2 Block diagram of solution to differential equation $y = f(x, y)$ by (a) continuous integration, and (b) Euler integration.

With this backward difference approximation, Equation 4.21 becomes

$$y_{n+1} = y_n + Tf_n + \frac{T}{2}(f_n - f_{n-1}) \qquad (4.23)$$

Applying the advance and delay formulas from Table 4.1, we obtain the transfer function of the trapezoidal integration formula:

$$\frac{Y(z)}{F(z)} = \frac{T}{2}\frac{(3z-1)}{z(z-1)} \qquad (4.24)$$

In a similar way it is possible to construct z-transform expressions corresponding to a variety of numerical integration formulas. Such expressions are given in Table 4.2.

Table 4.2 Transfer Functions Corresponding to Simple Integration Formulas

Formula	Difference Equation	z-Transfer Functions
1. Rectangular (Euler)	$y_{n+1} = y_n + Tf_n$	$\dfrac{Y(z)}{F(z)} = \dfrac{T}{z-1}$
2. Trapezoidal	$y_{n+1} = y_n + \dfrac{3T}{2}f_n - \dfrac{T}{2}f_{n-1}$	$\dfrac{Y(z)}{F(z)} = \dfrac{T}{2}\dfrac{(3z-1)}{z(z-1)}$
3. Two-step Adams-Bashforth	$y_{n+1} = y_{n-1} + 2Tf_n$	$\dfrac{Y(z)}{F(z)} = \dfrac{2Tz}{z^2-1}$
4. Two-step Adams-Moulton	$y_{n+1} = y_{n-1} + \dfrac{T}{3}(f_{n+1} + 4f_n + f_{n-1})$	$\dfrac{Y(z)}{F(z)} = \dfrac{T(z^2+4z+1)}{3(z^2-1)}$

With more complex single-step formulas, such as the Runge-Kutta methods, the z-transform method is still applicable, but the specific form of the right-hand side of the differential equation, Equation 4.17, must be specified. This will be illustrated in Section 4.5 where the solution of specific first- and second-order equations will be examined.

4.4 FREQUENCY DOMAIN CHARACTERISTICS OF INTEGRATION FORMULAS

To obtain a relationship between the new complex variable z introduced above and the frequency variable s, consider the process of sampling as

4.4/CHARACTERISTICS OF INTEGRATION FORMULAS

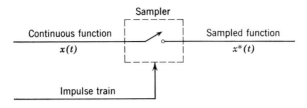

Figure 4.3 Transformation of continuous function to modulated-impulse train.

illustrated in Figure 4.3. In contrast to the description of the sampler given in Figure 4.1 where continuous functions were transformed into number sequences, we now view the sampler as a modulator. The impulse train is defined as the sequence of delta functions

$$p_\delta(t) = \sum_{n=-\infty}^{+\infty} \delta(t - nT) \qquad (4.25)$$

The areas of these delta functions are modulated by the information-carrying signal $x(t)$, resulting in the sampled function

or
$$\left. \begin{aligned} x^*(t) &= x(t)p_\delta(t) \\ x^*(t) &= \sum_{n=0}^{\infty} x(nT)\delta(t - nT) \end{aligned} \right\} \qquad (4.26)$$

Thus the areas or weights of the impulses form precisely the output sequence defined by Equation 4.1 earlier. This process of impulse modulation may be viewed graphically, as indicated in Figure 4.4. The height of

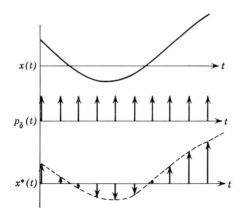

Figure 4.4 Sampling as impulse modulation.

each arrow symbolizes the weight of the corresponding impulse. In order to investigate the frequency domain characteristics of the impulse sequence (Equation 4.26), we can obtain the Laplace transform of each term of the sum:

$$X^*(s) = \sum_{n=0}^{\infty} x(nT)e^{-nTs} \qquad (4.27)$$

The frequency domain characteristics of the sampled signal $x^*(t)$ can now be obtained by replacing s by $j\omega$ in Equation 4.27.

To relate the s and z domains, consider now the definition of the z transform of the sequence $\{x(nT)\}$, as given in Equation 4.2 and repeated below:

$$X(z) = \sum_{n=0}^{\infty} x(nT)z^{-n} \qquad (4.28)$$

Comparison of the two relationships, 4.27 and 4.28, indicates that the relation between the complex variable z and the complex variable s is given by

$$z = e^{sT} \qquad (4.29)$$

This relation denotes a mapping from one complex plane to another. It is easy to verify by substitution in Equation 4.29 that the origin of the s-plane maps into the point $(+1, 0)$ in the z plane, that the entire negative real axis of the s plane maps into the line segment connecting the origin with the point $(+1, 0)$ in the z plane, and that each horizontal strip in the s plane of width $2\pi/T$ maps into the interior of the unit circle in the z plane, as illustrated in Figure 4.5. Each successive length of the $j\omega$ axis

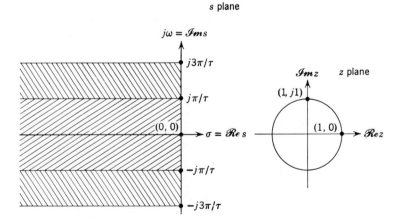

Figure 4.5 Mapping of s-plane into z-plane.

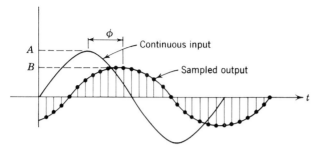

Figure 4.6 Frequency response of a system with continuous sinusoidal input and sampled output. The amplitude ratio is given by B/A and the phase shift by ϕ.

in the s-plane marking the right-hand boundary of one of the strips indicated in the figure maps into the circumference of the unit circle.

From the standpoint of stability considerations, the mapping relationship indicated above is particularly significant, since stable poles of closed-loop transfer functions of continuous systems must be located in the left half of the s plane. As a consequence of the mapping of Equation 4.29, stable poles of closed-loop transfer functions of data sequences (discrete systems) must be located on the interior of the unit circle in the z plane. Furthermore, as we shall see below, the substitution of $\exp(j\omega T)$ for z in z-transform expressions can provide insight into amplitude and phase behavior of sampled systems with sinusoidal inputs. This can be justified on the grounds of fitting a sinusoidal envelope over the samples of a system response, provided that the sampling rate is sufficiently high, as illustrated in Figure 4.6.[4,5]

In the following sections we shall apply the notions of stability analysis and frequency response to the study of numerical integration techniques, as applied to first- and second-order systems.

4.5 z-TRANSFORM ANALYSIS OF THE NUMERICAL INTEGRATION OF A FIRST-ORDER EQUATION

Consider the solution of the first-order linear equation

$$\frac{dy}{dt} + y = 0, \quad y(0) = y_0 \qquad (4.30)$$

which may be diagrammed as shown in Figure 4.7. Part (a) of this figure illustrates the analog solution; part (b) illustrates a numerical solution with the box labeled $I(z)$ representing the z transform of the appropriate numerical integration formula.

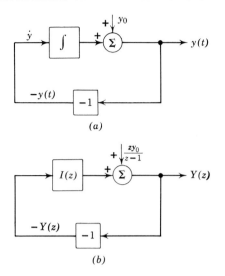

Figure 4.7 Solution of equation $\ddot{y} + y = 0$: (a) analog diagram, and (b) numerical (z-transform) diagram.

A. Rectangular Integration

From Table 4.2, the numerical integration formula is

$$I(z) = \frac{T}{z - 1} \tag{4.31}$$

From the block diagram of Figure 4.7 we directly obtain

$$Y(z) = -I(z)Y(z) + \frac{zy_0}{z - 1} \tag{4.32}$$

or

$$Y(z) = \frac{zy_0}{z - (1 - T)} \tag{4.33}$$

which represents the z-transform of the solution.

As pointed out above, for stability the poles of this expression must be located inside of the unit circle. In contrast to the conventional root-locus, which describes the locus of the poles of the closed-loop transfer function as the open-loop gain is varied, we shall plot the "*T*-locus," the locus of closed-loop poles as the sampling interval is varied from zero to infinity.

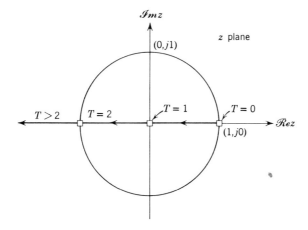

Figure 4.8 T-locus for solution of Equation 4.30 by rectangular integration.

The T-locus for Equation 4.33 is shown in Figure 4.8. This figure illustrates graphically that the solution of Equation 4.30 by rectangular integration is stable only so long as the sampling interval T is less than 2 seconds. The actual output sequence $x(nT)$ is easy to evaluate by long division, as discussed in Section 4.2, for any particular value of T. For example, for $T = 2$, the solution sequence becomes

$$\{y(nT)\}|_{T=2} = \{y_0, -y_0, +y_0, -y_0, \ldots\} \quad (4.34)$$

which is clearly oscillatory as may be expected from the fact that the root in Figure 4.8 is located on the unit circle.

In order to determine the seriousness of the truncation errors which are evident in the stability behavior illustrated above, consider an alternate approach, as suggested by Howe[1] and Gilbert.[2] The exact solution of Equation 4.30 is

$$y(t) = y_0 e^{-t} \quad (4.35)$$

A sequence which would agree exactly with the continuous solution at the sampling instants would then be

$$\{\hat{y}(nT)\} = \{y_0 e^{-nT}\} \quad (4.36)$$

Therefore, by using transform number 2 in Table 4.1, the z-transform of an exact solution would be given by

$$\hat{Y}(z) = \frac{zy_0}{z - e^{-T}} \quad (4.37)$$

To compare this expression with the approximate solution given by Equation 4.33, we can rewrite 4.33 as

$$Y(z) = \frac{zy_0}{z - e^{-\lambda T}} \quad (4.38)$$

where

$$-\lambda = \frac{1}{T} \ln(1 - T) \quad (4.39)$$

The degree to which the coefficient λ differs from unity is a measure of the error introduced by our numerical integration formula. By expanding in a series it can be shown that

$$\lambda = 1 + \frac{T}{2} + \frac{T^2}{3} + \frac{T^3}{4} + \cdots \quad (4.40)$$

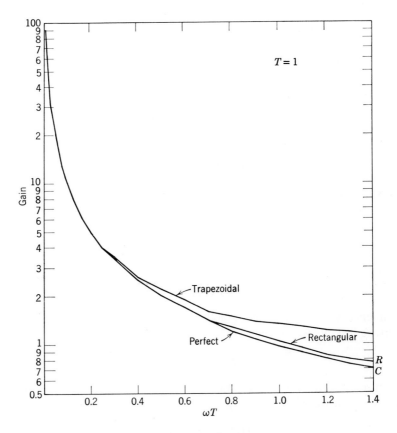

Figure 4.9 Gain vs. ωT for digital integrators.

Therefore, if T is sufficiently small, the error in the exponent λ is approximately $T/2$.

In order to investigate the frequency response behavior of the rectangular integrator, we can replace the variable z by its corresponding Laplace transform expression 4.29, and then let

$$s = j\omega \qquad (4.41)$$

With this substitution the rectangular integration formula can be written as

$$I(j\omega) = I(z)\Big|_{z=e^{j\omega T}} = \frac{T}{\cos \omega T - 1 + j \sin \omega T} \qquad (4.42)$$

With some algebraic manipulations we can derive the amplitude ratio and phase of this expression:

$$\left.\begin{aligned} |I(j\omega, T)| &= \frac{T}{\sqrt{2(1 - \cos \omega T)}} \\ \sphericalangle I(j\omega, T) &= -\tan^{-1} \frac{\sin \omega T}{1 - \cos \omega T} \end{aligned}\right\} \qquad (4.43)$$

Plots of these two expressions as a function of dimensionless frequency ωT are given in Figures 4.9 and 4.10. The amplitude and phase corresponding to a perfect integrator are also shown.

B. Trapezoidal Integration

From Table 4.2 the integration formula for a trapezoidal integrator is

$$I(z) = \frac{T(3z - 1)}{2z(z - 1)} \qquad (4.44)$$

Substitution of this expression in Figure 4.7b yields as the z-transform of the solution sequence to the differential Equation 4.30:

$$Y(z) = \frac{y_0 z^2}{z^2 + \left(\dfrac{3T}{2} - 1\right)z - \dfrac{T}{2}} \qquad (4.45)$$

It is interesting to note that the application of this formula has introduced a second root into the z-transform of the solution. Consequently, there are two branches in the corresponding T-locus, as shown in Figure 4.11. It is evident that one branch of the T-locus crosses the unit circle for $T > 1$. It is interesting to note that from a comparison of the two loci of Figure 4.8 and 4.11, trapezoidal integration is a worse formula than rectangular integration for the solution of this particular equation.

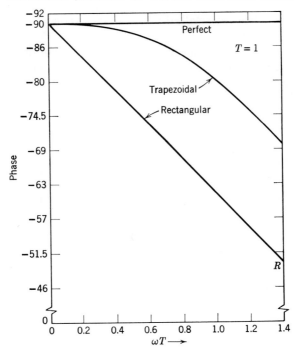

Figure 4.10   Phase vs. ωT for digital integrators.

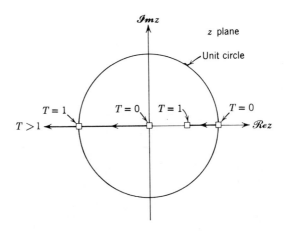

Figure 4.11   T-locus for solution of Equation 4.30 by trapezoidal integration.

The frequency response of the trapezoidal integration formula is included in the plots of Figures 4.9 and 4.10. An examination of these figures indicates that the gain characteristic of the rectangular integration formula is somewhat better than that of the trapezoidal formula, but that the reverse situation holds for the phase characteristic.

C. Runge-Kutta Integration

As a final example of the application of z-transform techniques to the solution of first-order differential equations by numerical methods, consider the application of a fourth-order Runge-Kutta integration formula, as defined in Chapter 3 by the expression

$$y_{i+1} = y_i + \tfrac{1}{6}(k_0 + 2k_1 + 2k_2 + k_3) \tag{4.46}$$

If this expression is applied to the solution of the general first-order differential equation

$$\frac{dy}{dx} = f(x, y), \quad y(x_0) = y_0 \tag{4.47}$$

Then the k's in Equation 4.46 are defined as

$$\begin{aligned} k_0 &= Tf(x_i, y_i) \\ k_1 &= Tf(x_i + \tfrac{1}{2}T, y_i + \tfrac{1}{2}k_0) \\ k_2 &= Tf(x_i + \tfrac{1}{2}T, y_i + \tfrac{1}{2}k_1) \\ k_3 &= Tf(x_i + T, y_i + k_2) \end{aligned} \tag{4.48}$$

It can be seen by inspection of Equation 4.46 and 4.48 that a transfer function for this numerical integration formula cannot be obtained in general. However, for the specific solution of Equation 4.30, i.e., a first-order differential equation with no forcing function, the application of the Runge-Kutta formula yields the expression

$$y_{n+1} = \left(1 - T + \frac{T^2}{2} - \frac{T^3}{6} + \frac{T^4}{24}\right) y_n \tag{4.49}$$

To obtain the z-transform of the solution of this equation, we apply relation number 6 from Table 4.1, which yields

$$Y(z) = \frac{z y_0}{z - \left(1 - T + \dfrac{T^2}{2} - \dfrac{T^3}{6} + \dfrac{T^4}{24}\right)} \tag{4.50}$$

As with the previous examples, the stability of the solution can be examined by plotting the T-locus of Equation 4.50, which is given in

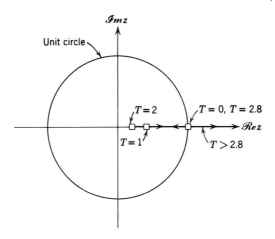

Figure 4.12 T-locus for solution of Equation 4.30 by Runge-Kutta integration.

Figure 4.12. It is interesting to note that in this case, as contrasted for example with the T-locus for the solution by rectangular integration, the locus reverses, and again crosses the point $(+1, j0)$ which corresponds to the solution at $T = 0$, that is, the exact solution. The solution becomes unstable only for $T > 2.8$, which is a significant improvement over the stability of either the rectangular or the trapezoidal formulas.

If we now evaluate the exponential constant corresponding to the sequence obtained by inverting Equation 4.50, it can be shown[2] to be

$$\lambda = 1 - \frac{T^4}{120} + \cdots \qquad (4.51)$$

Evidently, for small values of integration step T, this error is very small. The considerable superiority of accuracy of the Runge-Kutta formula as contrasted with the rectangular or Euler integration formula is clearly revealed. Of course, in any particular application the trade-off between accuracy and increased computation time must be examined carefully.

4.6 SOLUTION OF A SECOND-ORDER EQUATION

We now consider the solution of the second-order differential equation

$$\frac{d^2y}{dt^2} + 3\frac{dy}{dt} + 4y = x(t)$$
$$y(0) = y'(0) = 0 \qquad (4.52)$$

4.6/SOLUTION OF A SECOND-ORDER EQUATION   107

by numerical methods. It can be assumed that the input $x(nT)$ is obtained from the analog portion of the hybrid simulation, that the differential equation is solved on the digital computer, and the solution $y(nT)$ is to be returned to the analog computer through a digital to analog converter. In block diagram form, the solution of this equation by analog and digital methods is indicated in Figure 4.13. We shall apply both rectangular and trapezoidal integration formulas to the solution of this equation.

Substituting the transfer function for the rectangular integrator into the integration boxes $I(z)$, we obtain the closed-loop transfer function

$$\left[\frac{Y(z)}{X(z)}\right]_{\text{rect.}} = \frac{T^2}{z^2 + z(3T - 2) + (1 - 3T + 4T^2)} \qquad (4.53)$$

The roots of the denominator of this expression are located at

$$r_1, r_2 = (1 - 1.5T) \pm j(0.5\sqrt{7}T) \qquad (4.54)$$

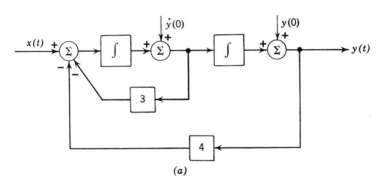

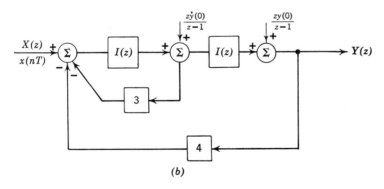

Figure 4.13 Solution of second-order equation $\ddot{y} + 3\dot{y} + 4y = x$: (a) analog diagram and (b) numerical (z-transform) diagram.

As before, a plot of the location of these roots as a function of T (the T-locus) gives an indication of the stability of this solution and provides an indication of the allowable range of sampling intervals. The T-locus is given in Figure 4.14. A second branch of the locus is located at corresponding negative imaginary values.

If trapezoidal integration is used for the two integrator transfer functions in Figure 4.13b, the closed-loop transfer function becomes

$$\left[\frac{Y(z)}{X(z)}\right]_{\text{trap.}} = \frac{T^2(3z-1)^2}{4z^4 + (18T-8)z^3 + (4-25T+36T^2)z^2 + (6T-24T^2)z + 4T^2}$$

(4.55)

The T-locus for the roots of the denominator of this expression is shown in Figure 4.15. Once again, only the roots in the upper-half plane are shown. As with the solution of the first-order equation, it can be seen that at least one root of the trapezoidal method leads to an unstable solution for smaller values of sampling interval than for the rectangular integration technique.

The frequency domain behavior of the two solutions can be investigated by examining Bode plots for the two transfer functions, 4.53 and 4.55.

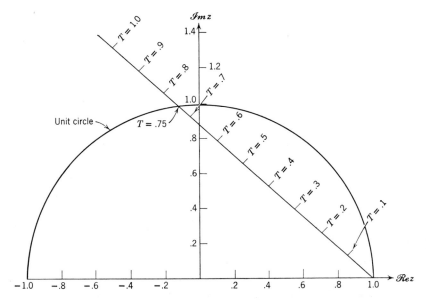

Figure 4.14 *T-locus for solution of Equation 4.52 by rectangular integration.*

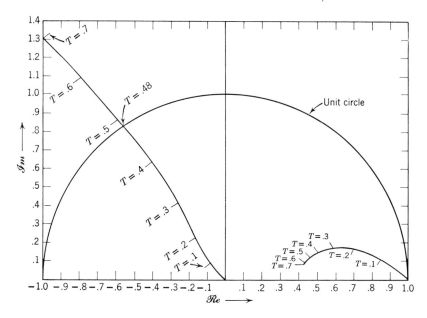

Figure 4.15 T-locus for solution of Equation 4.52 by trapezoidal integration.

The Bode plots are obtained by the substitution

$$z = \cos \omega T + j \sin \omega T \qquad (4.56)$$

as indicated previously. Bode plots showing the gain and phase of the solution by both trapezoidal and rectangular integration with $T = .5$, as well as the continuous (exact) solution of Equation 4.52, are given in Figures 4.16 and 4.17. Obviously, in general, families of such curves for different values of sampling interval (integration step size) may be required in order to study the properties of the solution. An examination of the sets of curves in these two figures shows results similar to those obtained for the first-order equation, namely, that trapezoidal integration has a better gain characteristic while rectangular integration has a better phase characteristic. This point is discussed further in Section 5.7.

4.7 SUMMARY

The purpose of this chapter was to demonstrate the utility of z-transform techniques in the study of numerical integration. The use of such transform methods provides new insight into the dependence of solution stability,

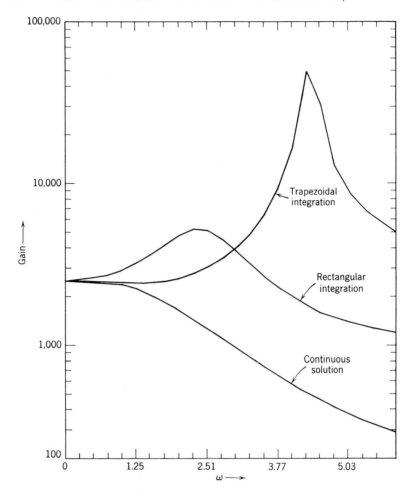

Figure 4.16 Solution of Equation 4.52: gain vs. frequency ω for $T = 0.5$.

truncation error, and the frequency domain behavior of solutions as a function of integration step size. The techniques of this chapter can be extended to the study of complete hybrid systems in which certain operations are performed on the analog computer and others on the digital computer. Since the computations required to obtain significant results by this technique are somewhat complex, they will not be reviewed here and the reader is encouraged to consult the work of Gilbert.[2] However, some results obtained by transform methods in the error analysis of hybrid flight simulations are discussed in Chapter 12.

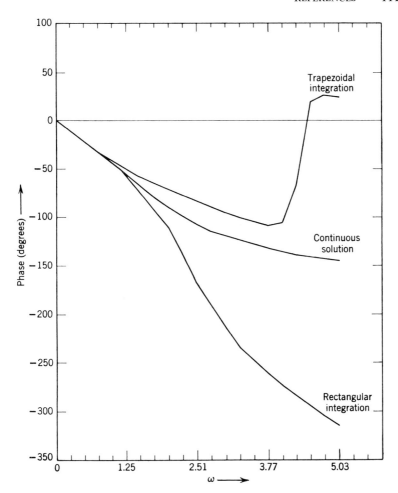

Figure 4.17 Solution of Equation 4.51: phase vs. frequency ω for $T = 0.5$.

References

1. Howe, R. M., "Notes on Error Analysis of Combined Analog-Digital Computer Systems," Unpub. Rpt., Univ. of Mich., 1964.
2. Gilbert, E. G., "Dynamic Error Analysis of Digital and Combined Analog-Digital Computer Systems," *Simulation*, **6**, 241–257, April 1966.

3. Jury, E. I., *Theory and Application of the Z-transform Method*, John Wiley, New York, 1964.

4. Kuo, B. C., *Analysis and Synthesis of Sampled-Data Control Systems*, Prentice-Hall, Englewood Cliffs, N.J., 1963.

5. Tou, J. T., *Sampled-Data and Digital Control Systems*, McGraw-Hill, New York, 1959.

6. Fowler, M. E., "Numerical Methods for Use in the Study of Spacecraft Guidance and Control Systems," *Proc. IBM Symp. on Digital Simulation of Continuous Systems*, IBM, White Plains, N.Y., pp. 47–69, 1967.

7. Gilliland, M. C., "A Spectral Stability Analysis of Finite Difference Operators," *IEEE Transactions on Electronic Computers*, **EC-15**, 849–854, Dec. 1966.

8. Henrici, P., *Discrete Variable Methods in Ordinary Differential Equations*, John Wiley, New York, 1962.

9. Monroe, A. J., *Digital Processes for Sampled-Data Systems*, John Wiley, New York, 1962.

10. Saucedo, R., and T. W. Sze, "Analog Simulation of Digital Computer Programs," *AIEE Transactions*, **80**, Part II, 703–709, Jan. 1962.

11. Wait, J. V., "State-Space Methods for Designing Digital Simulations of Continuous Fixed Linear Systems," *IEEE Transactions on Electronic Computers*, **EC-16**, 351–354, June 1967.

12. Hung, J. W., et al., "Investigation of Numerical Techniques as Applied to Digital Flight Control Systems," *U.S. Air Force Flight Dynamics Laboratory Technical Report*, AFFDL-TR-66-68, Feb. 1967.

5

Error Analysis of Hybrid Computer Systems

5.1 INTRODUCTORY REMARKS

The subject of error analysis is central to any meaningful discussion of hybrid computer systems. The specification of the characteristics of each subunit of the hybrid computer loop, as well as the topology of the computing components, must be governed to a large extent by the accuracy desired of the computer solution and by the effect of nonideal behavior in any of the system components upon overall solution errors. Inevitably the specification of a complex computer system involves a variety of engineering trade-offs and compromises, and an effective error analysis is the only meaningful approach to such engineering design problems.

The basic error-analysis problem can be stated in two ways:

1. Given a hybrid computer system consisting of a specific interconnection of computer components, each with a specified tolerance, what is the probable accuracy of the overall computer solution?
2. If it is desired to obtain a computer solution of a specified accuracy, what must be the accuracy or tolerance of each system component?

That it is unlikely that fully satisfying answers to the above questions will become available in the near future can be inferred from the present status of the error analysis in pure analog and digital computation. Although many sophisticated studies have been made of the errors inherent in the linear analog computer components and their effect on the solution of systems of linear ordinary differential equations, there have been very few extensions of these efforts to complex nonlinear differential equations. In most large analog simulations, it is virtually impossible to predict an overall solution accuracy. Similarly, in digital computations, the effects of accumulated truncation and round-off errors upon the solution of systems of differential equations are always difficult and usually impossible to predict accurately. If a system of differential equations is integrated over an appreciable number of time steps, the probable error in the solution is virtually impossible to estimate. A hybrid computer loop contains an analog computer, a digital computer, and linkage equipment. All the errors inherent in analog and digital computation are therefore present in the hybrid loop. Moreover the errors introduced by the linkage system must be taken into account. Nonetheless, it is of utmost importance that all available analytical and experimental tools for error analysis be brought to bear upon hybrid computation and that every effort be made to develop new theoretical approaches to the subject.

The hybrid computer system considered in this chapter is assumed to consist of a general-purpose digital computer, a general-purpose analog computer, a multiplexer, an analog-digital converter, a demultiplexer, a digital-analog converter, and sample-hold circuitry. In addition, as described in more detail later, it is assumed that some sort of prediction circuit is employed to compensate for serious errors due to time delay in the digital computer and in the analog-digital converter. The primary emphasis of this chapter is upon the errors introduced by the various components of the linkage system. To these should be added the errors introduced by the general-purpose computers. In considering hybrid computer systems, a distinction must be made between three classes of such systems:

1. Analog-computer oriented systems.
2. Digital-computer oriented systems.
3. Balanced linkage systems.

In an analog-computer oriented system, one or a number of continuous variables, usually generated within the analog computer, constitutes the final solution being sought. In that case, the digital computer is primarily used as a peripheral element for function generation, memory, or control. In digital-computer oriented systems, on the other hand, discrete variables usually generated by the digital computer constitute the final solution. In that case, the analog computer is the subsidiary element, usually performing a limited number of operations, including the generation of high-frequency components in control systems or permitting the utilization of actual system hardware in a hybrid simulation. In balanced linkage systems, both analog and digital outputs are of more or less equal importance. In performing an error analysis upon a hybrid system, it is of considerable importance to specify clearly to which of these three categories the system under analysis belongs. Errors are always defined as departures from a correct or ideal solution; a meaningful error analysis is impossible without a clear understanding as to which of the many dependent variables generated in the closed-loop hybrid system constitute the desired solution.

In this chapter the major types of errors occurring in a hybrid computer are first classified in a general way. A clear distinction is made here between component errors and solution errors. The errors inherent in the major components of the hybrid computer loop are then analyzed separately. The concept of sensitivity functions is then introduced to permit a discussion of the solution errors in the hybrid system. A description of several techniques for compensating the major sources of error in a hybrid computer loop concludes the chapter. Errors related to the approximation of continuous variables by discrete variables in the digital computer are also considered in Chapter 4.

5.2 CLASSIFICATION OF ERRORS

For present purposes the hybrid computer loop can be viewed as shown in Figure 5.1. Each of the major components of the hybrid system is shown in block form together with major sources of error introduced by the nonideal functioning of each component. Each block in this figure has one or a

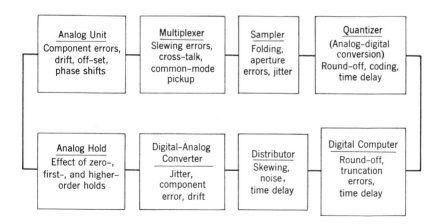

Figure 5.1 Hybrid loop including major error sources.

multiplicity of inputs and one or a multiplicity of outputs. For each unit there exists an "ideal" transfer characteristic or input-output relationship. As a result of the sources of error associated with each block, the input-output relationship departs from this ideal. Some of the error sources are in the nature of random noise and are completely unpredictable; others are systematic in nature and therefore lend themselves to some form of compensation. Both types, however, are instrumental in degrading the quality of the computer system solution and must be carefully controlled in designing and maintaining the computer system.

To date, most error studies pertinent to hybrid systems have been limited to investigations of errors in the transfer characteristics of the system components. Hybrid systems are usually built up from subunits furnished by different manufacturers; an input-output specification constitutes really the only satisfactory communication link between the designer of the system and the designer of the subcomponent. The system designer then hopes that if all subcomponents are conservatively specified, the overall solution accuracy will be within tolerable bounds. It is clear, however, that each specific utilization of the hybrid system will be influenced in a different way by the various error sources. Thus, in some computations, sampling errors can have very serious results, although they may be of minor importance in other applications. Similarly, an analog-digital converter using 6 bits may be sufficient to permit an overall solution accuracy of .1% in certain hybrid system configurations, while 10- or 14-bit word-lengths may be required in other problem areas. Not infrequently it is necessary to effect an engineering compromise or trade-off in component characteristics. For example, the longer the word in an analog-digital converter, the fewer outputs per second can be generated. It becomes clear that an effective utilization of hybrid computer systems demands the derivation of relationships between errors in the transfer characteristics of the system components and errors in the computer solution.

Most hybrid computer systems operate in a synchronous manner, information being transferred between the elements of the hybrid loop only at discretely and regularly spaced intervals governed by a precision clock. Between clock pulses the output of each of the subunits either remains at the value it had at the preceding clock pulse or is reduced to zero. Since the overall solution is therefore generated in a stepwise manner in the time domain, it is logical to inquire into the effect of each input-output error source on one step in the solution. That is, assume that no errors have been committed up to a certain point and that an input-output error is introduced at that point. As a result, the next step in the computer solution will be in error, this error then being a measure of the sensitivity of the overall solution to that specific input-output error. Per-step solution errors are relatively easy to derive and constitute an interesting and meaningful measure of the importance of each specific error source.

In actual hybrid system operation, errors are committed by each computer component during each computing cycle. The solution generated by the hybrid system therefore departs more and more from the correct value as time progresses. This may be viewed as being due to an accumulation of per-step errors. If a hybrid system is to be operated over a specified length of time, it is of greatest interest to know what will be the greatest error or greatest departure from the solution as the result of accumulated per-step errors. This constitutes the most difficult and most challenging problem in error analysis, and one which has by no means been successfully solved.

Summarizing the above discussion, an error analysis of hybrid systems involves a consideration of three types of errors:

1. Subsystem input-output errors
2. Per-step solution errors
3. Accumulated solution errors

Sections 5.3 through 5.7 are devoted to a discussion of the input-output errors of the major components shown in Figure 5.1. The remaining sections of this chapter consider the relation of these input-output errors to the per-step and the accumulated solution errors for analog-computer-oriented systems.

SUBSYSTEM INPUT-OUTPUT ERRORS

5.3 TIME-DELAY COMPENSATION

One of the most serious errors introduced into the hybrid loop arises because of the time delays inherent in analog-digital conversion and in digital computations. A digital computer operates in a serial manner, so that the computing time is directly related to the complexity of the computation. This is in contrast to analog devices in which all computations are performed in parallel so that the computing time is independent of problem complexity. In an analog-oriented hybrid computer system, an analog variable or a set of such variables is sampled in turn, converted into digital code, processed by the digital computer, reconverted into analog form, and distributed to the analog computer. The input to the analog system is therefore in the form of a set of staircase functions corresponding to the desired calculation but shifted in time. To make frequency domain analysis possible, let us assume that the digital computer solves a set of linear, constant coefficient differential equations. In that case, the operation performed by the digital computer can be represented by the transfer function $D(s)$. (It should be recognized, however, that the compensation schemes discussed in this section are generally applicable and do not depend on linearity). Recalling that time delay in the Laplace transform domain is equivalent to multiplying by $\epsilon^{-\tau s}$ where τ is the delay time, the time delays of the hybrid computer loop can be represented as shown in Figure 5.2. In this figure, τ_1 is the time delay inherent in analog-digital conversion and is usually related to the word-length of the digital output, τ_2 is the time required for the digital computations and is related to the complexity of these computations, and T is the time between successive samples. The analog output $X(s)$ is therefore related to the analog input $Y(s)$ by the equation

$$Y(s) = X(s)D(s)\epsilon^{-(\tau_1+\tau_2)s} \cdot \frac{1 - \epsilon^{-Ts}}{s} \tag{5.1}$$

The relationship between the analog input and the time delays is illustrated in Figure 5.3. Curve (a) represents the desired analog input, which would exist if no time delays were inherent in the hybrid computer loop. The staircase curve (b) is the output of the zero-order hold, that is, the actual analog input. Provided the sampling interval T is sufficiently small and the analog computer sufficiently insensitive to the high-frequency components of the staircase function, the output of the zero-order hold can be approximated by the dotted line (c). It is clear from Figure 5.3 that the total time delay introduced by the hybrid computer-loop is equal to the sum of the analog-digital converter time delay, the digital computer time delay, and one-half the sampling interval. Designating this total time delay by τ,

$$\tau = \tau_1 + \tau_2 + \frac{T}{2}$$

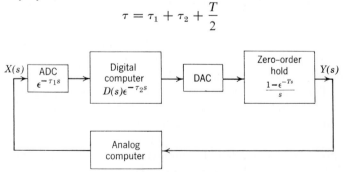

Figure 5.2   Time delays in hybrid loop.

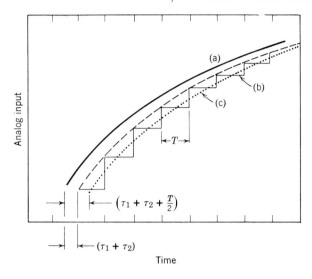

Figure 5.3 Time delay in hybrid loop due to conversion, digital computation, and zero-order hold.

Miura[1] has analyzed the effect of time delay upon the characteristic roots of systems of linear differential equations solved on a hybrid computer. The effect of this time delay can be illustrated in a simple fashion by considering the hybrid solution of the undamped second-order differential equation

$$\frac{d^2y}{dt^2} + \omega^2 y = 0 \tag{5.2}$$

This equation was treated on a hybrid computer by using the circuit shown in Figure 5.4. In this case, the digital computer was employed only to

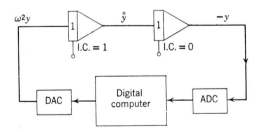

Figure 5.4 Test circuit for evaluating effect of time delay.

multiply by $-\omega^2$. The effect of time delay in this computation is shown in Figure 5.5. Curve (a) is the true solution, obtained by using pure analog techniques; curve (b) is the output of the system shown in Figure 5.4, and curve (c) is the difference between curve (a) and curve (b). It can be seen that the time delay is effective in introducing a progressively larger and larger solution error. Clearly, the oscillatory nature of this error is characteristic of the specific problem being solved. For other hybrid solutions, this error could be a monotonic or a more complex oscillatory function.

In most hybrid computations, the time-delay error is sufficiently serious to warrant continuous compensation. Since the basic compensation procedure is relatively simple and straightforward, such compensation circuits are very widely used. In considering errors inherent in the hybrid

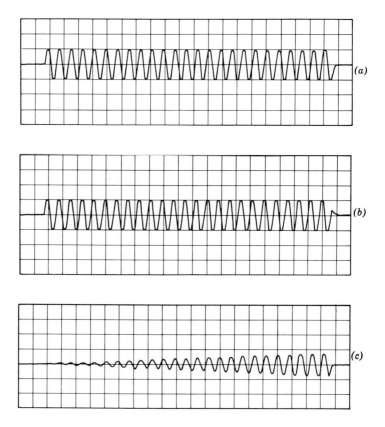

Figure 5.5 (a) *Correct solution of Equation 5.2.* (b) *Solution with time delay due to digital computation and conversion.* (c) *Error due to time delay.* (*Courtesy of E. Hartsfield.*)

computer loop, it therefore becomes necessary to discuss the errors introduced by these compensating circuits, that is, errors introduced by the incomplete or improper elimination of the time-delay error. Since the effect of time delay is to cause the analog input to arrive τ seconds too late, the obvious approach to the compensation of time-delay errors is to employ an approximate prediction scheme. In effect, it is necessary to insert into the hybrid loop an element with a transfer function which approximates the ideal predictor $\epsilon^{\tau s}$, in order to eliminate time-delay effects. The actual nature of such a predictive scheme can best be understood by considering a series expansion of the time-delay operator:

$$\epsilon^{-\tau s} = 1 - \tau s + \frac{1}{2!}\tau^2 s^2 - \frac{1}{3!}\tau^3 s^3 + \cdots \tag{5.3}$$

In most currently used compensators, it is assumed that the second term on the right side of Equation 5.3 is much larger than the higher-order terms. This is usually a reasonable assumption since τ is a very small number, so that higher powers of τ are very small indeed. Accordingly, to eliminate the major portion of the time-delay error term, it is necessary to process the analog output through a filter with the transfer function

$$F(s) = 1 + \tau s$$

Using the Taylor series 5.3,

$$e^{-\tau s}F(s) = 1 - \tfrac{1}{2}\tau^2 s^2 + \tfrac{1}{6}\tau^3 s^3 - \cdots \tag{5.3a}$$

It can be seen that use of the filter $F(s)$ has indeed eliminated the first-order term. However, since the operator τs represents differentiation, which is a noise-amplifying process, it is, of course, expedient to avoid the actual differentiation of the analog computer output if at all possible. There exist, in fact, three general approaches to time-delay error compensation:

1. Modification of the analog input by the addition of a voltage corresponding to $\tau s\ Y(s)$.
2. Modification of the analog output (the input to the linkage system) by the addition of a term corresponding to $\tau s\ X(s)$.
3. Modification of the output of the digital computer (obtained by digital calculations) by a term corresponding to $\tau s\ Y(s)$.

The choice of any one of the above methods is generally determined, at least to some extent, by the specific problem being solved. If the analog output is generated by an integrator, it is usually a relatively simple matter to obtain the derivative of the analog output from the analog computer unit generating the input of that integrator. This is illustrated in Figure 5.6 for the simple second-order equation, Equation 5.2. Note that the voltage $\dot{y}$ is multiplied by τ using a potentiometer, and the result after sign-change is added to the variable $-y$. This technique might be considered a pre-distortion method to take into account, in advance, the time delay inherent in the linkage system.

If the analog input is applied directly to an analog integrator, it is often more convenient to modify the output of the digital computer. This is illustrated in Figure 5.7a, again using Equation 5.2 as an example. Alternatively, the correction term can be applied to the input integrator through an input capacitor, as shown in Figure 5.7b.

A third approach to the compensation of time delay involves making the correction within the digital computer. The first derivative term is then approximated by a simple backward difference according to

$$\left(\frac{dy}{dt}\right)_n \simeq \frac{1}{T}(y_n - y_{n-1}) \tag{5.4}$$

where the subscripts $n-1$ and n identify the value of y at two successive sampling instants. In employing this method, it is necessary to store the digital outputs for one computing cycle so that these may be employed in calculating the first derivatives for the next step. The output of the digital computer is then modified by the addition of the term

$$\frac{\tau}{T}(y_n - y_{n-1})$$

This technique eliminates the need for analog differentiation but entails a number of other difficulties. Usually, in a hybrid computer-loop the digital

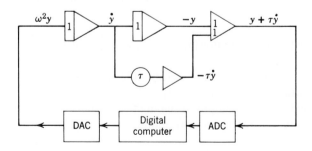

Figure 5.6 Method for compensating time-delay error by "predistortion."

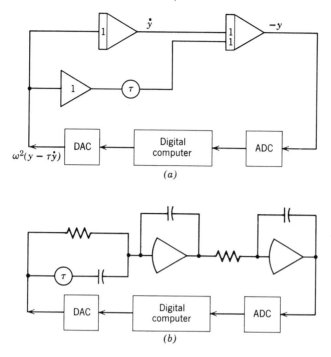

Figure 5.7 Alternative techniques for time-delay error compensation.

computer is the slowest computing element and therefore acts to limit the available bandwidth. This effect is accentuated if the additional calculations for the compensation of the time delay are performed in the digital computer. Also, if T is small, the calculation of the first derivative involves the subtraction of two almost equal numbers, and hence implies an appreciable loss of significant figures.

In evaluating the effect of shortcomings in the prediction scheme on the closed-loop hybrid solution, it is necessary to consider two types of errors:

1. Errors due to an incorrect specification of the loop time-delay τ.
2. Errors due to approximating Equation 5.3, using only the first two terms on the right-hand side.

The total time delay usually can be estimated with reasonable accuracy if the analog-digital converters are of the successive-approximation type and hence require the same number of iterative cycles regardless of the analog input and if the number of computations performed in the digital computer is invariant. Occasionally, problems are encountered in which the algorithm performed by the digital computer is modified in the course of the solution run. Also, if a compiler is used to program a digital computer or if the digital computer employs a magnetic drum or other serial memory, some uncertainty may exist as to the length of time required to generate the digital computer output. Errors due to the truncation of the series expansion in Equation 5.3 become significant if the analog variable read into the digital computer is rapidly varying so that the second and higher derivatives are not negligible. The control of these errors is discussed briefly in Section 5.9.

An example of the improvement possible with time-delay compensation is given in Chapter 12 in connection with digital function generation for flight simulation.

5.4 SAMPLER ERRORS

Sampling is unavoidable in a hybrid computer-loop, since at some point the continuous analog variables must be converted into a sequence of digital words. Sampling is nearly always a synchronous process, successive samples being equally spaced in the time domain. The choice of the sampling interval is usually dictated by the slowest elements in the computer-loop, that is, the analog-digital converter and the digital computer, as well as by the number of analog channels which must be processed in sequence. Some of the errors introduced into the hybrid loop by the sampling device are inherent in the sampling operation while others are due to shortcomings in the operation of the sampler itself.

Of fundamental importance in sampling theory is the sampling theorem usually associated with the name of C. E. Shannon. According to this theorem, the sampling frequency must be at least twice as great as the frequency of the highest Fourier component of the analog signal being sampled. If the sampling frequency is less than two times the frequency of the highest Fourier component, errors are introduced into the sampled signal; these errors cannot be removed by subsequent filtering or other manipulation. The reason for this is that a desampling or data-reconstruction device, even if operating in an ideal manner, is unable to distinguish frequency components of the analog signal which are ϵ higher than one-half the sampling frequency from components which are ϵ lower than one-half the sampling frequency. The higher-frequency components are therefore misinterpreted as lower-frequency components, and the desampled output is in error. Consider, for example, the sampling, at a sampling rate of 100 samples per second, of two sinusoidal signals, one having a frequency of 60 cycles per second and the other a frequency of 40 cycles per second.

It can be shown that at every sampling instant the magnitude of these two sine waves is exactly identical. The sampler will therefore misinterpret the 60 cycle per second signal and report it as a 40 cycle per second signal.

In applying the sampling theorem, it is important to recognize that the highest signal frequency, which determines the required sampling frequency, is not the highest frequency of interest but rather the highest frequency actually occurring in the analog signal. Usually, the need for an accurate reconstructed output dictates the use of a sampling rate far greater than twice the highest frequency of interest. Nonetheless, it is absolutely necessary to assure that unwanted frequencies resulting from noise pickup, cross-talk, or switching transients in the analog system are not permitted to enter the sampling mechanism. Accordingly, it is always desirable to precede the sampler with a signal conditioner in the form of a low-pass filter. If the analog-digital converter in the hybrid loop is preceded by a multiplexer, it must be recognized that the sampling operation is actually performed at the input to the multiplexer, so a separate filter must be provided for each multiplexer channel. If the filter is placed instead between the multiplexer output and the analog-digital converter input, the analog signals may be hopelessly degraded prior to filtering.

An error which generally arises as part of the sampling process is due to uncertainty as to the exact instant of time at which the analog signal is sampled. A sampler may be viewed as a switch which is closed for a brief instant of time, permitting the analog signal to be applied to the analog-digital converter. The sampling process is usually initiated by a signal from an electronic clock, a precision pulse generator. After the clock pulse is generated, some time is necessarily required for the sampling switch to become enabled. Uncertainties or variations in this length of time, termed sampler jitter, are a source of error in hybrid computer systems.

The sampling switch remains closed for a short but finite time interval, known as the aperture time. If the analog signal undergoes changes while the sampling switch is closed, some uncertainty exists as to the significance of the output of the analog-digital converter. Evidently, it is desirable to make the aperture time as small as possible, but technical considerations place a minimum upon this time interval.

5.5 QUANTIZING ERRORS

The analog-digital converter is instrumental in introducing a number of errors or perturbations into the hybrid computer-loop. Most modern converters consist of a closed loop, including at least one digital-analog converter, and one or several comparators. These devices are, of course, subject to the same component tolerance, zero off-set, and drift errors as other analog devices. In most high-class analog-digital converters, these errors are kept below .01 % of full scale by careful design and production techniques. Under these conditions, the analog errors introduced by the analog-digital converter are smaller than the errors introduced by the nonlinear operational elements within the analog computer.

A second form of error inherent in quantizers is due to the limited number of bits which are employed to represent the analog variable. Inevitably there is an uncertainty equivalent to $\pm\frac{1}{2}$ of the least significant bit of the converter. For example, if 10 bits are used, corresponding to slightly over 1000 quantization levels, the uncertainty in the value read out is $\pm.05\%$. That quantization is a highly nonlinear phenomenon is illustrated in Figures 5.8a and 5.8b. Figure 5.8a is a plot of the input-output relation

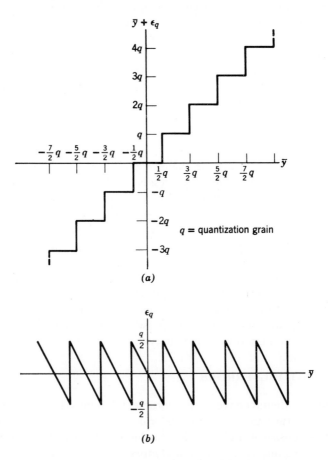

Figure 5.8 Per-step error introduced by quantizer.

of the quantizer; q is the distance between successive output levels and is known as the quantization interval or the quantization grain. Figure 5.8b illustrates the error ϵ_q as a function of the quantizer input. Quantization effects have been studied in considerable detail by Widrow.[2] It can be shown that quantizing is equivalent to the addition of random noise to the analog signal.

A final error which may be occasionally introduced by quantizers is due to the nature of the specific digital code employed. Consider, for example, that a pure binary code is employed and that the amplitude of the analog signal is changing from a magnitude of 3 to a magnitude of 4. This means that the output register of the analog-digital converter must change from 0011 to 0100. Suppose now that the flip-flops constituting this register are such that a change from the 0 to the 1 condition occurs more rapidly than a change from the 1 to the 0 condition. For a brief instant, therefore, the output register of the converter will be 0111, which is the binary equivalent of the number 7. This could produce an erroneous read-out of the converter. This type of error can be avoided by assuring that the "conversion complete" pulse emitted by the converter is timed so as to permit all flip-flops to set properly. Occasionally it is preferable to utilize a special binary code designed to eliminate this danger. Such a code is constructed so that the change from one number to the next smaller or larger number never involves a change of more than one binary digit. An example of such a code is the Gray code, widely used in certain hybrid systems.

5.6 MULTIPLEXING AND DISTRIBUTING

A multiplexing unit is employed in hybrid systems to permit the analog-digital converter and access to the digital computer to be time-shared among a number of analog channels. The input of the multiplexer is therefore connected to a large number of points within the analog computer while its output usually forms the input to the analog-digital converter. As explained previously, filters and sample-hold circuits may be employed as buffers at the input and output of the multiplexer. Two general classes of multiplexers are in wide use: electromechanical and solid-state.

Solid-state multiplexing units employ a field-effect transistor or a transistor switching circuit in each channel. Zero off-sets, noninfinite backward resistance, and nonzero forward resistance are then the major sources of error introduced by the solid-state switch into the hybrid loop. By careful design techniques, these errors can generally be kept below .01% of full-scale (usually 10 volts); these errors are not present in electromechanical multiplexers employing crossbar switches or reed relays.

Errors which are common to solid-state and electromechanical multiplexers are those due to inadequate isolation of each of the signal input lines. Coupling between adjacently located input lines results in "crosstalk," usually an unpredictable disturbance and therefore classified as noise. Not infrequently, switching transients are coupled from one line to another, leading to serious errors. A second source of error due to inadequate design results in "common-mode" signals. A common-mode voltage is defined as a potential which exists equally between each of two signal-carrying lines and ground. This is in contrast to signal voltages which appear only between the wire pair. Common-mode voltages occur in every large hybrid system and must be carefully controlled. Common-mode voltages usually result directly from an inadequate ground system. Even if the difference in ground potential at the analog computer, the multiplexer, and the analog-digital converter amounts to only a few millivolts, perceptible common-mode errors can arise. A final source of random noise in the analog signal lines is due to pickup from noise sources outside the system. This source of error is minimized by careful shielding of all lines.

An additional important source of errors in some multiplexed systems is due to the time delay between successive switch closures. If the multiplexer is employed to sample a sequence of analog variables so that these may be processed in the digital computer, it is generally assumed in programming the digital computer that all these samples are taken simultaneously, that is, that they refer to the same instant of time. For example, in the treatment of the Van der Pol equation, a digital computer may be employed to generate the term $y^2\dot{y}$. The voltages corresponding to y and to $\dot{y}$ are then obtained from the analog system and converted into digital form. As a result of time delay in the multiplexer unit, the data received by the digital computer actually refers to y and $\dot{y}$ at two different instants of time. This is known as "slewing," and slewing errors can cause a serious degradation of certain hybrid solutions. One technique for avoiding slewing errors involves the use of sample-hold units at the input lines to the multiplexer unit. All analog channels are then sampled simultaneously so that the data introduced into the digital computer all refer to the same instant of time, regardless of the number of analog channels or the multiplexer delay time. Alternatively, sensitivity analysis, as described in more detail in Section 5.9, can be employed to control the slewing error.

Demultiplexing, prior to digital-analog conversion, involves the distribution of digital information. Since digital signals are limited to "ones" and "zeros," there is little chance for perturbations or signal degradation as a result of electronic shortcomings. An error akin to the "slewing" error can arise in certain hybrid systems due to the time delay between successive outputs of the digital distributor. This error, sometimes termed "skewing," may introduce unwanted phase shifts into the analog system. On occasion it can be controlled by the introduction of additional sample-hold units.

5.7 DIGITAL-ANALOG CONVERSION AND DATA RECONSTRUCTION

The purpose of the digital-to-analog portion of the linkage system is to translate the sequence of words read out of the digital computer into continuous analog voltages. The main unit employed for this is, of course, the digital-analog converter. In some hybrid systems this converter or an array of converters serves also as an analog-hold; in other hybrid systems, separate sample-hold circuits are provided for each analog input channel. As in the case of sampling devices, two types of errors are encountered: (1) electronic errors due to the nonideal behavior of the analog components, and (2) errors inherent in the sampling-desampling process.

Electronic errors are usually due to zero off-set and component tolerances. Digital-analog converters are constructed by using a ladder network of weighted precision resistors. Most of these are usually kept in an oven to eliminate temperature drift. By careful design, errors in digital-analog conversion can be kept well within a .01% tolerance. Most linkage systems limit signal amplitudes to a range of -10 volts to $+10$ volts, whereas many analog computer systems have a -100-volt to $+100$-volt dynamic range. Accordingly, it is necessary to step up the output of the linkage system by a factor of 10 and to step down the input of the linkage system by a similar ratio. Special amplifiers must be provided for this purpose to minimize drift effects and the loading of critical points in the analog circuit.

A major source of error in the digital-analog portion of a linkage is due to the fact that the zero-order or first-order hold circuits which are used do not have the transfer functions which are ideally required. In effect, the data read out of the digital computer represent a continuous function sampled $1/T$ times a second. The amplitude of each sample corresponds to the amplitude of the continuous function at the corresponding instant of time, and the reconstruction device must perform the necessary extrapolation to generate a continuous function from these samples. It can be shown that if a continuous function is sampled with the sampling frequency $\omega_s = 2\pi/T$, the Fourier transform $F^*(\omega)$ of the sampled signal will be

$$F^*(\omega) = \sum_{k=-\infty}^{+\infty} C_k F(\omega + k\omega_s) \qquad (5.5)$$

where $F(\omega)$ is the Fourier spectrum of the continuous function being sampled. C_k are the Fourier coefficients and depend on the width of the sampling pulse. For impulse sampling, all the Fourier coefficients are equal.

The sampled signal therefore contains all the Fourier components of the original signal as well as so-called repeated spectra, repetitions of $F(\omega)$ as k ranges from $-\infty$ to $+\infty$. It is the purpose of the desampling device to separate the pure components, $F(\omega)$, from the repeated spectra. For this purpose, a low-pass filter is required—a filter which has zero attenuation within the desired frequency band and perfect attenuation outside. Such a filter is unfortunately physically unrealizable and can only be approximated by using conventional hold circuits.

In essence, the desampler or analog-hold unit is required to perform an extrapolation or prediction of the sampled signal. On the basis of the last received sample and possibly preceding samples, the analog-hold circuit is required to generate an output approximating as closely as possible the continuous function until the next sample is received. In a zero-order hold circuit, the analog output is merely updated each time a new sample is received and is held constant between sampling instants. The output of the zero-order hold is therefore a staircase function. In the case of the first-order hold circuit, the approximating function consists of a series of ramps, each starting with the magnitude of the latest sample and having a slope obtained by an equation similar to Equation 5.4. Typical outputs of zero- and first-order hold circuits are shown in Figure 5.9.

There are a number of ways of describing the error introduced by the analog hold unit. One widely used technique for describing linear systems involves the use of steady-state sinusoidal excitations. The amplitude and phase of the output sine wave are then compared with the data of the input sine wave. Figures 5.10a and b illustrate the amplitude and phase characteristics of zero-order and first-order hold circuits, respectively. The dotted line in each case represents the ideal response. Note that ideally the hold circuit should have a sharp cut-off at one-half the sampling frequency (i.e., at $\omega_s/2$ where $\omega_s = 2\pi/T$) in order to satisfy the sampling theorem. The humps present in the amplitude curves beyond $\omega_s = 2\pi/T$ are indicative of the fact that the hold circuit will pass a significant fraction of the harmonics introduced by the sampler (Equation 5.5). If we are concerned only with the amplitude distortion introduced by the hold circuit, then a linear variation of phase with frequency (which corresponds to a pure time delay) is acceptable. Unfortunately, attempts to use the phase and amplitude data of Figure 5.10 to predict the error of the sampler-hold operation are misleading, for several reasons. First, the curves of Figure 5.10 refer to the hold circuit alone, and not to the sampler-hold combination. A

5.7/DIGITAL-ANALOG CONVERSION AND DATA RECONSTRUCTION 131

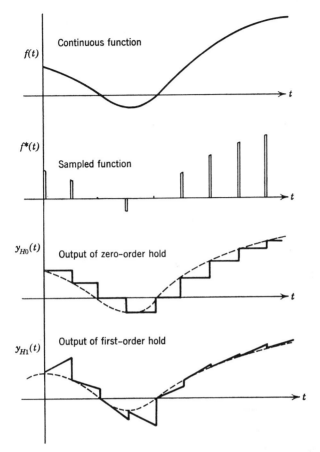

Figure 5.9 Outputs of zero-order and first-order hold circuits.

simple sinusoid of frequency ω_0, sampled and held, results in a wave-form containing higher harmonics of frequency $\omega_0 + n\omega_s$, $n = \pm 1, \pm 2, \ldots$. Therefore, the curves of Figure 5.10 could be used only to estimate amplitude and phase errors for the fundamental of the reconstructed wave (or any of its harmonics). If only the fundamental data are used, the following results are obtained: for a zero-order hold at 10 samples per cycle the amplitude error is 1.1% and the phase error 18%; for 30 samples per cycle the amplitude error is .7% and the phase error is 6.5%; for 50 samples per cycle the amplitude error is .1% and the phase error 3.6%. For the first-order hold and a sufficiently high sampling frequency the amplitude

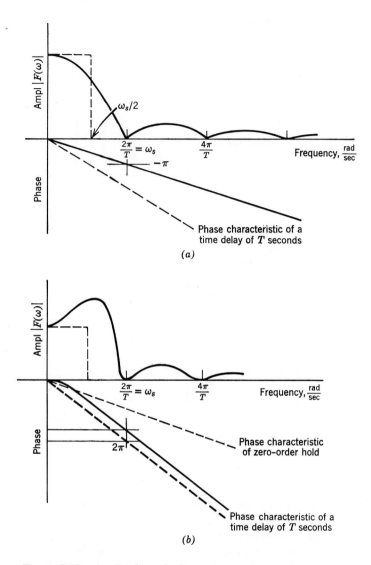

Figure 5.10 Amplitude and phase characteristics of commonly used sample-hold circuits: (a) zero-order hold and (b) first-order hold.

5.7/DIGITAL-ANALOG CONVERSION AND DATA RECONSTRUCTION

error is opposite in sign but of approximately the same magnitude, while the phase error is greatly reduced. These figures make it clear that it is necessary to sample continuous data far more frequently than required by Shannon's sampling theorem if accurate reconstructions are to be attained.

The use of sinusoidal excitations is misleading even if only the output fundamental is considered, because the amplitude response will lead to an estimate of the error in the sine wave at its maximum point (at 90°). The absolute error in the zero-order hold is a minimum in that point (it is a maximum at 0°), while the error in the first-order hold is a maximum at 90°. Accordingly, a comparison of the amplitude response of zero-order and first-order hold circuits is actually a comparison of the first-order circuit at its worst and the zero-order circuit at its best. A more realistic analysis is made by considering instantaneous error rather than amplitude and phase errors.

Consider that it is desired to reconstruct a continuous function

$$y(t) = A \sin \omega t \qquad (5.6)$$

which has been sampled with the sampling period of T, corresponding to p samples per cycle. For a zero-order hold, the reconstructed signal is held constant between sampling instants, so that the absolute error is zero at the sampling instant and varies according to the equation

$$\epsilon = A \sin \left[\left(n + \frac{\tau}{T} \right) \omega T \right] - A \sin n\omega T \qquad 0 \leq \tau < T \qquad (5.7)$$

The maximum instantaneous error occurs in the vicinity of the maximum slope of y in Equation 5.6, that is, in the vicinity of 0°. At the sampling point immediately following passage through zero, the second term on the right side of Equation 5.7, corresponding to the output of the hold unit, is zero. The error at that point is

$$\epsilon_{max} = A \sin T\omega = A \sin (2\pi\omega/\omega_s) \qquad (5.8a)$$

For a sufficiently high sampling rate,

$$\epsilon_{max} \cong AT\omega$$

$$\simeq \frac{2\pi A}{p} \qquad (5.8b)$$

The number of points per cycle, p, required in order to limit the instantaneous error to .1% of full-scale is, therefore, approximately 6280 for a zero-hold unit.

In the case of a first-order hold unit, the output function is in the form of a series of ramps, each ramp commencing at the value of the function at the preceding instant and having a slope proportional to the difference between the sampled function at the two preceding sampling points according to

$$y(\tau) = y(nT) + \frac{\tau}{T}\{y(nT) - y[(n-1)T]\}, \qquad 0 \le \tau < T \quad (5.9)$$

If the sine wave expressed by Equation 5.6 is to be reconstructed, the error is

$$\epsilon = A \sin\left[\left(n + \frac{\tau}{T}\right)\omega T\right] - A \sin(n\omega T) - A\frac{\tau}{T}\sin(n\omega T)$$

$$+ A\frac{\tau}{T}\sin[(n-1)\omega T] \quad (5.10)$$

where the first term on the right side represents the desired output and the other terms correspond to the output of the first-order hold. The error is maximum in the vicinity in which the second derivative of Equation 5.6 is maximum, that is, at 90° where

$$n = \frac{\pi}{2\omega T}, \qquad \tau = T \quad (5.11)$$

Substituting these values in Equation 5.10, the maximum error is found to be

$$\epsilon_{\max} = A\left[\sin\left(\frac{\pi}{2} + T\omega\right) - 2 + \sin\left(\frac{\pi}{2} - T\omega\right)\right] \quad (5.12)$$

which can be simplified to read

$$|\epsilon_{\max}| = 2A[1 - \cos \omega T]$$
$$\simeq A\omega^2 T^2$$
$$\simeq A\left(\frac{2\pi}{p}\right)^2 \quad (5.13)$$

In order to limit the instantaneous error to .1% of full-scale, 199 sampling points must be provided per cycle. This is seen to be appreciably less than for the zero-order hold.

A number of additional refinements are possible in the use of hold circuits in order to reduce the reconstruction error. Occasionally fractional-order holds, involving a compromise between the zero- and first-order hold, are employed. Comcor, Inc., has introduced a variation of the conventional hold circuit known as the "known-slope method." This approach is applicable if the function to be reconstructed has been pre-stored in a digital computer. For example, if the digital computer is employed to store a function to serve as the transient excitation for an analog system, the entire function is first placed in the digital computer memory. In this case, advantage can be taken of the fact that the succeeding sampling point, $n + 1$, is known in advance. In place of the preceding sample, the value of the sample at the succeeding sampling instant in the future can be employed in constructing the slope. Thus the output of the sampling circuit in the known-slope method is

$$y(\tau) = y(nT) + \frac{\tau}{T}[y[(n + 1)T] - y(nT)] \qquad (5.14)$$

In this case, the maximum instantaneous error occurs near the zero crossing as in the zero-order hold. The number of points per cycle required for a given accuracy then becomes approximately half that required for the first-order hold.

SOLUTION ERRORS

5.8 THE CONCEPT OF SENSITIVITY FUNCTIONS

In the preceding sections the errors introduced by each of the components of the hybrid loop are discussed. These errors refer to departures from the ideal of the input-output relationships for each of these units. While such errors in transfer characteristics are of considerable interest in themselves, they do not provide a direct insight into errors introduced into the solution of the problem being solved in the hybrid computer-loop. For this purpose it is necessary to develop relationships between the input-output errors and the solution errors. Clearly, such a relationship will depend on the specific problem being solved and the specific techniques employed for the computer mechanization. For this purpose a method originally introduced for the error analysis of pure analog systems appears to have considerable potential. This is the technique known as sensitivity analysis.

The sensitivity analysis approach was first presented by Miller and Murray,[3] applied to practical situations by Meissinger[4] and Bihovski,[5] and discussed in considerable detail by Tomovic.[6] The technique as originally developed by Miller and Murray concerns the sensitivity of the response of dynamic systems to variations or perturbations in fixed parameters. Consider the general second-order differential equation

$$F(\ddot{x}, \dot{x}, x, t, q_0) = 0 \qquad (5.15)$$

where q_0 is the parameter to be perturbed. The method also applies, of course, to equations containing a multiplicity of parameters. In the case of Equation 5.15, the solution has the form $x = x(t, q_0)$. It is now desired to study the effect upon the solution x, of a perturbation Δq in q_0, which causes Equation 5.15 to become

$$F(\ddot{x}, \dot{x}, x, t, q_0 + \Delta q) = 0 \tag{5.16}$$

A variable $u(t, q_0)$, termed the sensitivity coefficient, is now defined as

$$u(t, q_0) = \lim_{\Delta q \to 0} \frac{x(t, q_0 + \Delta q) - x(t, q_0)}{\Delta q} = \frac{dx(t, q_0)}{dq_0} \tag{5.17}$$

The sensitivity coefficient is therefore a measure of the influence of the perturbation Δq on the solution x. For systems in steady-state the sensitivity coefficient is a constant, while it is a function of time in transient systems.

The determination of the sensitivity function involves the solution of a sensitivity equation obtained from the equation governing the system under study. Differentiating Equation 5.15 with respect to q_0,

$$\frac{\partial F}{\partial \ddot{x}} \frac{\partial \ddot{x}}{\partial q_0} + \frac{\partial F}{\partial \dot{x}} \frac{\partial \dot{x}}{\partial q_0} + \frac{\partial F}{\partial x} \frac{\partial x}{\partial q_0} + \frac{\partial F}{\partial q_0} = 0 \tag{5.18}$$

Provided that x and $\dot{x}$ are continuous with respect to both t and q_0,

$$\frac{\partial \dot{x}}{\partial q_0} = \frac{\partial}{\partial t} \frac{\partial x}{\partial q_0} = \dot{u}, \quad \frac{\partial \ddot{x}}{\partial q_0} = \frac{\partial^2}{\partial t^2} \frac{\partial x}{\partial q_0} = \ddot{u} \tag{5.19}$$

Equation 5.18 takes the form

$$\frac{\partial F}{\partial \ddot{x}} \ddot{u} + \frac{\partial F}{\partial \dot{x}} \dot{u} + \frac{\partial F}{\partial x} u = - \frac{\partial F}{\partial q_0} \tag{5.20}$$

Equation 5.20 is the sensitivity equation which must be solved for $u(t, q_0)$. It should be noted that the sensitivity equation is a linear equation even if the original equation, Equation 5.15, is nonlinear. This must be true since the coefficients in Equation 5.20, that is, partial derivatives of F with respect to $\ddot{x}$, $\dot{x}$, and x, are not functions of u or its derivatives. Provided the initial conditions for Equations 5.15 and 5.16 are the same, all initial conditions of Equation 5.20 are equal to zero.

The solution of the sensitivity equation provides a function of time $u(t, q_0)$. The product of this function and Δq constitutes to a first-order approximation the difference between the perturbed solution $x(t, q_0 + \Delta q)$ and the unperturbed solution $x(t, q_0)$. The solution of the sensitivity equation by using computer techniques is a relatively straightforward process, since the structure of the sensitivity equation is always very similar

5.8/THE CONCEPT OF SENSITIVITY FUNCTIONS   137

to the structure of the original equation under study. Consider, for example, the second-order differential equation

$$\ddot{x} + \mu\dot{x} + \lambda x = f(t) \tag{5.21}$$

In order to determine the sensitivity with respect to perturbations of the parameter λ, Equation 5.21 is differentiated with respect to λ,

$$\frac{\partial^3 x}{\partial \lambda \, \partial t^2} + \mu \frac{\partial^2 x}{\partial \lambda \, \partial t} + \lambda \frac{\partial x}{\partial \lambda} + x = 0 \tag{5.22}$$

Letting

$$u = \frac{\partial x}{\partial \lambda}$$

Equation 5.22 becomes

$$\ddot{u} + \mu\dot{u} + \lambda u = -x \qquad u(0) = \dot{u}(0) = 0 \tag{5.23}$$

Similarly, if it is desired to obtain the sensitivity of the solution of Equation 5.21 to perturbations in the parameter μ, Equation 5.21 is differentiated with respect to μ:

$$\frac{\partial^3 x}{\partial \mu \, \partial t^2} + \mu \frac{\partial^2 x}{\partial \mu \, \partial t} + \frac{\partial x}{\partial t} + \lambda \frac{\partial x}{\partial \mu} = 0 \tag{5.24}$$

Letting

$$v = \frac{\partial x}{\partial \mu}$$

Equation 5.24 becomes

$$\ddot{v} + \mu\dot{v} + \lambda v = -\dot{x} \qquad v(0) = \dot{v}(0) = 0 \tag{5.25}$$

Note that the sensitivity equations 5.23 and 5.25 are identical in structure to Equation 5.21 except for the forcing terms. They can therefore be solved by using precisely the same analog set-up or the same digital computer program as is used to solve the original equation under study. The transient solutions of these equations then represent the transient sensitivity of the original equation to perturbations in the parameter. If the specific parameter perturbations $\Delta\lambda$ and $\Delta\mu$ are known, these terms can be multiplied by the transient sensitivity functions to provide a continuous error plot, or to compensate solution errors. This is illustrated in Figure 5.11. Meissinger[4] shows that for small perturbations the output of Figure 5.11 remains correct over a considerable integration interval. Eventually higher-order derivatives, which are neglected in the original derivation, introduce appreciable accumulated errors.

A number of extensions of the sensitivity concept have been introduced in past years. Time-varying parameters and time-varying perturbations

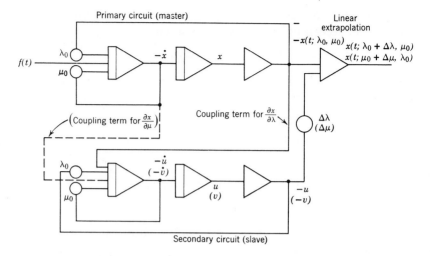

Figure 5.11 Primary (master circuit) and secondary (slave circuit) for second-order differential equation (H. Meissinger[4]).

can be handled under certain circumstances. For example, if the parameter μ in Equation 5.21 is a function of time, $\mu(t)$, and if this parameter is perturbed in a time-varying manner such that $\Delta\mu = \epsilon f(t)$, the sensitivity coefficient with respect to the perturbation is defined as

$$u = \left[\frac{\partial x}{\partial f(t)}\right]_{\epsilon=0}$$

The sensitivity equation then becomes

$$\ddot{u} + \mu(t)\dot{u} + \lambda u = -\dot{x}f(t) \qquad (5.26)$$

An operation akin to convolution is required to superpose the effects of the transient perturbations existing during each computing interval.

Bihovski[5] has described the application of the sensitivity approach to perturbations which are expressed statistically. He demonstrates that the standard deviation D_s in the solution is related to the standard deviation of the parameters D_r according to

$$D_s^2(t) = u^2(t) D_r^2 \qquad (5.27)$$

Bihovski further demonstrates the extension of this approach to non-stationary perturbations in which the parameter deviation D_r is a function of time.

5.9 EXTENSION OF SENSITIVITY ANALYSIS TO HYBRID COMPUTER SYSTEMS

Sensitivity functions provide precisely that link between subsystem errors and solution errors which is required for the specification and characterization of hybrid computer systems. Accordingly, it would be very desirable to develop sensitivity techniques to determine the influence of each of the error sources enumerated and described in the preceding paragraphs upon the solution generated in the analog and digital portions of the hybrid system. Such an extension of the sensitivity-analysis approach, which has been successfully applied in the past only to systems in which all the variables are continuous or all the variables are discrete, is subject to a number of difficulties:

1. Error sources such as quantization error, time-delay error, and reconstruction error do not appear as perturbations in the coefficients of the equation being solved in the computer.
2. Solution variables appear in the hybrid loop in discrete as well as continuous form, so that the nature of the dependent solution variable must be taken into account.
3. Differentiation can theoretically be performed only with respect to a continuous dependent variable, so that the concept of partial differentiation with respect to the perturbed parameter must be modified.

The general technique[7] presented in this section represents a first tentative attempt to employ sensitivity functions to characterize the relationship between error sources in hybrid systems and solution errors. Though cumbersome to apply to complex systems, it does facilitate the specification of tolerances on the subunit input-output relations so as to maintain specified error bounds on the overall solution.

In accordance with modern system theory, state-variable notation is employed. That is, it is assumed that the differential equations under study have been reformulated as a set of n simultaneous first-order differential equations. Consider then that the system of equations to be solved on the hybrid computer is expressed in vector form as

$$\dot{Y} = f(Y, P, t) \qquad (5.28)$$

where
$$Y = \{y_n\}\, n = 1, \ldots, N$$
$$P = \{p_m\}\, m = 1, \ldots, M$$

The elements of P are all the parameters involved in the description of the system 5.28, including its initial state. The vector P thus includes all initial conditions as well as all the coefficients affecting either the variables or the forcing functions.

To represent a hybrid computer implementation of 5.28 along with specific hybrid effects such as sampling, quantizing, and time delays, it is necessary to extend the description of the system in order to accommodate the particular transfer function of each element. It may be assumed that this extended system can still be described by a continuous equation. Accordingly, the system is modified by the addition of supplementary time-varying forcing functions, so that

$$\dot{Y} = f(Y, P, t) + H(Y, P, t)Q \qquad (5.29)$$

where

$$Q = \{q_j\}; \quad j = 1, 2, \ldots, J$$
$$H = \{h_{nj}\}$$

$H(Y, P, t)$ is an $N \times J$, Jacobian matrix of functions, the elements of which express the relationship between each individual error source, identified by the subscript j, and each state equation identified by the subscript n. The unperturbed system is characterized by

$$Q = 0$$

in which case Equations 5.28 and 5.29 become identical.

Consistent with the classical sensitivity approach, it is assumed that there exists a relation of the type

$$\Delta Y = U(Y, P, t)Q \qquad (5.30)$$

where it should be noted that U is a $N \times J$ matrix of sensitivity functions:

$$U = \{u_{nj}\}$$

with

$$u_{nj} = \frac{\partial y_n}{\partial q_j} \qquad (5.31)$$

Equation 5.30 implies the additivity of the effect of each individual perturbation upon the solution; that is, it supposes that the perturbations take place in a linear region imbedded within the nonlinear system 5.29. This implies in turn that the resulting state-variable deviations must remain sufficiently small.

A differential equation in U may be obtained by differentiating 5.29 with respect to Q and permuting the differentiation symbols. The latter operation is permissible if the solution is continuously dependent on Q, an assumption already made implicitly in 5.30. Therefore,

$$\dot{U} = \left[\frac{\partial f(Y, P, T)}{\partial Y} + \frac{\partial H(Y, P, T)}{\partial Y} Q\right]_{Q=0} U + H(Y, P, t) \quad (5.32)$$

or

$$\dot{U} = G(Y, P, t)U + H(Y, P, t) \quad (5.33)$$

where the $N \times N$ matrix G is defined as:

$$G(Y, P, t) = \left[\frac{\partial f(Y, P, t)}{\partial Y}\right]$$

Equation 5.33 is a linear matrix differential equation with time-dependent coefficients and forcing functions (specified by the Jacobian matrices G and H).

The initial conditions associated with 5.33 are always zero, since the introduction of $Q \neq 0$ does not affect the initial conditions of the original system 5.29. Each elementary solution u_{nj} relates the perturbation of the state-variable y_n to the error source q_j. If Equation 5.33 is solved on the computer, Equation 5.30 may be applied to obtain the state-variable deviation vector resulting from all error sources. Since 5.30 is a correction equation, moderate accuracies suffice.

It should be noted that for each state-variable the individual equations of the system 5.33 have the same homogeneous part, characterized by

$$\dot{U} = G(Y, P, t)U \quad (5.34)$$

Using standard computer methods, a plot of u_{nj} versus t for all state-variables of interest is generated over the specified integration interval. Next, the specified tolerances of these state-variables are employed to determine the maximum permissible values of the corresponding multipliers q_j. These bounds on q_j are then translated into engineering specifications of the subunits comprising the hybrid loop.

In summary, the application of sensitivity techniques to the specification and characterization of hybrid computer systems takes the following steps:

1. All sources of error which are capable of significantly affecting the state-variable of interest are identified.
2. The effect of each of these error sources upon the system state-equations is expressed by an additive forcing function $h_{nj} \cdot q_j$.
3. A sensitivity function u_{nj} is associated with each of these forcing functions such that $u_{nj} = \partial y_n/\partial q_j$.
4. The corresponding sensitivity equations are solved on the computer to provide a separate plot of each u_{nj} over the entire time domain of interest.

5. These plots are employed to determine the maximum q_j which can be permitted without excessive perturbations Δy_n in the state-variable of interest.
6. The bounds on q_j are translated into engineering specifications for each subunit.

Most of the hybrid sources of error indicated in Figure 5.1 may be analyzed using the method described in the above paragraphs. To illustrate a typical application of the method, the errors due to time delays in the analog-digital converter and the digital computer will be considered, with special reference to Van der Pol's equation,

$$\ddot{y} - (1 - y^2)\dot{y} + y = 0 \tag{5.35}$$

which becomes in state-variable form:

$$\dot{y}_1 = y_2$$
$$\dot{y}_2 = (1 - y_1^2)y_2 - y_1 \tag{5.36}$$

The two differential equations are integrated by the analog computer, while the term $(1 - y_1^2)y_2$ is formed by the digital computer.

For each variable, the digital treatment introduces a time delay defined as the time elapsed between the sampling of each variable (y_1 and y_2) and the application of the product $(1 - y_1^2)y_2$ to the analog computer. The time delay thus includes the digital-computing time, the time required for multiplexing and analog-digital conversion, as well as delays introduced in reconstruction and filtering. The symbols τ_1 and τ_2 designate those delays affecting y_1 and y_2, respectively.

If no compensation scheme is used, the most significant portion of the time-delay error is given by the first-order terms of a Taylor series expansion of the time-delay operator $\epsilon^{-\tau s}$, so that

$$y_1^*(t) = y_1(t - \tau_1) = y_1(t) + \dot{y}_1(t)q_1$$
$$y_2^*(t) = y_2(t - \tau_2) = y_2(t) + \dot{y}_2(t)q_2 \tag{5.37}$$

where the time-delay error terms q_1 and q_2 can be considered as elements of an error vector,

$$Q = \begin{pmatrix} q_1 \\ q_2 \end{pmatrix} = \begin{pmatrix} -\tau_1 \\ -\tau_2 \end{pmatrix}$$

If first-order compensation (Section 5.3) is used, the error can be approximated by the second-order terms $\ddot{y}_1 q_1$ and $\ddot{y}_2 q_2$, where

$$Q = \begin{pmatrix} \tau_1^2/2 \\ \tau_2^2/2 \end{pmatrix}$$

The first case is considered here, and Equation 5.37 is used to generate the forcing functions by substituting Y^* for Y in the product. This procedure yields the Jacobian matrices:

$$G(Y, P, t) = \begin{bmatrix} 0 & 1 \\ -(1 + 2y_1 y_2) & (1 - y_1^2) \end{bmatrix}$$

$$H(Y, P, t) = \begin{bmatrix} 0 & 0 \\ -2y_1 y_2^2 & -(1 - y_1^2)(y_1 - y_2 + y_1^2 y_2) \end{bmatrix}$$
(5.38)

The four corresponding sensitivity functions were computed on an analog computer. The resulting curves are periodic functions of continuously growing amplitude due to error accumulation. The plots of u_{11} and u_{12} are shown in Figures 5.12 and 5.13 and demonstrate the sensitivity of the state-variables y_1 and y_2 to a delay τ_1 affecting y_1.

It may be seen from those curves that if six solution cycles are to be obtained with an error y_1 not exceeding .02 (or 1% of the solution amplitude), the maximum value of τ_1 is given by

$$\tau_1 \leq 1.6 \text{ msec}$$

Similar curves have been obtained for u_{21} and u_{22}, for which

$$\tau_2 \leq 1.1 \text{ msec}$$

It is concluded therefore that, if the time delays exceed 1 millisecond, first-order compensation should be included in the hybrid loop.

5.10 COMPENSATION OF PER-STEP ERRORS

Consider that a hybrid computer system is to be employed to solve a set of nonlinear equations represented in vector form as

$$F(Y, t) = 0$$
(5.39)

where it is assumed that the equations are formulated or reformulated in state-variable notation, so that F contains only first-order differential operators. The dependent variable vector Y is an n-dimensional vector $Y = (y_1, y_2, \ldots, y_k, \ldots, y_n)$. If the system under analysis is an analog computer oriented system, it can be assumed that all dependent variables y_i appear as continuous voltages in the analog computer, and that it is this set Y which is the solution being sought.

Since conventional hybrid systems operate synchronously, transfer of information between the computers can occur only at times corresponding to $t = 0, T, 2T, 3T, \ldots, kT, (k+1)T, \ldots, KT$. The total observation interval is $(0, KT)$ and the sampled variables are kept constant during

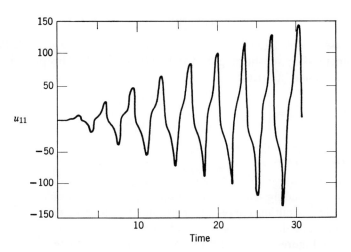

Figure 5.12 Sensitivity function u_{11} for Van der Pol's equation

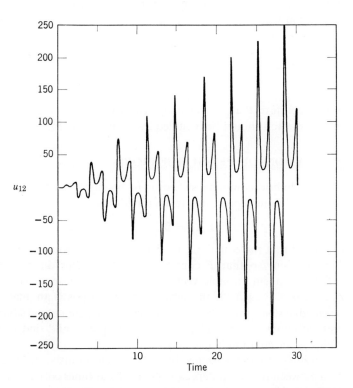

Figure 5.13 Sensitivity function u_{12} for Van der Pol's equation.

any sampling interval $kT < t < (k+1)T$. To emphasize the division of labor between analog and the digital computer, Equation 5.39 can be rewritten as

$$A(Y, t) + D(Y, t) + G(t) = 0 \qquad (5.40)$$

where A and D represent the operations performed by the analog and digital computers respectively, and $G(t)$ is the set of specified forcing functions. Since not all of the dependent variables $y_1 \cdots y_n$ are processed by the digital computer, indices 1 to m are reserved to designate those dependent variables which are converted into digital form for digital computer processing. It is convenient to define a new vector X to describe the output of the digital computer

$$X = D(y_1, y_2, \ldots, t)$$

Typically, the dimension of this vector will be neither m nor n. It is the vector X which is applied to the analog computer through the hybrid loop, so that Equation 5.39 can be rewritten as

$$\phi(Y, X, t) + G(t) = 0 \qquad (5.41)$$

As a result of errors contributed by the units comprising the hybrid computer-loop as well as by errors inherent in sampled-data systems per se, the solutions generated by the analog and digital computers deviate from their correct values. This deviation can be expected to become larger and larger as time progresses and errors accumulate. For the purposes of this discussion, the solutions generated by the analog and digital computer in the presence of significant errors are designated as $\bar{Y}$ and $\bar{X}$, respectively. The system of equations actually processed by the analog computer is then

$$\phi(\bar{Y}, \bar{X}, t) + G(t) = 0 \qquad (5.42)$$

The hybrid computer system under discussion can be presented as shown in Figure 5.14. The analog computer generates the n-dimensional solution vector $\bar{Y}$. The first m components of this vector are sampled simultaneously at each sampling instant, and the instantaneous values of these variables are stored in the set of sample-hold units SH_0. If it is desired to study the effect of slewing errors, these sample-hold units can be omitted. The sampled values are converted in turn into digital code by using an analog-digital converter. These converted values are introduced into the digital computer which, in turn, generates the vector $\bar{X}$. The digital computer output is reconverted into analog form and applied to a series of sample-hold units SH_i. Thus the solution proceeds in the continuous manner within the analog system, but in discrete or step-wise fashion in the rest of the hybrid loop.

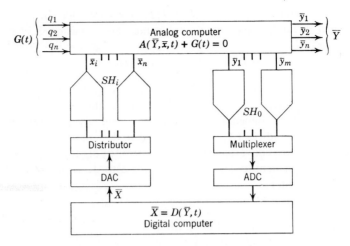

Figure 5.14 Block diagram of closed-loop hybrid computer system.

From the point of view of error analysis, the effect of major error sources in the hybrid loop can be represented as shown in Figure 5.15. With reference to this figure, assume that the hybrid system has been operating in a closed loop fashion for kT sampling instants and that, as a result, accumulated errors exist in both the analog and digital computers. The additional errors introduced in the solution during the $(k + 1)$th step

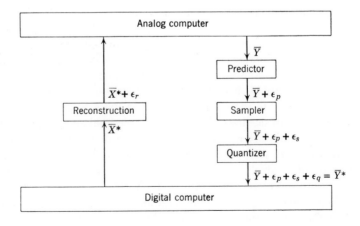

Figure 5.15 Major error sources in hybrid computer-loop.

due to some of the major error sources can then be expressed as

$$\bar{Y}^*(k+1) = \bar{Y}(k) + \epsilon_p(k) + \epsilon_s(k) + \epsilon_q(k) \tag{5.43}$$

where ϵ_p, ϵ_s, and ϵ_q represent the predictor error, the sampler error, and the quantizer error during the kth step, respectively. $\bar{Y}^*$ is employed to indicate that the per-step error has been added to the accumulated error. The digital computer now operates on erroneous data, so that output vector $\bar{X}$ is likewise modified by an additional per-step error and is designated by

$$\bar{X}^* = D(\bar{Y}^*) \tag{5.44}$$

This erroneous vector $\bar{X}^*$ is read into the analog computer and causes an additional departure of the analog solution from the true solution. The additional per-step error in the analog solution is designated by $\Delta \bar{Y}^*$. It is the purpose of error compensation schemes to minimize this error as far as possible at each step in the calculation. This automatically minimizes the accumulated error.

The application of sensitivity analysis to the characterization and compensation of the errors introduced by the linkage system involves the recognition that as far as the digital computer is concerned, the output vector of the analog computer $\bar{Y}^*$ is a forcing function for the digital computer; similarly, the output of the digital computer $\bar{X}^*$ acts as a forcing function for the analog computer. Errors introduced by the linkage system can therefore be regarded as perturbations in the forcing function. Since the terms $\bar{Y}$ and $\bar{X}$ are kept constant during each computing interval (between sampling instants), these terms can be considered as parameters. This approach has been applied successfully to the compensation of quantization errors.[8,9]

5.11 SENSITIVITY ANALYSIS APPLIED TO SAMPLING RATE

Recent studies[10,13] have applied sensitivity analysis to the determination of the influence of variations in sampling interval on the solution of difference equations. The motivation for the work stemmed from studies of discrete and sampled-data systems and led to the development of a technique for the design of adaptive sampled-data systems where sampling rate is adjustable. When applied to hybrid computer systems, the technique makes possible a study of the dependence of the digital (or iterative-analog) solution on the particular sampling rate used.

The following vector difference equation defining a discrete system is used as a starting point:

$$Y(k+1, T) = F[Y(k, T), k, X(k, T)] \tag{5.45}$$

In this expression Y is the R-dimensional state vector, k represents the particular sampling instant, and X is the 1-dimensional (scalar) input. The dependence of the solution upon perturbations in the sampling interval, T, is expressed by a vector sensitivity function $u_T(k)$ according to

$$u_T(k) = \lim_{\Delta T \to 0} \frac{Y[k,(T+\Delta T)] - Y(k,T)}{\Delta T} \tag{5.46}$$

If $Y(k, T)$ is a variable in the analog portion of the system, it is continuous in T, so that Equation 5.46 becomes

$$u_T(k) = \frac{\partial Y(k, T)}{\partial T} \tag{5.47}$$

Provided that certain restrictions are satisfied, the sensitivity equation can be obtained by differentiating Equation 5.45 with respect to T:

$$\frac{\partial}{\partial T} Y[(k+1), T] - A(k, T) \frac{\partial}{\partial T} Y(k, T) = a(k, T) \frac{\partial X(k, T)}{\partial T} \tag{5.48}$$

where

$$A(k,T) = \begin{vmatrix} \frac{\partial F_1(k,T)}{\partial y_1} & \cdots & \frac{\partial F_1(k,T)}{\partial y_R} \\ \vdots & & \vdots \\ \frac{\partial F_R(k,T)}{\partial y_1} & \cdots & \frac{\partial F_R(k,T)}{\partial y_R} \end{vmatrix}, \quad a(k,T) = \begin{vmatrix} \frac{\partial F_1(k,T)}{\partial X} \\ \vdots \\ \frac{\partial F_R(k,T)}{\partial X} \end{vmatrix} \tag{5.49}$$

In accordance with Equation 5.46, Equation 5.48 can be written as:

$$u_T(k+1) - A(k,T)u_T(k) = a(k,T)\frac{\partial X(k,T)}{\partial T} \qquad u_T(0) = 0 \tag{5.50}$$

The solution of this equation yields the desired sensitivity functions $u_T(k)$. Knowledge of the sensitivity function can be used in a number of ways. If $u_T(k)$ is large for certain values of k (indicating that the solution is very sensitive to small changes in sampling frequency), it may be appropriate to rerun the problem with a higher sampling rate. In addition, knowledge of $u_T(k)$ makes it possible to extrapolate to neighboring solutions by use of the first two terms of a Taylor series, i.e., given a solution $Y(kT_0)$ obtained with a sampling period T_0, the solution with a new sampling period $T_0 + \Delta T$ can be obtained from:

$$Y[k(T_0 + \Delta T] \simeq Y(kT_0) + \Delta T u_T(k)$$

5.12 THE METHOD OF CORRECTED INPUTS

An interesting alternative approach to the compensation of sampling, quantizing, and time-delay errors in analog-computer-oriented hybrid systems was presented by Gelman.[11] Assume, as before, that analog and digital solutions are to be performed in closed-loop fashion as expressed in Equation 5.29. To minimize a number of the errors which occur in the closed loop, an analog simulation $d(Y,t)$ of the digital calculation $D(Y,t)$ is provided, and this analog circuit is connected in a closed loop with the major analog computer performing $A(Y,t)$. The digital computer also performing $D(Y,t)$ is employed to update or correct any errors in $d(Y,t)$ committed by the analog simulator. This makes it necessary, of course, that the digital computation lag the analog computation by one computing interval. At the end of each computing interval, the digital computer compares its own calculation of $D(Y,t)$ with that generated by the analog equivalent; a correction signal is then sent across the hybrid linkage to eliminate this error in the analog loop. The technique has been applied to a number of problems by Connelly[12] and has had very favorable results.

References

1. Miura, T., and J. Iwata, "Effects of Digital Execution Time in a Hybrid Computer," *Proc. Fall Joint Computer Conference*, Las Vegas, pp. 251–266, 1963.
2. Widrow, B., "A Study of Rough Amplitude Quantization by Means of Nyquist Sampling Theory," *IRE Transactions PGCT*, **III**, 266–276, 1956.
3. Miller, K. S., and F. J. Murray, "A Mathematical Basis for an Error Analysis of Differential Analyzers," *J. Math. and Physics*, No. 2 and 3, 136–163, 1953.
4. Meissinger, H., "Parameter Influence Coefficients and Weighting Functions Applied to Perturbation Analysis of Dynamic Systems," *Proc. Third International Congress, AICA*, Opatija, Yugoslavia, pp. 207–216, 1961.
5. Bihovski, M. L., "Sensitivity and Dynamic Accuracy of Control Systems," *Engineering Cybernetics*, **2**, (6), Dec. 1964.
6. Tomovic, R., *Sensitivity Analysis of Dynamic Systems*, McGraw-Hill, New York, 1964.
7. Karplus, W. J., and J. Vidal, "Characterization and Evaluation of Hybrid Systems," *IFAC, Proc. Symposium on System Engineering*, Tokyo, Japan, Aug. 1965.
8. Vidal, J., W. J. Karplus, and G. Keludjian, "Sensitivity Coefficients for the Correction of Quantization Errors in Hybrid Computer Systems," *IFAC Symposium on Sensitivity Analysis*, Dubrovnik, Yugoslavia, 1964.

9. Vidal, J., and W. J. Karplus, "Characterization and Compensation of Quantization Errors in Hybrid Computer Systems," *IEEE International Convention Record*, New York, Part III, 236–241, 1965.

10. Bekey, G. A., and R. Tomovic, "Sensitivity of Discrete Systems to Variation of Sampling Interval," *IEEE Transactions on Automatic Control*, **AC-11,** 284–287, April 1966.

11. Gelman, R., "Corrected Inputs—A Method for Improved Hybrid Simulation," *Proc. Fall Joint Computer Conference*, Las Vegas, pp. 267–276, 1963.

12. Connelly, M. E., and O. Federoff, "A Demonstration Computer for Real-Time Flight Simulations," *Simulation*, **5,** 191–200, Sept. 1965.

13. Tomovic, R., and G. A. Bekey, "Adaptive Sampling Based on Amplitude Sensitivity," *IEEE Transactions on Automatic Control*, **AC-11,** 282–284, April 1966.

6

Complete Hybrid Computer Systems

6.1 INTRODUCTORY REMARKS

This chapter is devoted to the considerations underlying the acquisition and operation of complete hybrid computer systems. The factors entering into the specification of hybrid computers are first enumerated. This is followed by a brief historical survey of the evolution of hybrid computer systems. In succeeding sections the general philosophies of operation, training of staff members, and administration of hybrid computer centers are discussed. The chapter closes with a description of three modern hybrid computer systems.

6.2 CONSIDERATIONS IN SPECIFYING A HYBRID COMPUTER SYSTEM

The specification of a complete hybrid computer system for a given application requires a number of basic and far-reaching decisions. Upon the wisdom and technical insight exercised in making the initial decisions depends, to a very large extent, the overall cost of the computations, the accuracy and dependability of the results, and, therefore, the success or failure of the computer system. No doubt the most basic question which must be answered at the outset is whether a hybrid simulation or computation is sufficiently advantageous relative to a pure digital or a pure analog approach to justify a major expenditure of time and funds for the design and purchase of linkage equipment. Assuming that a hybrid approach is justified, it is necessary to select the major components of the hybrid computer-loop and to specify the topology of the entire system.

To assure satisfactory operation, the design of the hybrid computer system must be governed from start to finish by the demands of the specific applications for which the system is intended. The computer operating characteristics required for the many and diverse problems arising in engineering vary to such an extent that it is most unwise to plan a "general-purpose" hybrid computer system. It is even more dangerous for a computer laboratory to construct a hybrid computer-loop by linking their major analog and digital units and then to await customers. The following rule of thumb has been found useful by a number of computer engineers: the principal problems for which the hybrid computer is designed must be of sufficient importance to justify by themselves the complete cost of the hybrid computer system.

Unfortunately, designers of hybrid computer systems rarely are able to start from scratch. Usually an analog computer or a digital computer or both are already on hand, and economic considerations dictate their use in a hybrid computer-loop regardless of their suitability. Sometimes, the objectives of the applications for which the hybrid system is intended change while the design is in progress. The organizational structure of the computer laboratory, particularly the manner in which digital and analog computer groups are administered, introduces "political" and psychological constraints beyond the control of the designer of the hybrid system. The considerations outlined below can therefore be regarded only as very general guidelines. The basic alternatives in specifying complete hybrid computer systems can be summarized in the following questions:

1. Is the system to be analog-computer-oriented, digital-computer-oriented, or a balanced hybrid computer system?
2. What size of digital computer is to be employed? Is it to be a large, a medium-sized, or a small machine?
3. What type of digital computer is to be employed? What should be its minimum word length? Are such features as priority interrupt and floating-point hardware mandatory? What is the minimum requirement for core memory and auxiliary bulk storage?
4. What should be the bandwidth and accuracy of the analog computer?
5. What set-up and control functions are to be vested in the digital computer?
6. Is real-time operation required? What external system hardware or human operators are to be connected into the hybrid computer-loop?
7. What input-output devices will be used? Is paper-tape I/O adequate, or will card punches and readers, graphical displays, light pen systems, or analog and digital plotters be needed?

6.2/CONSIDERATIONS IN SPECIFYING A HYBRID COMPUTER SYSTEM   153

8. How many analog-digital and how many digital-analog conversion channels are required?
9. What speed and accuracy characteristics can reasonably be demanded from the linkage equipment?
10. What should be the topology of the linkage system? Should control of the linkage system be vested in the digital computer or should it be controlled by a separate clock? What units are to be time-shared by multiplexing?
11. What software is required for digital operation and for hybrid operation?

As indicated in Chapter 1 and illustrated in Figure 1.4, most hybrid computer systems are used for the simulation of physical systems. In analog-computer-oriented hybrid systems, the digital computer is employed primarily for mode control, potentiometer setting, and the generation of special functions; in balanced hybrid systems, on the other hand, integrations and other essential computing operations are performed by both analog and digital computers; in digital-computer-oriented systems, the analog portion plays a relatively minor role. For the most part, the hybrid computer systems constructed or acquired by large industrial and university laboratories are either analog-computer-oriented or balanced hybrid systems. Balanced computer systems require, of course, a more powerful digital computer as well as a more sophisticated linkage system and, therefore, are considerably more expensive. The decision as to which of the three types of hybrid systems is to be acquired is therefore one of key importance and will govern all succeeding logical decisions.

Usually, in all hybrid systems, it is the digital computer which determines the operating characteristics of the overall system in the hybrid mode. The basic choice here is between the so-called large digital computers, the so-called medium-size computers, and the so-called small digital computers. Large digital computers include such systems as the IBM 360-75 and the Control Data 6400. Medium-sized computers include the Scientific Data Systems Sigma-7, the Scientific Data Systems 9300, and the IBM 360/44. Small computers include the Scientific Data Systems Sigma-2 and 930, the Digital Equipment Corporation PDP-8, the Honeywell (CCC) DDP-116, the IBM 1800, and the Control Data 1700. All of these machines have provision for online operations, a characteristic indispensable for hybrid computations. Most often, the choice between the three classes of digital computers is determined by economic considerations, since the lease cost for a large machine can easily exceed $100,000 per month compared with less than $2,000 per month for the more modest small computers. Although it is frequently feasible to purchase and maintain a small or even medium-size digital computer for the sole use of the hybrid computer facility, such an arrangement is almost always out of the question for large digital computers. If a large digital computer is used in the hybrid computer-loop, it is virtually a foregone conclusion that this computer will be utilized for tasks other than hybrid computation for a major portion of its running time. The use of a large digital computer therefore may introduce serious scheduling problems, problems which become particularly severe if the digital computer facility is run on a "batch-processing" rather than on a "time-sharing" basis. For this reason, in the early 1960's a definite trend toward the use of small computers took place, notwithstanding their lower memory capability and lower speed. With the advent of the so-called third generation digital computers, especially adapted to time sharing, this trend may gradually reverse.

With very few exceptions, most present-day hybrid computer applications make use of the electronic analog computer in the slow or one-shot mode. Even for slow analog computation, digital computers are often hard pressed to keep up. However, high-speed or repetitive operation is extremely useful in digitally oriented hybrid systems, where the analog computer is used as a high-speed subroutine generator. Therefore, in the late 1960's, more and more bandwidth is being demanded of analog systems. Whereas computing bandwidths of 50 kHz were considered adequate in the early days of hybrid computing, by 1966 numerous users of hybrid computing equipment demanded analog bandwidths of 500 kHz or more. One advantage of such large bandwidths is that the analog portion of a hybrid computer system can be set up and checked out in the repetitive mode, permitting rapid adjustment and optimization of potentiometer settings. Once the system is checked out, the analog system is switched into the real-time mode for coupling to the digital computer.

6.3 EVOLUTION OF HYBRID COMPUTER SYSTEMS

The first major efforts to link large analog and digital computers were undertaken in 1955 and 1956 by Ramo-Wooldridge Corporation (subsequently renamed TRW Systems Group) and Convair Astronautics. The primary application for both systems was the solution of three-dimensional trajectory problems associated with intercontinental ballistic missiles. The simulation was to be carried out in real-time and to include a realistic representation of the control system. Actual missile hardware items, including the autopilot, were to be used in the simulation. In the time that elapsed between the original specification of the hybrid system and the delivery and acceptance of the conversion equipment, it was established that the guidance and control problems associated with missile flight are not closely coupled and can be studied separately. Consequently the basic problem for which the hybrid computing system was designed vanished. Nonetheless, both the Ramo-Wooldridge and the Convair systems have been successfully applied to a number of important problems, some of which are reviewed elsewhere in this book.

The Convair hybrid system originally utilized an IBM 704 digital computer while the Ramo-Wooldridge system was designed for a Univac 1103A. In other respects, however, these first two major hybrid systems were very similar. Both made use of Electronic Associates' PACE analog computing equipment, and both employed the Addaverter conversion system developed by Epsco, Inc. These linkages contained fifteen channels of analog-digital conversion and fifteen channels of digital-analog conversion, all channels operating simultaneously without multiplexing. The maximum sampling rate was 10,000 samples per second, and a 10-bit word-length was employed. The conversion time was 90 microseconds for analog-to-digital conversion and slightly less for digital-to-analog conversion. The equipment was housed in nine racks containing several thousand vacuum tubes. No doubt many of the operational difficulties associated with the Addaverter system were a result of the fact that the design antedated by some years the introduction of compact and reliable solid-state hybrid system components.

In 1960, the TRW Systems laboratories decided to redesign their hybrid computer system so that a Packard-Bell 250 digital computer could be used in place of the 1103A. This involved the transition from a large parallel machine to a small serial computer. Numerous timing and control problems were encountered in this redesign. From 1958 to 1962, a number of other major hybrid computing systems were constructed. Most of these employed linkage systems constructed by Electronic Associates under the trade name Addalink and by Packard-Bell under the trade name Multiverter.

A major breakthrough in the design of hybrid systems occurred in 1964 when Comcor/Astrodata introduced the first 100-volt solid-state analog computer for hybrid operation. Up to that time, the realization of transistorized 100-volt output stages had been impractical. Soon thereafter, the other major manufacturers of analog and hybrid equipment, including Electronic Associates, Inc., Applied Dynamics, Inc., and Beckman Instruments, Inc., turned completely to transistorized systems. In the mid-1960's, most linkages were to small or medium-sized digital computers, including, particularly, the Scientific Data Systems 930 and 9300, the Computer Control Corporation's DDP-24, and Digital Equipment Corporation's PDP series. During this era, virtually every large company in the aerospace/aircraft field acquired a hybrid computing system. Many of these systems were designed to permit the simulation of space flights and to give human operators simulated flight experiences.

In 1966, Electronic Associates Incorporated introduced its 680, a high-precision analog computer with a 10-volt dynamic range. To that time, 100-volt dynamic ranges in analog computers had been preferred to the exclusion of all others because of the relatively more favorable signal-to-noise ratio and because of their higher accuracy, particularly in performing nonlinear operations. Advances in solid-state technology finally permitted a reduction of the dynamic range to 10-volts with a relatively minor deterioration in the static accuracy of the computer. By reducing the dynamic range by a factor of ten, it became possible to expand bandwidth by a factor of at least four. It is to be expected that in the late 1960's and early 1970's, the trend towards reduced dynamic range and increased bandwidth will continue.

6.4 DIFFERENCES IN PHILOSOPHIES OF OPERATION

The development of increasingly larger and more complex computer systems has resulted in a gradual specialization of analog and digital facilities in most large organizations. In particular, as the cost per hour of rental of a large digital computer has increased, it has become increasingly necessary to provide specialized personnel for its operation and, consequently, the problem originator may find it difficult or impossible actually to run his own problem on such a computer. By contrast, even where very large analog machines are employed, the problem originator is frequently heavily involved in actual machine operation. It is here that a fundamental difference in philosophy appears. To examine the differences, consider first the differences between "open-shop" and "closed-shop" operation.

"Open-shop" operation refers to an installation where the user is free to program and run his own problem. Most analog computer facilities operate on either an "open-shop" basis or a "modified open-shop" basis. In the latter mode of operation, some programming assistance and some operating assistance are provided to the problem originator, but generally his constant participation during checkout and running of the problem is required. Closed-shop operation implies that the originator is not allowed to run his own problem but must turn it over to a group of specialists. Several degrees of closed-shop operation are possible. In a completely rigid closed-shop installation, the problem originator must submit the statement of the problem in writing to the computation center. It is first processed by a programming group which prepares the necessary computer programs and then may pass it on to an operating group which will actually place the problem upon the computer. In analog computer facilities, complete closed-shop operation is rare. In large digital installations, however, it is the most common mode of operation, and several layers of processing are involved between the problem originator and the actual machine operation. Typically, his problem may go from its initial statement to a mathematical-analysis or numerical-analysis group which will recommend methods of solution; the problem then may pass to a programming group which will prepare the required programs in an interpretive language or in machine language, depending on the problem and the computer involved. From the programming group it may be submitted to another group for keypunching the required cards. The program and required values of coefficients, initial conditions, and so forth, properly punched on cards, will then be submitted to an operation group which will actually run the problem on the machine, returning the cards and output listings to the programmer. The programmer will then examine the results in the listing or obtain a plot, correct any errors, and finally submit the results to the problem originator. Clearly, such a mode of operation provides for the maximum use of skilled specialists but, on the other hand, introduces considerable human processing and consequently the opportunity for errors.

It should not be assumed from these statements that all digital computer facilities operate on a closed-shop basis. Many companies have available smaller, moderately priced digital computers to which engineers have direct access. In general, these machines can be programmed by means of a symbolic language such as **FORTRAN** which makes it convenient for the originator to prepare his own machine program. In many university laboratories, a completely open-shop operation with small or medium-size digital computers is also possible. It is interesting to note, however, that the University of Michigan studies on the use of computers in education have indicated that where large numbers of students are to be trained in the use of digital computers, the use of large computers operating in essentially closed-shop operation provides considerable benefit, both in speed and in resulting reduction of total cost. In such operations the problem originator (in this case the student) prepares his own computer program, but he does not run the problem. Machine operation is reserved to a crew of specialists. The development in the later 1960's of powerful time-shared systems with remote consoles promises to change this situation to a large extent.

In summary, then, there tends to be a fundamental difference in philosophy of operation between large digital computation centers and large analog computation centers. Typically, in the digital computation center there is a greater distance between machine and problem originator than there is in the analog center. It is important to keep these differences in mind when planning for hybrid computation.

6.5 DIFFERENCES IN BACKGROUNDS OF THE STAFF

Differences in the philosophy of operation also imply differences in the backgrounds of the personnel involved in computer operation. Where a large digital computer operates on a closed-shop basis, the machine operator is so removed from the problem itself that his training in mathematics and engineering need not be very extensive. In fact, to the man who loads cards into the machine, moves tape decks, and operates the machine controls, all problems are alike since he is completely removed from the mathematical statement. Consequently, in many cases, technicians are used for machine operation. Programmers and mathematical analysts generally have training in mathematics, but they too have relatively little physical insight into the nature of the problem being solved.

On the other hand, because of the greater involvement of the analog computer operator with his problem, some degree of engineering sophistication is required on the part of the staff of analysis-computer facility. That is, the man operating an analog computer must have some understanding of differential equations, and he must have a sufficient background in engineering to visualize the simulation process. In a number of companies the analog computation center serves as a training ground for young engineers who use it to gain experience in the mathematical formulation of problems and to gain insight into the dynamic behavior of the systems being studied. Since the analog operator is typically an engineer, he can move from the computation center staff to a number of other engineering positions such as those involving control systems, guidance, or structural dynamics. On the other hand, the digital computer operator or programmer tends to be a specialist for whom future positions will lie within the computation center staff itself. It should not be assumed that there are no analog personnel for whom problem solution, development of computation techniques, and operation of the computation center in themselves represent worthwhile goals, but such personnel are in a minority in industry. While digital computation center staffs are almost invariably part of a service or staff organization, the operation of the analog computer and solution of analog problems is often done by "line" engineers rather than personnel of a service organization.

The differences among operating personnel outlined above are due in part to the differences in the levels of reliability and maintenance practices used with the two types of computers. In the use of modern digital computers, daily maintenance practices, diagnostic routines, and self-checking programs are employed. The problem originator and even the programmer can therefore remove themselves from the machine hardware. On the other hand, the analog computer operator cannot completely abstract himself from computer hardware and the details of its operation. He cannot assume that details of construction, bandwidth limitations, and machine noise are problems he need not consider in the process of formulating a machine diagram. On the contrary, even the best of modern analog computers usually cannot be considered to be ideal mathematical devices. In general, the man who prepares an analog computer diagram must keep in mind both the dynamics of the problem he is solving and also the dynamic limitations of the elements with which he is performing the solution. He must consider the bandwidths of his amplifiers, their noise level, their output current limitations, and so forth. The modified open-shop mentioned in the preceding section provides a way of combining the greater knowledge of machine components of an analog operator with a greater insight into the problems possessed by the originator, but each still requires some knowledge of the other's specialty.

The differences in staff background outlined above should not be considered universal, but they form a sufficiently typical pattern so that they must be considered seriously in the attempt to organize and staff a hybrid computation center. The question of whether analog personnel should be taught digital programming or whether digital personnel should be taught programming and operation must be raised seriously. Numerous organizations have found that there is considerably greater reluctance on the part of the highly specialized digital computer staffs to learn analog operation. On the other hand, the increasing use of digital elements in analog computing has made analog computation center personnel much more interested in becoming acquainted with the programming and operation of digital computers.

6.6 OPERATING-COST CONSIDERATIONS

Cost studies of the operation of computation centers are often extremely difficult to make. The picture is clouded by the fact that practices vary from installation to installation. In many cases, a significant percentage of the operating cost of the analog or digital machine is borne by the company as "overhead." In other cases, attempts are made to include the entire operating cost in hourly charges made against the particular group or contract employing the computer. Even in these cases, however, charges may vary, and a careful inquiry generally reveals that significant costs are not included in the rate computation. Therefore, it is extremely difficult to make fair comparisons between operating costs per hour on an analog and on a digital computer. The difficulty is further compounded by the great difficulty in evaluating the usefulness of one hour of computation time. The differences in the uses of the machines are such that no figure of merit equating a particular number of amplifiers to a particular set of specifications, say, add-time on a digital computer, is possible.

In spite of these limitations, it is reasonable to state that in a number of large companies the charges for one hour of analog computation time will be about 5 to 10% of the charge for one hour of computation time on a large digital computer of the class of the IBM 7094 or 360/75. Whether or not this charge includes provisions for depreciation, maintenance, administrative personnel, operation on second and third shifts, and so forth, cannot be ascertained in general but must be investigated in any particular situation. As will be noted later, the important point in considering the differences in rate for the utilization of the computers is that the rate may play a significant part in administrative decisions concerning the establishment of hybrid computation facilities.

It should also be noted that, in general, charges for rental of a digital facility include all necessary maintenance. In many cases, the manufacturer will establish a permanent field engineering installation at the customer's plant and provide what services are required to keep the facility operating. On the other hand, large analog computer facilities with several hundred operational amplifiers generally require not only a permanent maintenance staff but also one or more engineers. The engineers have the responsibility for the design of circuit modifications and improvements to the analog computers in addition to providing the necessary supervision over maintenance services. In order to keep a modern analog computer facility in proper operation, it has been estimated that the cost for maintenance may run as high as 15 to 20% of the initial cost of the machine per year.

6.7 ADMINISTRATIVE CONSIDERATIONS

In addition to the technical factors just outlined, a number of significant administrative factors must be considered. It has been the experience of several organizations that decisions made on the basis of purely technical factors have resulted in considerable administrative difficulty. Two major considerations emerge:

1. Use of a large digital computer generally implies that hybrid computation occupies only a portion of the total time available on the machine. The cost and versatility of the large digital computer are such that the

total time allotted to hybrid problems may in fact constitute only a small percentage of its total use. This operation implies that some provisions for interrupting the digital program and transferring control to the analog computer must be provided. Such time sharing has been implemented in a number of laboratories, but it should be noted that while an interrupt provision presents few technical problems, it presents at least the following administrative problems:

(a) How is priority determined?
(b) Should interrupt be effective any time upon command, or should it depend on the remaining computation time of the current digital problem? In some installations, 20 minutes is the maximum running time on a large digital computer. Should the analog computer be required to wait for up to 20 minutes? Or should there be a maximum of 2 minutes? How is the decision made?
(c) During what times of the day, that is, during which shifts, should hybrid computation be allowed?
(d) Who will be responsible for hybrid computation when it does take place? Will the responsibility be in the hands of the digital computation center, the analog computation center, or should a new organization be established with joint responsibility for both analog and digital computers?

Some of these questions are particularly critical during the initial checkout phase of the combined system. It should be realized that the checkout of a hybrid installation requires a considerable expenditure of time on the part of both personnel and computers. During this period, no useful computation will be performed and yet significant blocks of digital computer time may be required. In the process of establishing a combined simulation operation, provisions should be made for the financing of the checkout period and for the establishment of a system of priorities that will make it possible to obtain an adequate amount of digital computer time to complete the checkout successfully. The checkout of a large hybrid system may in fact consume many hours of computer time and consequently represent a significant proportion of the total budget for the establishment of the laboratory.

2. The use of a small or medium-sized digital computer in a price class compatible with that of an analog computer represents an entirely different administrative problem. It is now possible to utilize significant proportions of the computer time for hybrid computation purposes for two reasons. First, the computer itself will generally be slower and, consequently, longer periods of computer time will be required. Second, the more moderate price of the computer makes it easier to justify the expenditure

of large periods of time in checkout and preliminary runs when required. Consequently, the use of a small digital computer raises the following questions:
(a) Will it be possible within the structure of the organization for the analog computation center to own and operate a digital computer?
(b) Will the use of the computer be shared with other organizations and for other purposes than hybrid computation?
(c) How will priorities be determined?
(d) How will charges be made?
(e) Who will perform the programming if the computer is placed in the hands of an organization specialized in analog computation?

In summary, then, it can be seen that if the digital computer selected for hybrid computation possesses by virtue of its size considerable speed and versatility, it may create a significant administrative problem in the selection of mode of operation, cost, and assignment of priorities. On the other hand, if the computer lacks these qualities, the administrative problems of its operation may be significantly reduced but at the expense of reduced efficiency in the hybrid computation itself.

6.8 PLANNING FOR HYBRID COMPUTATION

Planning for hybrid computer installation presents unique problems in each installation. However, the following is a suggested checklist of some factors to be considered in the process of planning for the establishment of a hybrid computer laboratory.

Is There a Genuine Need? This is perhaps the most crucial of the questions to be raised. Hybrid computation represents too substantial an expenditure of money and effort to be justified simply on the grounds that "it is a good idea" or "other companies are doing it." Within each organization, the need must be established on the basis of carefully selected problems whose solution not only *could* be done but in fact *would* be done by means of hybrid equipment. Furthermore, these problems must be anticipated for a sufficiently long period in advance of the completed installation so that a backlog of problems can be anticipated for a period of perhaps 6 months to a year following installation. After significant problems have been identified and a definite need established, the allocation of portions of the problem to the analog or digital computer must be made. Careful studies of error, bandwidth, accuracy, flexibility, solution time, programming convenience, etc., are required.

6.8/PLANNING FOR HYBRID COMPUTATION

How Will the Digital Computer be Used? In some installations, the digital computer is used for hybrid computation only a portion of the time with the remaining time being used for pure digital computation. In such cases, the availability of a high-level FORTRAN compiler, floating point hardware, and a flexible priority interrupt structure need to be considered with great care. The instruction repertoire, the word-length, and the nature of the input-output equipment also affect the versatility of the digital computer. For example, a 16-bit machine with paper tape input may lend itself very well to certain hybrid problems in which one program will be used for long periods of time. On the other hand, pure digital utilization would be slowed by double-precision computations, by software floating point computation, by a limited compiler, and by the lack of a card punch and reader. Similar requirements would of course apply to a hybrid system with a dedicated digital computer where rapid turn-around is desirable.

On the other hand, the acquisition of a digital machine with too many optional features and with more than the necessary computing power may defeat the objective of obtaining a computer sufficiently economical to permit it to be dedicated completely to hybrid computation.

How Will Charges for Hybrid Computer Services be Determined? This question includes a number of subsidiary ones such as the cost of the checkout period, the considerably higher cost which may be incurred in the solution of the first few problems due to the inadequate experience of the staff, and the training time required.

What Personnel Will Program and Operate the Equipment? As outlined earlier, the background of the personnel on the analog and digital computer staffs is often quite different. If analog computer personnel are given the responsibility of programming and operating the digital portion of the system, it is reasonable to assume that this programming will be less effective and less efficient than if it were carried out by experienced digital computer programmers. On the other hand, cooperative efforts among teams of analog and digital specialists may present serious administrative difficulties. Consequently, it is recommended that provisions be made for the training of analog personnel in order to establish a separate group, within the organization, charged with the responsibility of programming for hybrid simulations.

How Will Priorities for Utilization of the Equipment be Established? It is important to emphasize that this question should be settled, at least in a preliminary fashion, before any equipment is actually ordered. The experience of many organizations has been that inadequate planning in the area of priority assignment has resulted in "political" problems of far-reaching consequence.

Who Will be Responsible for Procurement, Installation, and Checkout of the Equipment? The significance of this problem becomes evident if it is considered that, sometimes, neither the analog computer manufacturer nor the digital computer manufacturer may be willing to assume responsibility for system operation. If three different manufacturers are involved (an analog computer manufacturer, a digital computer manufacturer, and a conversion equipment manufacturer), the problems of compatibility and system engineering necessary to insure adequate performance become formidable. If the organization attempting to establish the laboratory keeps the problems of system engineering for itself, it must also make adequate provisions for engineering personnel to devote significant amounts of time to coordination with the manufacturers and, furthermore, it must provide sufficient funds for any necessary modifications in the standard equipment available. This problem is, of course, essentially alleviated when it is possible to assign system responsibility to one manufacturer.

The above list of problems is by no means exhaustive, but it does serve to indicate the seriousness of some of the questions which must be answered before hybrid computation can be justified. The experience of many installations shows very clearly that this is one area where foresight is better than hindsight.

6.9 TYPICAL MODERN HYBRID COMPUTER SYSTEMS

By 1967, more than ten manufacturers were engaged in the commercialization of analog computers for hybrid computer use. The number of manufacturers of online digital computers suitable for hybrid computation was even larger. Since each manufacturer generally has available a number of different models, the total number of combinations of analog and digital computers which can be employed to fashion a hybrid computer system is therefore very large. The purpose of this section is not to present a comprehensive summary to these combinations, but rather to describe in some detail three hybrid systems, all of which became operational in 1967 and each of which has played, in some respects, the role of "pace setter."

The first of these systems involves the interconnection of a Comcor Ci-5000 analog computer and a Control Data Corporation 6400 digital

6.9/TYPICAL MODERN HYBRID COMPUTER SYSTEMS 165

computer. Since, in 1967, the CDC 6400 constituted the largest available digital computer designed specifically for time-sharing applications, the system may be considered as making maximum use of newly emerging digital computer technology. The second hybrid system, the Electronic Associates, Inc., 8900 system, represents the largest fully integrated system commercialized by a single manufacturer. The final example, the Electronic Associates 690 system, was the first large-scale American system utilizing a 10-volt dynamic range, rather than the more conventional 100-volt range. In the discussion that follows, component complements and characteristics are included to provide the reader with an insight into the kind of equipment available. However, it should be recognized that the analog portion of hybrid systems and the available software vary markedly from system to system and that, in fact, large hybrid computer systems are often "tailor made" for specific applications.

The Comcor-CDC System. The first system to be described was designed by Comcor, Inc., and delivered to Lockheed Missile and Space Division in April, 1967. At that time, this represented the largest single procurement of a hybrid computer, costing more than $4,000,000. The complete system consists of four large analog computers, Comcor Model Ci-5000/5; two high-speed intercommunication systems, Comcor Model Ci-5100 (INTRACOM); a large digital computer, Control Data Corporation Model 6400; a comprehensive hybrid software system; and an assortment of analog and digital peripheral equipment. A photograph of a model of the complete hybrid computer system is shown in Figure 6.1.

The Comcor Ci-5000/5 is a solid-state, 100-volt, high-speed, iterative differential analyzer capable of being completely controlled by a digital computer. The quantity and specification of the major components contained in each of the four Ci-5000/5 analog computers are presented in Table 6.1. Figure 6.2 is a photograph of a typical analog computer unit. The INTRACOM linkage system, shown in Figure 6.3, provides all necessary conversion and control devices required to facilitate information transfer between the analog and digital computers. The quantity and specifications of major components contained in each INTRACOM are listed in Table 6.2. The digital computer is a Model 6400, manufactured by Control Data Corporation, and the digital equipment provided in the hybrid installation is listed in Table 6.3. A complete and comprehensive software system designated "CLASH" was supplied with the hybrid computer hardware. Table 6.4 lists the major programs supplied.

The EAI 8900 Hybrid Computer System. The Electronic Associates 8900 Scientific Computing System was, in 1967, the largest hybrid computing

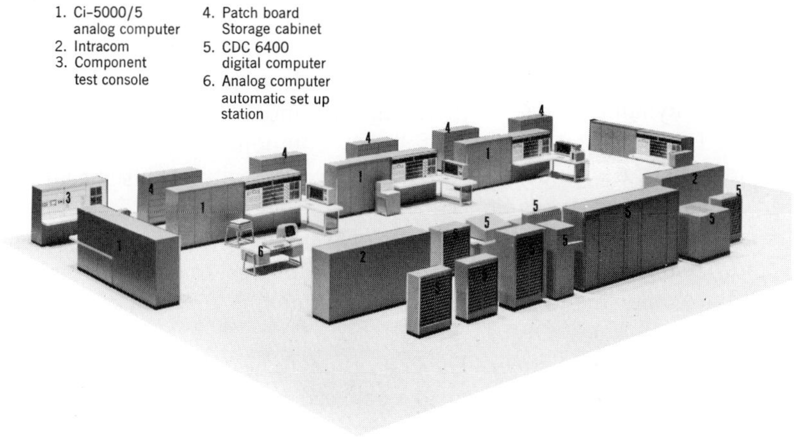

166   COMPLETE HYBRID COMPUTER SYSTEMS/6

1. Ci-5000/5 analog computer
2. Intracom
3. Component test console
4. Patch board Storage cabinet
5. CDC 6400 digital computer
6. Analog computer automatic set up station

Figure 6.1   Model of complete Comcor-CDC hybrid computer system installed at Lockheed Missile and Space Company, Sunnyvale, California.

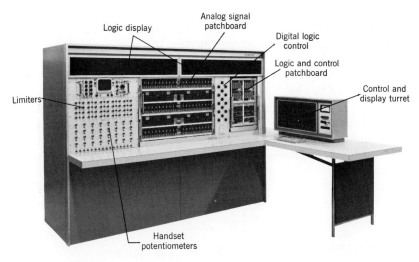

Figure 6.2   Comcor analog computer model Ci-5000/5.

Table 6.1 Complement of the Comcor Ci-5000/5 Analog Computer

Quantity	Component	Specification
400	Amplifier	±100 v, 50 ma DC gain: >2.5 × 10⁷—chopper stabilized Drift: <20 μv/day Bandwidth (summer): −3 db, 100 kHz, SCI load Total dynamic error @ 1 kHz (summer): <0.8% of 100 v
70	Quarter square multiplier	Static error: ±.015% of 100 v Total dynamic error: 0.9% of 100 v
48	Hybrid operated digital attenuator device (HODAD)	Setting time: 7 μs Band-width: −3 db, 230 kHz Phase error: <0.4° Total dynamic error @ 1 kHz: 0.85% of 100 v
22	Card set interpolative function generator ($y = f(x)$)	Data entry: standard 80-column data processing card Data points: 31 Slope: 200:1 maximum Static error: 100 mv Band-width: 15 kHz Total dynamic error @ 100 Hz: 0.55% of 100 v
1	Card set surface generator ($z = f(x, y)$)	Data points: 31 × 12 matrix Other specifications similar to card set interpolative function generator.
6	Resolver	Static error: Forward resolution: 0.06% of 100 v Inverse resolution: 0.12% of 100 v Operational bandwidth: Continuous mode: 1000 radians/second Total dynamic error @ 10 Hz: Forward mode: 0.2% of 100 v Inverse mode: 0.25% of 100 v

system developed and fabricated by a single supplier. The system consists of two major subsystems, the 8400 scientific digital computer and the 8800 analog computer. These are linked via a high-speed conversion system and an extensive control and information network. A photograph of a typical system configuration is shown in Figure 6.4.

The EAI 8800 analog computer has a 100-volt dynamic range and is modularized to permit ready expansion from a medium-sized to a very

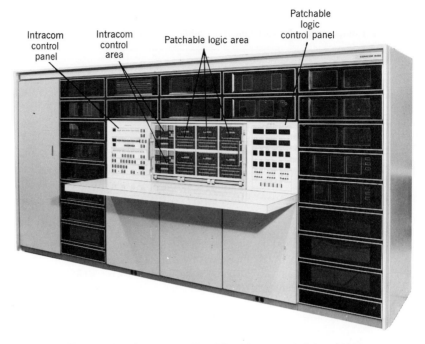

Figure 6.3   Comcor INTRACOM System Model Ci-5100.

Figure 6.4   Electronic Associates, Inc. 8900 hybrid computer system.

6.9/TYPICAL MODERN HYBRID COMPUTER SYSTEMS   169

Table 6.2 Complement of the Comcor Ci-5100 Interface

Quantity	Component	Specification
64	Multiplexer channels	Error: ±0.01% of 100 v Switching time: 2 μs
16	Sample and hold channels	Error: ±0.02% of 100 v Settling time to 0.01%: 50 μs Drift: 200 mv/sec
1	Analog-to-digital converter	Error: ±0.01% of 100 v, ±$\frac{1}{2}$ LSB Conversion time: 10 μs Resolution: 14 bits + sign
40	Multiplying digital-to-analog converters	Overall DAC error: ±0.014% of 100 v Settling time: (200-v step) 50 μs to 0.01% of 100 v Multiply error: ±0.024% of 100 v Loading time: 5 μs
1	Precision interval generator	Resolution: 1 μs Range: 1 μs to 524 seconds
24	Interrupts	
68	Sense lines	
68	Control lines	
300	Free patchable logic elements (flip-flops, counters, delay flops, gates, etc.)	

Table 6.3 Digital Equipment Provided with the Hybrid System Shown in Figure 6.1

Quantity	Item
1	CDC 6400 with 10 peripheral processors (each with 4K 12-bit core storage)
32K	60-bit word core storage
37 × 10^6	12-bit word disk storage
2	Magnetic tapes, 75 IPS density
1	Card reader, 1200 CPM
1	Card punch, 250 CPM
5	Display consoles

Table 6.4 Software System for the Comcor-Control Data Hybrid Computer

Program	Description
Hybrid Monitor: (modified scope 2.0) (CLASH)	Provides all necessary software control for simultaneous computation of up to seven jobs, two of which can be real-time hybrid programs. Programs are processed on a priority basis, real-time hybrid programs always operating with highest priority. Controls all processing of interrupts.
Display Monitor (DIS 211)	Processes information to and from five portable display entry consoles. Provides user interaction in the areas of: (1) source modification, and (2) FORTRAN variable display and modification during execution.
Hybrid Input/Output (HIO)	Provides all necessary routines for real-time data intercommunication between analog and digital domains.
Analog Input/Output (AIO)	Provides all necessary routines for nonreal time set-up, monitoring and control of the analog computers by the digital computer.
Automatic Problem Verification (APV)	Automatic static check of analog computer problem set-up.
Preventive Maintenance (PM)	Automatic operational test of all major computing elements contained in the four analog computers and two Intracoms.

large system. The total complement of a fully expanded system is presented in Table 6.5. Two program patch-panels are provided. The analog patch-panel terminates analog voltage signals only, while the logic patch-panel terminates logic and control switching elements only. The 8900 system interface was designed to be controlled primarily by the digital computer. Since the analog portion of the hybrid program can include up to six consoles, the control interface must provide the capability for fanning out control and data information to all of these consoles. The complement of a linkage system which would be required under these circumstances is shown in Figure 6.5.

The EAI 8400 digital computer has a 36-bit word-length. Of these 36 bits, 32 are used for data or instruction, 2 for parity, and 2 for control to provide memory protection. Memory size ranges from $8K$ to $64K$ words. The EAI 8400 operating system incorporates both digital and hybrid programming features into a single integrated software system

6.9/TYPICAL MODERN HYBRID COMPUTER SYSTEMS

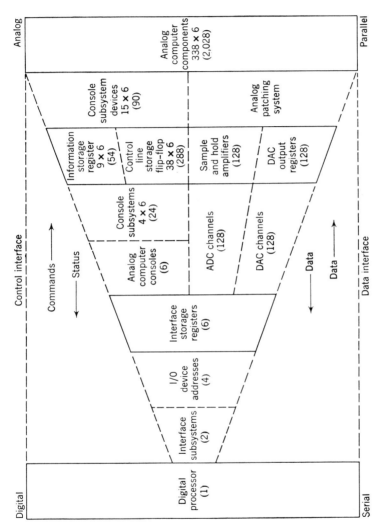

Figure 6.5 Electronic Associates, Inc. 8900 hybrid interface data and control fan-out.

Table 6.5 Complement of the EAI 8800 Analog Computer

Quantity	Item
	ANALOG COMPUTING COMPONENTS
60	Summer-integrators
6	Integrators (assigned to electronic rate resolvers)
60	Summers (with or without track-and-store networks)
48	Inverters
20	Inverters (with high-gain mode)
18	Inverters (assigned to electronic resolvers)
48	Other amplifiers (assigned to electronic resolvers)
240	Coefficient attenuators
72	Electronic multipliers
6	Sine-cosine generators (assigned to electronic resolvers)
6	Electronic resolvers
30	Ten-segment arbitrary diode function generators
30	Limiter networks
15	Passive element resistors (one megohm)
15	Passive element capacitors (with three or six time scales)
15	Analog function switches (SPTT)
1	Dual-range noise generator
	LOGIC CONVERSION COMPONENTS
30	Digital-to-analog switches
30	Analog (voltage) comparators
24	Function relays (DPDT)
15	Interpanel logic trunks (voltage-limited)
	PARALLEL LOGIC ELEMENTS
100	General-purpose logic gates
15	Quad registers (with preset switch)
30	Comparator flip-flops
8	Logic differentiators
4	Monostables (with delay selector)
9	Digital function switches
—	General-purpose indicators
	EXTERNAL TRUNK LINES
96	Analog input trunks (monitored)
96	Analog output trunks (nonmonitored)
50	Logic input trunks
50	Logic output trunks
	PERIPHERAL EQUIPMENT
1	Stored-program input-output system
4	X-Y plotters
1	Line printer
3	Recorders
1	Display unit
1	Oscilloscope

Table 6.5 (*continued*)

COMPUTER CONSOLE	
Mode Control and Timing System: Mode and time-scale selector (with rate and static test busses) Selector indicator panel Two megacycle system clock Clock mode and frequency selector Repetitive operation counter Slave selector Quad interval timers (3) *Input-Output System:* Input keyboard Input-output selector Address register (with automatic sequencer and selective scan) Transistorized voltmeter Address and value display Precision voltage divider Servo amplifier I/O system interface buffer Comparator indicator panel Logic selector and display Logic display panel	*Program Patching System:* Dual patch-bay (with latching mechanism) Analog pre-patch panel (4080 terminals) Control logic pre-patch panel (2400 terminals) Analog patching kits Control logic patching kits *Overload Indicator System:* Overload indicator panel (with storage option) Overload indicator bus (with automatic HOLD option) Audible overload alarm *Manual DFG System:* Set-up panel Dual set-up amplifier (with automatic nulling circuit) *Power Distribution System:* Bus-bar distribution matrix (with overload indicators) Monitor voltmeter ±100 volt reference supply (3 ampere)

which is described in Chapter 7. This system operates under the control of an executive monitor which accommodates a variety of programming languages. An idea of the magnitude of a complete programming system for hybrid computer purposes may be gained from Figure 7.1 in the following chapter.

The EAI 690 Hybrid Computer System. The Electronic Associates 690 is a medium-sized hybrid computer making use of a 10-volt dynamic range. It includes the EAI 680 analog computer, the EAI 640 digital computer, and the EAI 690 hybrid interface system. Figure 6.6 is a photograph of this system.

The EAI 680 analog computer was the first relatively large American analog computer to employ a 10-volt dynamic range. The reduction of dynamic range from 100-volts to 10-volts permitted the designers to expand

Table 6.6 Complement of the EAI 680 Analog Computer

Quantity	Item
	ANALOG EQUIPMENT (STANDARD PATCH-PANEL)
156	Amplifiers
18	Variable diode function generators (10 segment)
24	Multipliers (quarter-square)
120	Pots (servo-set)
12	Pots (hand-set)
30	Combination amplifiers (integrator/summer)
12	Sine/cosine DFG's
6	Log DFG's
12	Feedback limiters
12	Hard zero limiters
36	Loose diodes
	Loose resistors
48	Gain 1 (100k)
36	Gain 10 (10k)
6	Gain 100 (1k)
	LOGIC AND INTERFACE EQUIPMENT
24	Electronic comparators
24	D/A electronic switches
24	D/A relay (DPDT)
12	Track/store units
6	Four-bit general-purpose registers
42	AND gates
6	Monostable timers (one-shots or pulsers)
6	Differentiators
3	BCD counters (2-decade, bi-directional)
	AMPLIFIER COMPLEMENT
	The complement of 156 amplifiers includes the following:
30	Combination amplifier (may be used as integrator, summer, or high-gain amplifier)
24	Summer (may also be used as high-gain amplifier)
42	Inverter/high-gain amplifier (may be used as inverter or as output amplifier for nonlinear equipment; may also be used as multi-input summer by patching additional resistors to amplifier junction)
24	Output amplifier for quarter-square multiplier (may be separated from its associated multiplier, allowing independent use of multiplier and amplifier)
36	Amplifiers associated with variable diode function generators (when DFG not in use, the output amplifier is available as an inverter)
156	Total

6.9/TYPICAL MODERN HYBRID COMPUTER SYSTEMS   175

Table 6.7 Specification of the EAI 640 Digital Computer

COMPUTER WORD

Type	Fixed
Length	17 bits including memory protect

MEMORY

Basic	4096 or 8192 17-bit words
Expansion options	to 8192, 16,384, or 32,768 words
Addressing	8 addressing options available by combining 3 groups:
1 Absolute:	512 words directly addressable
or relative:	±256 words directly addressable relative to present location of program
2 Indexing:	before or after indirect addressing
3 Indirect addressing:	Multilevel indirect addressing through all core locations

SPEED

Core memory cycle time	1.65 microseconds
Input/Output:	
Single word mode	93,000 words/second
Record mode	600,000 words/second

INSTRUCTION REPERTOIRE

62-unduplicated hard-wired instructions

Instruction Class:	Number
Arithmetic	8
Transfer	4
Logical	5
Shifts	8
Control	7
Interrupt	2
Protect	2
Exchanges	4
Jumps and skips	4
Skip on register condition	6
Skip on condition code	1
Input/output	11

INDEX REGISTERS

One hardware register: indefinite number of index counters using core memory

INTERRUPTS

Internal priority	7 levels
External priority	5 levels

INTERFACES

Interval timer interfaces	4

Figure 6.6   Electronic Associates, Inc. 690 hybrid computer system.

the bandwidth of the analog computing elements to 500 kHz. This bandwidth exceeds at least by a factor of two the performance of 100-volt systems. The complement of the analog portion of the system is presented in Table 6.6. The EAI 640 digital computer system is a general-purpose stored program digital computer with a 17-bit word including 1 memory protection bit. Its characteristics are itemized in Table 6.7.

The EAI 690 hybrid interface contains two levels of monitor and control activity. The first is hard-wired, requires no patching, and performs all functions required for automatic set-up, checkout, and general operation of up to six EAI 680 consoles. All these functions are under direct program control, using the standard instruction repertoire of the EAI 640. The second level provides logic control through patchable buffered control, sense, and interrupt lines. These provide single-bit logic communication between the EAI 640 and the EAI 680 computers. The system provides 8 sense-bits and 8 interrupt-bits. The EAI 690 hybrid programming system consists of programs for problem preparation, for set-up and checkout, and for operation. The problem preparation programs include a special hybrid **FORTRAN**, a symbolic assembler with linkage instructions, and a **HYTRAN** operations interpreter. For set-up and checkout, a hardware diagnostic program and a hybrid de-bug program are provided. Hybrid programming routines include a program for function storage and playback as well as a multivariable function generation program.

7

Software for Hybrid Computers†

7.1 INTRODUCTION

When a new digital computer is announced and marketed, the manufacturer of the equipment is expected to deliver, as part of his product, automatic programming language systems and various programming aids. The original need for these program packages, with written procedures describing their use, developed when it was found at an early stage that the serial nature of the digital machine presented inordinate difficulties for the user. He is primarily concerned with getting an answer, getting it quickly, and without great effort. Such a task can be accomplished by either the user knowing the machine language or the machine understanding the user's. Experience has taught us that the latter represents the only feasible alternative.

Coding, program testing, and interpreting large volumes of digital results proved most tedious, time consuming, and expensive. Some very clever techniques have since been employed to reduce these undesirable activities. By using the computer to help in the preparation of programs (automatic programming), by taking advantage of previously written codes (library or canned routines), though automatic monitoring of a program during their execution (dumping, tracing, dynamic de-bugging), and by having the computer handle the needs of many users during a single time period (monitors, executive routines), an efficiency has been achieved through the creation of a *programmed interface* between the engineer and

---

† This chapter was contributed by Man T. Ung, Manager of Education and Training, Electronic Associates, Inc.

the machine. This interface, an indispensible adjunct to the hardware, is usually referred to as "software." This software development, therefore, by making digital computer programming easy for the noncomputer expert, is responsible for almost universal acceptance of digital computers in the scientific field.

The history of early hybrid computation attempts contains a number of failures in which the lack of software unfortunately played a principal role, because it is the hybrid software that makes the computer system into a usable unit. Nowadays, sophisticated software has made it possible to increase computer utilization, to gain wider use of computers with minimum training of personnel, and to reduce duplication of programming effort for programs of general utility.

On the other hand, total dependence on automatic programming has the disadvantage of isolating the problem analyst, and even the programmer himself, from the computer. The analyst is restricted from communicating with his computer model while computation place. Similarly, the programmer is often limited in taking full advantage of the computer's special features.

Who should be responsible for the software—the user or the analog or digital manufacturer? This question contains so many factors that it cannot be responded to with a simple answer. There are three distinct facets of the software: design, installation, and maintenance. Naturally, the initial design of the software represents a major investment, perhaps, more than half a million dollars for a large general-purpose system. An installation cost arises since the software must function on different systems with different peripheral equipment, core memory size, and often diversified features. Software maintenance frequently escapes the user's consideration when he is in the market for a computer. Yet it amounts to a sizeable expenditure accrued over the years to come. This expense is necessary to improve progressively the original package for future use, to correct some "bugs" not detected at the beginning, and to keep up with hardware modifications as suggested by the manufacturer. In general, we cannot expect software to be transferable, even with a carefully defined system such as FORTRAN. Utmost prudence should be practiced when contemplating the purchase of a "one-shot" configuration. Now let us get back to the question of who will have to supply the software. If the analog manufacturer were in charge of software, we should make sure that he is intimately knowledgeable about the particular machine being used. Although the software operates on analog and linkage systems, it is still a digital program and requires digital experts for its design. On the other hand, the user may have reservations regarding hybrid software written by the digital manufacturer, for understanding the analog and especially the linkage operation entails a full-time effort. Ideally, hybrid computers should be constructed by experts in hybrid computation. A tutorial description of hybrid computer software is given in an article by McGhee and Lew.[4]

7.2 SOFTWARE REQUIREMENTS FOR HYBRID COMPUTATION

Hybrid computation has a number of unique characteristics which impose special demands upon the hybrid software. These characteristics depend upon the environment within which the software must operate and can be divided into four categories.

A. Problem Characteristics

In many problems there is a minimum speed constraint arising from the need to communicate with another hardware system, external to the computer, which must operate on a real-time basis. Consequently the hybrid program must be capable of accepting continuous or discrete inputs from external systems, performing the necessary computations, and responding with the required information within a predetermined time interval. This is generally accepted as "real-time" computation. Real-time computation covers a wide spectrum of applications such as aircraft and space simulators for system analysis and pilot training. Also, chemical process control studies on actual pilot plants and real-time on-line data reduction demand this type of operation.

B. Program Preparation

The software requirements for hybrid program preparation fall into two general areas: (1) programming languages including subroutine libraries for the digital portion and (2) analog computer programming and checking aids. It is of paramount importance that, whatever language the program is coded in, the entire hybrid problem must run as a compatible unit. To accomplish this, the hybrid program must match the sequential, time-dependent operation of the digital computer to the parallel properties of the analog computer. The programming languages with subrouting libraries must provide features to handle the following unique situations of hybrid programs:

1. Synchronization of the analog and digital computer operation.
2. Optimization of computation time when necessary.
3. Communication of data and control information between the analog and digital programs.

C. Operating Procedures

The overall objective of a hybrid software system is to gain the greatest efficiency in utilization, measured only by the ratio of results per dollar. To achieve this goal, hybrid software should facilitate the checkout, analysis of results, and introduction of changes, and assume the most effective use of the hardware. Again, the characteristics of the problem enter into consideration; for a time-critical problem with external constraints, the overriding consideration is to satisfy these constraints.

The property of interaction becomes of greater importance when dealing with hybrid than with digital software. The users who have become accustomed to the standard batch-processing operation of most large digital installations would benefit little from the conversational ability of the computer. Analog-oriented programmers, on the other hand, have grown up with the ability to interact freely with their problem, and they prefer the same features to be included in the hybrid software.

D. Hardware Testing

The last major class of hybrid software requirements is special diagnostic programs to test the operation of the hybrid computer, especially the analog components. Any help to the technicians to improve the up-time of the system will eventually reflect in the results/dollar balance sheet. Therefore, hardware testing should be given equal status compared to other types of diagnostic software.

There exist, at the moment, several software packages available commercially and no two of them are exactly alike. Thus, rather than discuss software in general, the standard package provided by Electronic Associates for the EAI 8900 hybrid computing system will be singled out and described in detail. Other hybrid computer manufacturers supply similar programs which should contain the same basic features as the EAI package. It is hoped that, through this maze of information, the reader will draw his own conclusions and generalization. From time to time other software names will be mentioned when applicable.

Hybrid computer software can be classified as depicted by Figure 7.1:

Program Language Processors
Control Systems
Program Libraries
Utility Programs
Diagnostic Programs

In the following sections of this chapter, a number of the most important routines will be examined in detail.

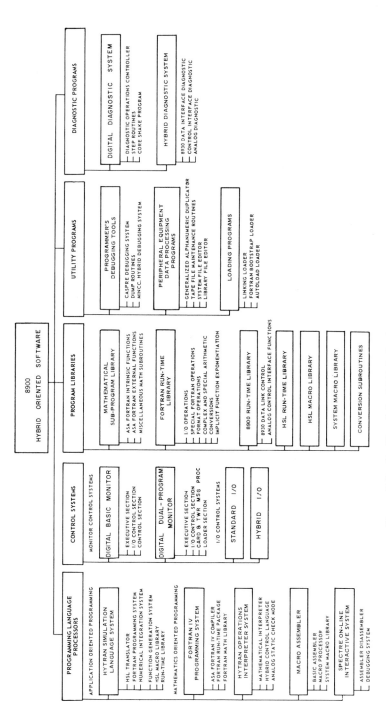

Figure 7.1 Profile of the 8900 hybrid oriented software package.

7.3 FORTRAN COMPILER

Because FORTRAN has been the standard programming language for mass digital scientific computation since its acceptance by the American Standards Association (ASA), the inclusion of a FORTRAN compiler in a hybrid computation software package is mandatory. Not only is it necessary to have available a programming language with which most programmers are familiar, but many of the other programs in a software package use a "FORTRAN-like" language and, in fact, usually produce a FORTRAN program as a result of their processing, and this secondary program must be input to the standard FORTRAN compiler.

To be entirely objective, however, it must be stated that a FORTRAN compiler, however efficient it may be, cannot produce object coding which is as efficient as that written directly in machine-assembly language. For some applications, this will not pose any restrictions; for others, a few problems will arise. The EAI 8900 FORTRAN compiler has an "in-line assembly" feature which offers to programmers the convenience of FORTRAN input-output and the increased efficiency and speed of assembly-language coding.

This in-line assembly feature most commonly finds applications in problems in which timing within a loop is of critical importance, but for which, at termination of the hybrid phase, large amounts of calculated data must be output in printed form. Although such a problem would be instructive, it would be prohibitively large for use here as an example.

A more tractable sample program, as illustrated by Figure 7.2, demonstrates the in-line assembly feature. This program reads six characters from the magnetic tape unit B, sorts them into month, day, and year, and then outputs them on the typewriter. Statements 10 and 30 illustrate the convenience of using FORTRAN for what would be tedious to program in assembly language. The statement immediately following 40 is an assembly-language instruction. The slash in the first column is one of the two ways in which the assembly-language mode may be entered. The effect of the instruction is to cause the machine to pause and transfer control to FORTRAN statement 50 when the operator presses the "EXECUTE" button. If several machine-language statements appear consecutively, the use of the "ASSEMBLE" card permits more efficient compiler operation. The coding sequence following the ASSEMBLE declaration is a program to sort out the date in a format such as 12/25/68 for type-out in the statement 80. The meaning of each mnemonic operation mode is not important and, therefore, will not be discussed. The appearance of the "FORTRAN" card causes return to the compile mode. With particular reference to hybrid applications, the "mixed-mode" assembly permits efficient setting,

7.4/HYTRAN OPERATION INTERPRETER (HOI)

```
C          THIS PROGRAM ILLUSTRATES THE IN-LINE ASSEMBLY
C          FEATURE OF THE EAI 8900 FORTRAN COMPILER.
           DIMENSION DATE(2)
10         TYPE 20
20         FORMAT(27H MOUNT DATA TAPE ON UNIT 8. )
30         TYPE 40
40         FORMAT(23H OPERATOR ACTION PAUSE.)
/          HJ       50S
50         REWIND 8
C          READ THE DATE OFF THE TAPE
60         READ (8, 70) DATE
70         FORMAT(A4,A2)
C          DATE IS SIX HOLLERITH CHARACTERS.
C          CONVERT IT TO THREE SETS OF TWO CHARACTERS.
ASSEMBLE
           ECA      DATE
           ST       /MONTH
           EROT     16
           ST       /DAY
           CA       DATE+1
           ST       /YEAR
FORTRAN
C          PRINT OUT THE DATE
80         TYPE 90,MONTH,DAY,YEAR
90         FORMAT(18H DATA TAPE CREATED ,2X,3(A2,1X))
C          OTHER
C             DATA
C                PROCESSING
C                   ASSUMED
C                      TO
C                         OCCUR
500        STOP
           END
```

Figure 7.2 Sample FORTRAN mixed-mode listing.

testing, and read-out of sense lines, conversion channels, and external interrupts, and the FORTRAN input/output (I/O) routines offer convenience and simplicity of use. The overall result is to free the hybrid programmer from some of the tedium of coding and permits concentration on the important aspects of the problem under consideration.

7.4 HYTRAN OPERATION INTERPRETER (HOI)

This on-line interactive language was created to minimize the bookkeeping chores associated with the preparation, set-up, control, monitoring, checking, execution, and documentation of the analog program. HOI has benefited from the design and the reported experience of earlier systems

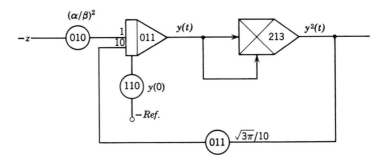

Figure 7.3 Circuit diagram for analog solution of Equation 7-1.

such as JOSS and FORTRAN. All the language generality of FORTRAN exists in this interpreter, plus many improvements for ease of man-machine communication. In the past, the burden of pot-setting calculation static-checking, and actual reading of static-check values fell on the programmers. These not very challenging tasks sometimes required days on the desk calculator. To illustrate few prominent features of the interpreter, let us consider a small portion of an analog program.

Assume we are given the equation:

$$\dot{y} + \sqrt{3\pi}\, y^2 = \left(\frac{\alpha}{\beta}\right)^2 z \qquad (7.1)$$

where $\quad z = .5$

$\qquad\quad \beta = 5$

$\qquad\quad \alpha \leqslant 4.5$

and $\quad y(0) = .3820$

The computer diagram for Equation 7.1 is shown in Figure 7.3.

The suggested HOI listing is found in Figure 7.4. Each statement is associated with a number called an identifier. Under program control, the steps are executed in numerical order (2.1000 precedes 2.2000, and 2.2000 precedes 3.4, etc. . . .) irrespective of their order of entry. This system of identification permits simple modification of an existing program by writing the changes at the bottom of the coding sheet.

—The first statement is a comment.
—Statement 1.0100 calls for the input DATA required to be read in by the CARD reader.
—Step 1.0200 demands the value of ALPHA (from the card reader as indicated above).

—Statements 1.0300 to 2.3000 follow FORTRAN conventions with the exception of the square root (SQR) subroutine. These statements are used to calculate the analog value at each point in the circuit diagram, Figure 7.3.
—Statement 3.1000 places the analog computer in pot set mode, and statement 3.2000 causes all pots entered earlier to be set, namely C010, C011, and C110 (C = coefficient).
—Step 3.3000 puts the analog computer in initial condition.
—Step 3.5000 reads all components in the symbol table such as A011, A213, and D011 (D = derivative or summing junction of integrators). The newly obtained values are then verified by statement 3.4000 against the calculated value, in steps 2.1000 to 2.3000, to a tolerance of 2 parts per 10,000 or .0002. If discrepancies are detected, these values will be printed out.
—Step 4.1000 prints all coefficients. The above discussion suggests that "reprogramming" is not a difficult job when using HOI. Besides the stored program mode, HOI could be utilized on-line in a sequence of request-response. In this mode of operation the user will type a command and the computer will act upon it. If the command is unacceptable a warning will be given. As an example, warnings may take the following forms:

UNDEFINED SYMBOL

FORMAT ERROR IN 2.5000

or

ITERATE ERROR IN 1.6201

It is evident that HOI contains some of the features of the APACHE (Analog Programming and Checking) language published by the Euratom

```
1.0000) ΔEXAMPLE OF HOI USAGEΔ
1.0100)   √DATA, √CARD, INPUT;
1.0200)   ALPHA ←
1.0300)   BETA = 5,
1.0400)   Z = .5,
1.1000)   C010 = (ALPHA/BETA)**2,
1.2000)   C011 = SQR(3*3.1416)/10,
1.3000)   C110 = .3820,
2.1000)   A011 = +C110,
2.2000)   A213 = A011*A011,
2.3000)   D011 = (C010*Z) - (C011*A213*10),
3.1000)   √PS, MODE;
3.2000)   $COEF, SET;
3.3000)   √IC, MODE;
3.4000)   .0002, VERIFY;
3.5000)   $COMP, READ;
4.1000)   $COEF;
```

Figure 7.4 HOI listing.

Computation Center at Ispra, Italy.[1] In particular, both HOI and APACHE compute potentiometer settings and static-check values. However, since HOI is part of a hybrid program package, it can be used to "close the loop" and set the coefficients after they have been computed.

7.5 HYTRAN SIMULATION LANGUAGE (HSL)

Hytran Simulation Language is an algebraic programming language for a digital computer which makes extensive use of the FORTRAN IV programming language. Many of the HSL special features are indeed valid FORTRAN statements. HSL provides the user with the power to solve large sets of simultaneous differential equations with a minimum of coding effort. This feature alone provides the hybrid programmer with a means of dynamic checking, that is, running an all-digital "test case" of the hybrid problem. Naturally, the solution using HSL takes longer, but it is a "test" to be executed only once.

The availability of a block-oriented programming language is an invaluable part of a hybrid software library. Such languages as MIMIC (written at the U.S. Air Force Wright-Patterson Air Force Base Computer Center) and the IBM Continuous System Modeling Program (CSMP) are other examples of such languages. Excellent surveys of digital simulation languages are found in the references 2 and 3.

Since the HSL program actually passes through the FORTRAN compiler, most of the FORTRAN features and library are available. HSL also contains many numerical integration algorithms available at the user's option. These range from a simple first-order technique (Euler's method) to higher-order predictor-corrector routines with variable step control The structure of an HSL program follows that of a typical simulation program and contains an initialization section, integration loop, and a terminal section in the order indicated in the HSL demonstration program listed below The initialization section (INITIAL) reads input data, lists the input data, and prints titles The DYNAMIC section contains the derivative evaluation routines; it is here that the differential equations are solved. In the TERMINAL region the programmer is free to do anything he wishes before returning to the INITIAL region. For instance, it may contain such tasks as testing for convergence, parameter updating, error sensing, etc.

As an example, let us consider a set of two coupled second-order differential equations:

$$(FM1)\ddot{X}_1 = -(G1)\cdot(X_1 - X_2) - (Q)\cdot(\dot{X}_1 - \dot{X}_2) \quad (7.2)$$

and

$$(FM2)\ddot{X}_2 = (G1)\cdot(X_1 - X_2) + (Q)(\dot{X}_1 - \dot{X}_2) + (G_2)\cdot(G - X_2) \quad (7.3)$$

Let us further define a few more terms:

T   Time
TZ   Initial time, $t = 0$
TMAX   Maximum time when integration is stopped
DELT   Integration step size Δt
X1D, X2D $\dot{X}_1, \dot{X}_2$
X1DD, X2DD $\ddot{X}_1, \ddot{X}_2$

An HSL program to solve Equations 7.2 and 7.3 is shown in Figure 7.5. The meanings of the important statements are individually explained later.

The INITIAL region (statement 1 to statement 40 inclusive) needs no elaboration because it includes only FORTRAN-like coding.

```
        PROGRAM≠EXAMPLE OF COUPLED DIFFERENTIAL EQUATIONS≠
        INITIAL
1       CONTINUE
        READ 10, TZ,TMAX,DELT,Q,G1,G2,G,FM1,FM2
10      FORMAT (9E8.0)
        PRINT 20
20      FORMAT(1H1)
        PRINT 30, TZ,TMAX,DELT,Q,G1,G2,G,FM1,FM2
30      FORMAT (2(6E20.8/))
        PRINT 40
40      FORMAT (1H1,11X1HT,18X1HG,18X2HX1,*
        19X3HX1D,18X2HX2,17X3HX2D//)
        END
        DYNAMIC
        DEFINE [TZ,TMAX,DELT,Q,G1,G2.G,FM1,FM2]
        PRINT 100, T,G,X1,X1D,X2,X2D
100     FORMAT (1X,6E20.8)
        INTAL [RK4]
        COMDEL [64*DELT]
        CALF [64]
        INITVAL [TZ]
        TERMVAL [TMAX]
        DERIVATIVE
            W = G1*(X1-X2) + Q*(X1D-X2D)
            X1DD = -W/FM1
            X2DD = (W - G2*(X2-G))/FM2
            X1D = INTEG [0, X1DD]
            X2D = INTEG [0,X2DD]
            X1 = INTEG [0,X1D]
300         X2 = INTEG [0,X2D]
        END
        END
        TERMINAL
        GO TO 1
        END
        END
```

Figure 7.5   HSL listing.

The DYNAMIC region immediately follows the INITIAL region and ends at statement 300:

> DEFINE [TZ, TMAX ... etc.] indicates to the translator that those variables listed are actually defined in different sections of the program, e.g., input data.
> INTAL [RK4] chooses 4th order Runge-Kutta as integration algorithm.
> COMDEL [64* DELT], communication interval, causes a printout at intervals of 64*Δt, i.e. after each 64 integration steps.
> CALF [64] expresses that there are 64 integration steps between each printout (COMDEL). Since COMDEL = 64*DELT the integration steps is DELT, or Δt by previous definition.
> INITVAL [TZ] instructs the integration routine to start at TZ read in earlier.
> TERMVAL [TMAX] sets the upper limit of the independent variable t, i.e., to stop integrating when $t =$ TMAX.
> X1D = INTEG [0, X1DD] instructs HSL to integrate $\ddot{X}_1$, with initial condition of zero (0) to obtain $\dot{X}_1$.

The TERMINAL section in this problem merely returns control to statement 1, which is the beginning of the INITIAL region. When compiled, loaded into memory and run, the program will read a data card to obtain all the parameters TZ, DELT, Q, G1, G2, G, *FM*1, and *FM*2 and solve the said equations. It will print out T, G, X_1, $\dot{X}_1$, X_2, $\dot{X}_2$ at time intervals of 64 Δt from TZ to TMAX.

7.6 SET-UP AND CONTROL SOFTWARE

A. The Monitor

Hybrid systems demand a special monitoring system to control their operation. As a result, the monitor must be quite different from the normal digital system monitor since it must allow direct on-line control of program action at all times. The highly interactive nature of hybrid computation requires that operator interrogation of programs be possible during program run-time. It is also necessary that the monitor provide the programmer direct access to all computer I/O facilities; otherwise, system delays will result, preventing true real-time operation of the hybrid system. Furthermore, it permits him to interrogate randomly the operation of the system, without being delayed by the operation of certain programs and without worry of destroying the control information vital to a program's continued execution. These requirements contrast with most large batch-processing systems where the monitor system takes full control of a program and runs it to completion before meaningful access is returned to the operator. To further extend the control flexibility needed in a hybrid system, the Monitor allows operating programs to initiate most of the control directives that are available to the programmer via control input statements. Thus an operating program may initiate a Macro Assembly as an intermediate step in its total operation, or it may request additional peripheral equipment or interrupts, and then release them while the operating program is being executed, without operator intervention. Such an ambitious Monitor should be efficient; otherwise, it will have to spend a considerable amount of time worrying about what to do next and not doing productive work during this period.

Specifically, upon user request, the 8900 Monitor is capable of performing any of the following functions:

1. Load a processor from the system tape and control its processing through several source files. This processor may create files which will be used by a subsequent processor. The Monitor will maintain control over intermediate files created by one processor, if they are on magnetic tape, in order that they be properly positioned for the subsequent processor. The processors, which are recognized by the Basic Monitor at this time, are the following:

    HYTRAN Simulation Language
    Macro Assembler
    FORTRAN
    Linking Loader
    Post Mortem Dump
    Operations Interpreter

2. Save user's program (with its registers) on magnetic tape, and restore it on command.
3. Label, rewind, reposition, and write end-of-file marks on magnetic tape.
4. Change a word in core memory at a location given in absolute form.
5. Control the core memory limits used by the various processors to allow a real-time program or de-bugging package to coexist in core with the processors.
6. Assign a physical device to a logical unit designation.
7. Execute a program at a location, given either in absolute or symbolic form. This function will also complete any load process in progress by loading from the subroutine library on magnetic tape.
8. Start, continue, or abort a job.
9. Transmit operating instructions to the operator as either a comment or a pause.

B. Analog-Digital Interface

The analog-to-digital interface can be divided into two broad categories: control interface and data interface. Under the control interface software falls a large number of routines:

> Test sense lines†
> Set and reset control lines†
> Input and output binary data from logic patch panel
> Read the values of analog components such as multipliers, amplifiers, trunk lines, and resolvers
> Handle external interrupts
> Control analog modes of operation and time constants
> Set potentiometers
> Control the analog printer
> Keep track of the present status of the analog console(s).

The data interface represents functions such as:

> Digital-analog conversion
> Analog-digital conversion
> Multiplexing and track/store amplifiers control
> Digital-analog multiplier

All the routines are FORTRAN compatible. On the other hand, reentrant features should be left to the personal choice of each programmer. If SIN is a reentrant subroutine, it can be interrupted during its execution. Before servicing the interrupt, it stores all registers and intermediate calculation results in some specified locations called "stack." This is done since, during the interrupt, the SIN subroutine may again be utilized. When the program finishes servicing the interrupt, control is returned once again to the reentrant SIN subroutine. This routine must restore all registers and intermediate calculations and reinitiate, starting at the location where it exited before. For the reason cited earlier, reentrant subroutines are not as efficient as their nonreentrant counterparts. Ideally, both kinds should be available to the programmer so he can use either one at his discretion whenever the need dictates.

The nature of interface software can be best illustrated in a small boundary value problem.

---

† Sense lines are flip-flops controlled by logic signals on the analog console that can be interrogated by the digital computer. Control lines represent flip-flops located on the analog-logic panel that can be set and reset by the digital machine.

Given the following equation:

$$\frac{d^2 T}{dx^2} = KT^4 \tag{7.4}$$

where $T(0) = T_0$ is a known constant and (7.5)

$$\left.\frac{dT}{dx}\right|_{x=L} = \dot{T}(L) = 0 \tag{7.6}$$

To solve Equation 7.4, we must have $T(0)$ and $\dot{T}(0)$. The latter condition is not at the programmer's disposal. Hence, we need to use an iteration scheme to satisfy Equation 7.6 such as:

$$\left.\dot{T}(0)\right|_{n+1} = \left.\dot{T}(0)\right|_n - (\text{Gain}) \cdot \left.\dot{T}(L)\right|_n \tag{7.7}$$

The analog diagram for Equation 7.4 is shown on Figure 7.6. When one executes the program, it will request from the operator an initial guessed value of $\dot{T}(0)$. Once the operator types in a number, the program proceeds

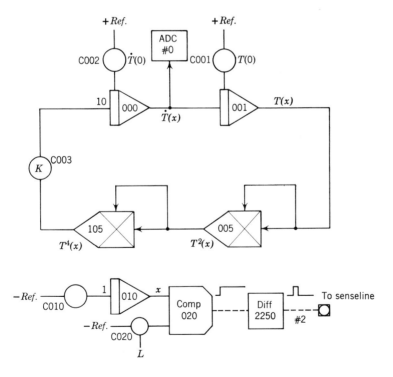

Figure 7.6 Interface exercise analog diagram.

by solving Equation 7.4 repetitively and varying $\dot{T}(0)$ according to Equation 7.7. But $\dot{T}(0)$ is the coefficient of potentiometer C002. Consequently, the rep-op (repetitive operation) cycle is composed of four separate phases:

>PS: during which C002 is set to a new value
>IC: allow all capacitors to change
>OP: from $x = 0$ to $x = L$
>Hold: for $x > L$

The cycle repeats itself once a new value of $\dot{T}(0)$ is obtained. In practice, this scheme should *not* be used since a digital-analog multiplier (multiplying DAC) can replace C002 much more efficiently and the PS period can be completely eliminated from the re-op cycle. Potentiometer C002 is used in this problem solely for illustrating how an attenuator can be set digitally by using interface software. A FORTRAN listing for this problem is found in Figure 7.7. There are two conditions that cause a termination of the run:

$$\dot{T}(L) < \text{Tolerance}$$
$$\text{Number of iterations (ITERNO)} > 100$$

The program in Figure 7.7 will not be explained, only the interface subroutines will be analyzed in detail.

90 CALL QSFST0 (1, ERROR): the numeral "1" represents analog console No. 1. This statement selects capacitor size of 0.1 μfd for the integrators. On the 8900 system, second-slow (QSSLO0), second medium (QSMED0), and second fast (QSFST0), choose 10 sec, 1 sec, and 0.1 sec, respectively. These time constants correspond to 10 μfd 1 μfd, and 0.1 μfd in that order. When the argument ERROR $= 2$, it indicates that the analog console is disconnected from the digital machine. It is necessary to push a button to put the analog console under digital control. Therefore, when someone is working on the analog panel he cannot be disturbed if he does not want to be. Every time that the analog console is referred to for the first time, we should look at the ERROR argument as indicated by the following statement in the program in Figure 7.7. This "safety" procedure is to make sure that the hybrid mode is performed only whenever the analog computer(s) is (are) ready.

110 CALL QPS0 (1, ERROR) places analog console No. 1 in pot set mode. There is no need to test ERROR because if the analog computer is disconnected, it would be detected earlier.

7.6/SET-UP AND CONTROL SOFTWARE   193

```
              LOGICAL  SENSELINE
              INTEGER ERROR, ADDERR
              DATA POT/4HC002/
      C  RECEIVE INPUT DATA FROM CARD READER AND TWR
              READ 50, GAIN, TOLERANCE
       50     FORMAT (2E10.5)
              TYPE 70
       70     FORMAT (10X, 32H ENTER A GUESS FOR TDOT(0) =     )
              ACCEPT 80, TDOT
       80     FORMAT ( E10.5 )
      C  TIME CONSTANT = SEC + FAST = .1 SEC
       90     CALL QSFS10 ( 1, ERROR )
              IF (ERROR.EQ.2)  GO TO 280
      100     ITERNO = 1
      C  RESET SENSELINE NO. 2 BEFORE STARTING
      105     CALL QTEFF0 (1,2,SENSELINE,ERROR)
      C  ENTER ITERATION LOOP,  PLACE CONSOLE IN POTSET MODE.
      110     CALL QPS0 ( 1, ERROR )
      112     CALL QSTPT0 ( 1, POT, TDOT, ERROR )
              IF ( ERROR.GT.1 )  GO TO 300
      C  PLACE CONSOLE IN INITIAL CONDITION MODE
      120     CALL QIC0 ( 1, ERROR )
      C  ENTER DELAY SUBROUTINE FOR 1 MILLISEC
              CALL SDLY
              CALL TDLY(1)
      C  SWITCH TO OPERATE MODE
      130     CALL QOP0 ( 1, ERROR )
      C  TEST TO SEE IF X = L
      132     CALL QTEFF0 (1,2,SENSELINE,ERROR)
              IF (.NOT.SENSELINE ) GO TO 132
      C  PUT ANALOG CONSOLE IN HOLD MODE
      150     CALL QHOLD0 ( 1, ERROR )
      170     CALL QADCV0 (0,1,TDOTFINAL, ERROR, ADDERR )
              GO TO (190, 320, 340) , ERROR
      190     TDOT = TDOT - GAIN*TDOTFINAL
              ITERNO = ITERNO + 1
              IF (ITERNO.GT.100) GO TO 220
              IF (TDOTFINAL.GT.ABS(TOLERANCE)) GO TO 110
      C  ITERATION COMPLETED - EXIT ROUTINE
      220     TYPE 240, TDOTFINAL, ITERNO
      240     FORMAT(10X,12HTDOTFINAL =    ,E10.5,10X,20HNO. OF ITERATIONS=    ,I3)
              STOP
      280     TYPE 290
      290     FORMAT (10X,24H CONSOLE DISCONNECTED   )
              GO TO 90
      300     TYPE 310, TDOT
      310     FORMAT (10X,35H POT C002 FAILED TO SET, COEFF. IS    ,E10.5)
              GO TO 112
      320     TYPE 330
      330     FORMAT (10X, 15H ADC OVERLOAD    )
              GO TO 170
      340     TYPE 350
      350     FORMAT (10X, 27H NON EXISTENT ADC CHANNEL    )
              GO TO 170
              END
```

Figure 7.7   Interface exercise listing.

112   CALL QSTPT0 (1, POT, TDOT, ERROR) selects C002 (POT) on console No. 1 and sets it to the value TDOT. If ERROR = 1, no error; ERROR = 2, invalid address; ERROR = 3, coefficient greater than unity; ERROR = 4, console disconnected; ERROR = 5, null failure, i.e., pot fails to set within 2 second. With a single "computed GO TO" statement, all five conditions mentioned above can be tested and a message can be printed. In case a block of pots are to set, the following calling sequence is necessary:

CALL QSTBK0 (3, C400, C411, 6, COEFF, ERROR) the above line seeks out analog console No. 3. It sets a block of potentiometers, starting with C400 and ending with C411. There are 6 attenuators in all (C400, C401, C402, C403, C410, and C411). No attenuator carries the addresses from C404 to C409 inclusive. Potentiometers are set to the values contained in the COEFF array, i.e., the value in COEFF is meant for C400, COEFF + 1 for C401, and so on. ERROR indications are identical to the case of setting one single attenuator.

132   CALL QTEFF0 (1, 2, SENSELINE, ERROR) tests senseline No. 2 on console 1 (TEFF = test flip-flop). It should be clear that a senseline is nothing but a flip-flop, as indicated earlier. If the flip-flop was set at the time of interrogation, then the routine sets SENSELINE = .TRUE. and the flip-flop is cleared immediately after interrogation. Statement 105 insures that the program starts with senseline No. 2 in the reset position.

170   CALL QADCV0 (0, 1, TDOTFINAL, ERROR, MCHAN) starts an analog-to-digital conversion on ADC channel No. 0, stops after one conversion is finished, and places the result in location TDOTFINAL. The variable named TDOTFINAL is deliberately made longer than 6 letters for easier recognition of its function. As far as the FORTRAN compiler is concerned, only the first 6 letters, TDOTFI, are considered the actual name, the rest is ignored.

To appreciate the complexity of only one of the subroutines contained in the Set-Up and Control Software Library, the reader is referred to a description of the potentiometer setting system given by McGhee and Chandler.[5] A list of the Set-Up and Control subroutines supplied by EAI for the 8900 system is given in Table 7.1.

7.7   CHECKOUT AND DE-BUGGING SOFTWARE

A.   Digital De-bug

Digital de-bug software (nicknamed CASPRE) is an interactive program used as an aid in diagnosing program malfunctions. The typewriter is

7.7/CHECKOUT AND DE-BUGGING SOFTWARE

Table 7.1 EAI Set-Up and Control Library

Data Interface Subroutines	Purpose
1. QDALD0	To load a continuous block of DAC channels
2. QDAJM0	To load both registers (load and initiate conversions simultaneously) of a double-buffered DAC
3. QDAXR0	To initiate simultaneous conversions on one or more DAC channels
4. QDASV0	To read out a DAC control word
5. QDARE0	To restore a DAC control word
6. QDACR0	To reset DAC registers
7. QDATS0	To test status of DAC devices
8. QADCV0	To execute analog-to-digital conversions on one or more continuous channels
9. QTRCK0	To place a contiguous block of track/store amplifiers into the "track" mode
10. QSTRE0	To place a contiguous block of track/store amplifiers into the "store" mode
11. QADSV0	To read out and store an ADC control word
12. QADRE0	To restore DAC control register to its earlier status
13. QADTS0	To test status of ADC device

Control Interface Subroutines	Purpose
1. QTEFF0	To determine the status of a specific senseline flip-flop
2. QSEFF0	To set a specified control-line flip-flop
3. QREFF0	To reset a specified control-line flip-flop
4. QSYNC0	To provide synchronization with respect to a specified sense-line
5. QLGOT0	To output a 16-bit data word to the analog computer
6. QLGO80	To output an 8-bit data word to the analog computer
7. QLGIN0	To input a 16-bit data word to the digital computer
8. QRDAL0	To read a selected analog component
9. QSTPT0	To set a selected potentiometer to a desired value
10. QPRNT0	To print address and output of a sequence of analog components on the analog printer (not to be confused with the system line-printer)
11. QSTBK0	To set a block of potentiometers
12. QSTWD0	To read in an analog computer status word
13. QMOWD0	To read in a hybrid monitor word
14. QFLWD0	To read fault condition in analog computer

(Continued)

Table 7.1 (*continued*)

Control Interface Subroutines	Purpose
15. QANMD0	To determine current analog computer mode
16. QTCST0	To determine current analog computer time constant selection
17. QLGMD0	To determine current logic mode selection on analog computer
18. QENIT0, QDIST0	To enable or disable general-purpose external interrupts

Computer Mode Selection Subroutines	Purpose
1. QHOLD0	Hold
2. QIC0	Initial condition
3. QRT0	Rate test
4. QPS0	Pot set
5. QSTST0	Static test
6. QRUN0	Run ⎫
7. QSTOP0	Stop ⎬ Applicable to parallel logic only
8. QCLR0	Clear ⎭

Time Constant Selection Subroutines	Purpose
1. QSSL00	Seconds—slow (10 mfd capacitor)
2. QSMED0	Seconds—medium (1 mfd capacitor)
3. QSFST0	Seconds—fast (.1 mfd capacitor)
4. QMSL00	Milliseconds—slow (.01 mfd capacitor)
5. QMMED0	Milliseconds—medium (.001 mfd capacitor)
6. QMFST0	Milliseconds—fast (.0001 mfd capacitor)

used as a control source from where the operator can command actions from all I/O devices. CASPRE has the capability to provide snapshot printouts of specified memory location(s) while testing a program. This includes the ability to open selected memory locations for inspection in decimal, octal, or hexadecimal format. For example, the operator may type:

    F!   15000A   15700Z   2D   (carriage return)

7.7/CHECKOUT AND DE-BUGGING SOFTWARE

In response to this the operator will have:

F!	The content of each location is treated as floating point data. For instance, .02679 will be outputted as 2.679 E-2.
15000A	First location to be opened is 15000.
15700Z	Last location inspected is 15700.
2D	2 = device 2 = line printer by convention. D = dump

In summary, the contents of memory locations 15000 to 15700 will be printed out on the line printer in floating point format.

CASPRE also allows the user to insert breakpoints at strategic points in the program without destroying it. These breakpoints halt the program to allow an inspection of the forward progress. In addition, CASPRE can do many other tasks, a few of which are:

—Save the contents of hardware registers so that calculations can be resumed later without loss of continuity.
—Move and relocate a block of words. Naturally all references to the moved block must be readjusted.
—Read an I/O and verify input against preselected block.

B. Hybrid De-bug

One important diagnostic package used in the 8900 system is called MINCC (Miniature Control Compiler). MINCC was designed:

1. To help the hybrid programmer de-bug his program *on-line*. For this purpose he can add a short hybrid program, modify the main program, change any constants, or command the analog console, all from the typewriter.
2. To provide the maintenance staff an easy means of communicating with one or more analog console during trouble shooting. The typewriter is utilized again.

Because the operator is usually an engineer, he is not expected to be an accurate typist. If an error is made and he wants to ignore the unfinished line, it suffices to back space. Even if an error is already entered, MINCC will type a question mark (?) when a command is invalid.

When using MINCC, all operations are referred to in mnemonic forms. It contains a number of macro commands that call complicated subroutines even though the user types only a single line. Suppose we want to insert an output sequence at location 30000. This sequence will type out all amplifiers starting at A000 and it should not displace the main program because there is nothing located at 30000 itself. First, we type

$$R(Sp)30000$$

where (Sp) = space. The above line instructs MINCC to "Retain" the newly created program at 30000. Upon receiving this line, MINCC will type

$$30000$$

and wait for the operator to type in the instruction, say, Cl. MINCC then closes the bracket behind Cl and starts typing the next location for the next input. The resulting listing on the typewriter is shown below. Remarks are added only for an explanation purpose, and are *not* part of the actual program:

R(Sp)30000	
[30000 Cl]	Select analog console 1
[30001 A000]	Select A000
[30002 RAA]	Read analog address
[30003 D]	Delay (waits for address selection)
[30004 RDVM]	Read digital voltmeter (DVM)
[30005 D]	Wait for the DVM
[30006 T]	Type both address and value
[30007 D]	Wait for typewriter
[30010 STEP]	Step to next legal address
[30011 WADR]	Wait for the new address
[30012 J 30002]	Jump back to 30002

MINCC is far more powerful than the above example depicts. Here is a partial list of its capabilities:

—Control and monitor analog mode and time constant.
—Set potentiometers.
—Control external interrupts, senselines, control lines, and logic communications with the analog computers.

It should be stressed again that there is no need for compilation if the program is written in MINCC. This feature alone gives the operator intimate control over all analog consoles even though he may be physically remote from the analog pushbuttons.

C. Hardware Diagnostic

The unique requirements of hardware diagnostics for a hybrid system divide into two categories: (1) comprehensive tests for all portions of the hybrid interface and (2) complete analysis of the static and dynamic operation of the analog computer. These programs must permit a service engineer to determine if all portions of the systems are operating properly on a go-no-go basis, with appropriate documentation of any failures to do so. The complexity of de-bugging a hybrid program demands that the system be operating correctly so that the programmer can assume that his difficulties are due only to the program.

A typical list of test routines which should be included in the linkage diagnostic software are:

1. Analog addressing.
2. Digital voltmeter conversion and readout.
3. Analog mode selection.
4. Analog value selection and potentiometer setting.
5. Logic mode selection.
6. Analog time constant selection.
7. Fault testing.
8. Discrete logic line communication (sense line, control line).
9. Analog-to-digital and digital-to-analog conversion.
10. Analog interrupts and mask control.
11. Track/store drift rate
12. Analog status readout.

Interactive program control should be provided to repeat any one or more of these tests independently of the others. To the extent available in the hardware, the test routines should isolate the fault to a particular hardware component that has failed and the error messages should indicate that component.

The analog diagnostic test program works with a standard analog patch-panel to test each analog computer component in both a static and dynamic mode of operation. This must include both the analog components such as amplifiers, potentiometers, and multipliers and the patchable logic components. Again, isolation of the fault with appropriate documentation is essential.

7.8 UTILITY LIBRARY

A. Integration Library

The real-time integration package provides a convenient facility for integrating sets of state variables by using any of these algorithms:

1. Euler
2. Simpson's Rule
3. Adams Fourth-Order Predictor
4. Improved Euler
5. Third-Order Runge-Kutta
6. Fourth-Order Runge-Kutta
7. Gill
8. Second-Order Predictor-Corrector
9. Milne Predictor-Corrector
10. Adams-Moulton Predictor-Corrector

The 8900 integration routines have been specifically organized to allow change in both step size and integration algorithm, without reassembly or recompilation of the program, and are designed for real-time computation which demands fixed step size and minimal execution time. Special features of this package of particular interest are:

(a) The routines are reentrant so that multispeed integration may be accomplished with one routine in core.
(b) Algorithm selection is controlled by one statement in the calling sequence, thus the user can easily experiment to determine the simplest algorithm that will satisfy the job requirements.
(c) Only the algorithms of interest need be included in memory at execution time, thereby minimizing core storage requirements.
(d) The routines are self-initializing.
(e) Coding of the routines is structured so that addition of other algorithms by the user can be accomplished with relative ease.
(f) The routines have a built-in mode detection system which alters the state variables according to system mode specified at the entry point as follows:

IC	Set state variables to their initial values
H	Leave state variables unchanged
OP	Update state variables according to selected integration algorithms

Derivative information is supplied to the integration package in the form of single-precision floating-point numbers, except for the step size which must be double precision. Additionally, state variables are accumulated internally as double-precision quantities, but the resultant values are returned to the calling program as rounded single-precision floating-point numbers.

These routines find their chief application in situations in which the analog hardware is fully utilized and additional integration requirements must be met, or when the number of state variables is excessively large. Another common use is for "slowly" changing variables which, when using analog integrators for a prolong period, deteriorates the signal-to-noise ratio due to integrator drift. Full explanation of the calling sequences is beyond the scope of this chapter and regretably must be omitted.

B. Function Generation

The function generation processor was developed to prepare function tables and to provide values of the functions for n arguments where n can vary from one to three. Many problems, especially in aerospace applications, include functions of more that one argument. As a result, the analog computer is unable to perform this task with its traditional diode function generator.

The initial task of the function generation system is to prepare function tables in either floating-point format or fixed-point format and a binary point that can be selected by the user. The remainder of the system consists of subroutines which are used by the user for argument normalization and function interpolation, and a function display package which provides a visual display of the functions as they would appear in memory during program execution.

Consider the function in Figure 7.8. X is the slow varying argument and Y the fast varying. The user provides the function generation system with a description of the arguments (variable or constant breakpoints) and a table

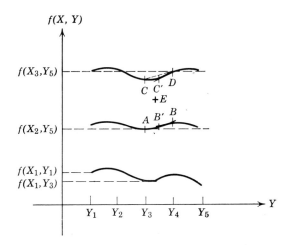

Figure 7.8 A function of two variables.

of numbers which are usually obtained while performing actual experiments:

$$f(X_1, Y_1), f(X_1, Y_2), f(X_1, Y_3), \ldots f(X_1, Y_n)$$
$$f(X_2, Y_1), f(X_2, Y_2), f(X_2, Y_3), \ldots f(X_2, Y_n)$$
$$f(X_m, Y_1), f(X_m, Y_2), f(X_m, Y_3), \ldots f(X_m, Y_n)$$

During program execution, subroutines are used to evaluate the function for the prescribed argument values which have previously been normalized. The normalization subroutine is called upon to normalize each new argument value into an integer and fractional part. The integer part is the number of discrete points contained in the argument and the fractional part the distance to the next discrete point.

Consider a function X and Y, defined for the following values of X,

$$0, 10, 20, \ldots 50$$

The normalized value for $X = 35$, would be 3 with a fractional part of .5. The starting position in the function table is computed by the function "generate routine" as follows:

Starting position = 3 (number of points per curve)
  + (the integer value of the normalized Y argument)

The starting position for each value of a given set of arguments needs to be computed once. Thereafter, the LOOK-UP entry of the subroutine can be used to perform the interpolation, thus resulting in faster computation times. To determine the value of the function at point E, for example, B' and C' are determined by a linear interpolator on AB and CD followed with an interpolation on the segment $B'C'$ itself.

As an example, consider the drag coefficient C_D of a missile which, under certain conditions, can be treated as consisting of two components; one, a function of thrust and Mach number, and the other a function of altitude and thrust. In this example, the sea-level component which is a function of thrust and Mach number will be considered:

$$\text{CD0} = f(\text{THRUST, MACH}) \qquad \text{Sea-level component of drag coefficient}$$

The following information (see top of opposite page) in the format indicated is needed by the function generation system to prepare the tables.

In this example, THRUST is defined as a variable breakpoint argument (VARBPT) that assumes values from 0.0 to 14220. The processor requires the value of the argument at each breakpoint to be specified for variable

THRUST	VARBPT	0.0, 14220.0
	DATA	0.0, 2720.0, 14220.0
	ENDVAR	THRUST
MACH	CONBPT	0.0, 4.0, .5
	INPUT	CD0, (THRUST, MACH)
	OUTPUT	CD0, (THRUST, MACH), .FIX
FDATA	CD0	
POINTS	$f(THRUST_1, MACH_1), f(THRUST_1, MACH_2),$	
	$\quad f(THRUST_1, MACH_3), \ldots$	
	$\quad f(THRUST_1, MACH_8), f(THRUST_2, MACH_1)$	
	$\quad f(THRUST_2, MACH_2), f(THRUST_2, MACH_3), \ldots$	
	$\quad f(THRUST_3, MACH_8).$	

breakpoint arguments. In this case, there exist three breakpoints whose corresponding arguments are 0.0, 2720.0, and 14220.0.

MACH is defined as a constant breakpoint (CONBPT) argument that varies from .0 to 4 in steps of .5.

CD0, a function of Mach and Thrust, is the input function provided by the user (FDATA). The function that will be prepared by the processor is fixed-point (.FIX) and has the same arguments as the input function. The programmer can define different arguments for the output function as long as the total number of points for the input functions are the same. The above information is all that is required by the processor to prepare tables with the correct headers so that the standard subroutines can be called upon by the user for argument normalization and linear interpolation.

Examples of subroutine calling sequences as they are written in FORTRAN are:

1. Argument Normalization
   CALL MACHV (MACH, NMACH)—the subroutine, MACHV, is used to normalize the argument MACH. The normalized value is stored in location NMACH.
2. Function Generation
   CALL CD0G (NTHRUST, NMACH, CD0)—the subroutine CALL CD0G determines, for the function CD0, the starting address in the function table, for the particular value of the normalized arguments thrust, (NTHRUST), and Mach (NMACH), and also determines the value of CD0. The function value is stored in CD0.
3. Function Look-up
   CALL CD0L(CD0)—this call is made if the starting address in the function table was determined by a previous call to the "generate routine" of a function of the same arguments. The interpolated value for the function is stored in CD0.

The parameters (MACH, NMACH, and NTHRUST) used in the above example are treated as real variables by declaration at the beginning of the subprogram.

A further discussion of function generation in flight simulation is given in Chapter 12.

C. Card-Programmed Diode Function Generation (CPDFG) Routine

Operationally, the card-program diode function generator differs from the conventional manual DFG only in the set-up mode where its breakpoints and outputs are automatically set by punched card. Thus CPDFG just follows the trend toward automating the set-up and preparation of an analog program. The CPDFG program is designed to receive the description of an arbitrary function as K pairs of numbers, usually experimental results. From these raw data it fits the curve with straight-line segments, a tedious job that the analog programmer is usually stuck with. It then punches out a new card complete with new breakpoints and slopes to be placed on the diode function generator.

D. Variable Time Delay

The representation of an analog transport delay is a special case of function storage and playback in which the table is continually filled and readout in a circular loop. A subroutine written by a user is shown first, followed by the description of the standard time delay subroutine.

The suggested subroutine can effect the delay of three analog variables,

```
            DIMENSION F(3,100), FIN(3), FOUT(3)
            DO      100     I =1, 3
100         FOUT(I) = F (I,100)
            CALL QDAJM0 (0, 3, FOUT, IERR)
            DO      200     J =99, 1, −1
            DO      200     I =1,3
200         F(I, J +1) = F(I,J)
            CALL QSTRE0 (0, 3, IERR)
            CALL QADCV0 (0, 3, FIN, IERR)
            DO      300     I =1,3
300         F(I,1) = FIN (I)
            RETURN
            END
```

Figure 7.9 Delay subroutines. (Program written by H. Sherrill, Lockheed Electronics, NASA/MSC, Houston).

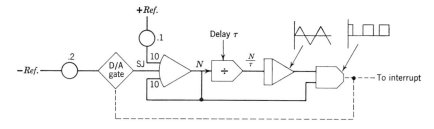

Figure 7.10 Analog circuit for generation of interrupt.

the delay time being a function of another computer variable. This subroutine utilizes three tables, each of which are 100 locations in length, to accomplish a delay. Each time the subroutine is executed, the numbers in the last location of each table are outputted by the digital computer. Then each preceding location is transferred to its adjacent lower location until the value in the first location is placed into the second (refer to Figure 7.9 for the program listing). Finally, the new values of the analog variables are placed into the first locations in the tables. It can be seen that after 100 executions, the values read into the digital on the first execution of the subroutine are then transferred back to the analog. This simulation utilizes the external interrupt system to communicate the desired frequency, i.e., it enters the subroutine once for every interrupt. The interrupt is generated by the analog program shown on Figure 7.10. If T is its period, then:

$$\text{Time Delay} = 100T$$

A precautionary measure must be taken to determine that the subroutine execution time does not exceed the smallest time increment that will be observed between interrupts during a run. A simple analog circuit generates a saw-toothed function, the frequency of which is determined by the program computed delay time.

A typical calling sequence for hybrid transport delay software would be:

```
CALL    FSPBIN (INPMED, TABLE, INITIAL, COUNT)
CALL    TDLY (TABLE, POINTR, DAC, ADC, COUNT,
        ERROR)
```

The first statement serves as initialization (Function Storage and Play-Back INitialization). The second instruction actually performs the delay (Time DeLaY). Here is the significance of the arguments:

INPMED   a code specifying the input media from which the data is read in.
TABLE   starting address of the dynamic function table.
INITIAL   starting address of array containing the initial values. This array is provided for dynamic initialization.
COUNT   an integer specifying the TABLE size. This is best understood if one relates it to the DIMENSION statement in FORTRAN.
POINTR   a cell used to sequence through the table. The user needs not worry about this except for resetting it to zero initially.
DAC   address of the digital-to-analog converter used to output the delayed signal.
ADC   address of the corresponding analog-to-digital converter.
ERROR   an error return as discussed earlier in the chapter.

A single transport delay channel would appear to the hybrid computer user as shown in Figure 7-11. The interrupt 0, by convention, causes all defined functions to be initialized. This involves loading the dynamic function table with the initial data table. Also, the first data point of the table is loaded into the appropriate DAC to establish the initial output. The time delay is equivalent to the COUNT times the period of the interrupt N. One sample is taken each time the interrupt N is high.

7.9 CONCLUSION

The design considerations for a software programming system to fulfill the requirements of the hybrid environment have been described. Where appropriate, existing software designs have been extrapolated to indicate means for most effectively satisfying the existing requirements. New features in hybrid computer hardware as well as changes in the operating environment will result in many further changes and additions to the hybrid software designs in the near future. More and more specialized packages, such as power spectral density or optimization programs, will undoubtedly be available to the user.

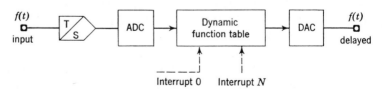

Figure 7.11 Schematic representation of a single transport delay channel.

References

1. Green, C., H. D'Hoop, and A. Debroux, "APACHE—A Breakthrough in Analog Computing," *IRE Transactions on Electronic Computers*, **EC-11**, 699–706, Oct. 1962.
2. Linebarger, R., and R. D. Brennan, "A Survey of Digital Simulation," *Simulation*, **3**, 22–37, December 1964.
3. Tiechroew, D., J. F. Lubin, "Discussion of Computer Simulation Techniques and Comparison of Languages," *Simulation*, **9**, 181–190, Oct. 1967.
4. McGhee, R. B., and A. Y. Lew, "Software for Hybrid Computers," *Simulation*, **5**, 367–374, Dec. 1965.
5. Chandler, W. J., and R. B. McGhee, "Digital Computer Control of Analog Computer Potentiometers," *Simulation*, 14–18, Jan. 1965.
6. Shaw, J. C., "The Joss System," *Datamation*, **10**, 35–42, Nov. 1964.
7. Electronic Associates, Inc., "HYTRAN Operation Interpreter," Publ. No. 07 800 0008–1, 1967.
8. Electronic Associates, Inc., "HYTRAN Simulation Language Programming Manuals," Publ. No. 07 800 0006 0, 1967.
9. Kovacs, J., and J. Strauss. "An Approach to Hybrid Programming Language." SCi Third Annual Simulation Software Meeting, St. Louis, Mo.

Part 2

Applications

8

Hybrid Computer Solutions of Field Problems

8.1 GENERAL REMARKS

The solution of problems governed by partial differential equations constitutes a special challenge both to the analog computer and to the digital computer.[1] Analog methods are complicated by the fact that such problems have two or more independent variables, whereas analog computers are restricted to one independent variable: time. Moreover, field problems are usually formulated as boundary value problems, whereas analog computers are particularly useful for initial value problems. The application of digital computer techniques demands stepwise integration in the time domain and most often leads to uneconomically long computer runs if reasonable accuracies are required.

This discussion begins with a brief survey of conventional analog and digital techniques for treating transient field problems characterized by partial differential equations; several hybrid computer methods are then described in some detail. These techniques offer some important advantages over conventional methods and indicate the potentialities of computers in which analog and digital hardware are interconnected.

Probably the most widely occurring partial differential equation is Laplace's equation:

$$\nabla^2 \phi = 0 \qquad (8.1)$$

where ∇^2 is the Laplacian operator, which in two Cartesian dimensions takes the form

$$\nabla^2 = \frac{\partial^2}{\partial x^2} + \frac{\partial^2}{\partial y^2} \tag{8.2}$$

Equation 8.1 is one of the family of elliptic partial differential equations whose solutions are not time dependent.

The parabolic partial differential equation

$$\nabla(\sigma\nabla\phi) = S\frac{\partial\phi}{\partial t} \tag{8.3}$$

occurs in the study of transient heat transfer, the flow of fluids through porous media, and a wide variety of diffusion phenomena. σ and S are field parameters which may be functions of time, space, or the potential function ϕ. For constant parameters and in one space dimension, Equation 8.3 becomes

$$\frac{\partial^2 \phi}{\partial x^2} = \frac{1}{\alpha}\frac{\partial \phi}{\partial t} \tag{8.4}$$

Fields governed by Equation 8.4 are usually characterized as initial-value problems. In addition to the partial differential equation, the specifications include an initial condition $f(x, 0)$ for all points in the field at $t = 0$ and two boundary conditions, $f_1(0, t)$ and $f_2(X, t)$, applying to the two extremities of the field for $t > 0$.

Other partial differential equations that are of frequent interest include the hyperbolic wave equation

$$\nabla^2\phi = \frac{\partial^2 \phi}{\partial t^2} \tag{8.5}$$

and the biharmonic equation

$$\nabla^4\phi = \frac{\partial^2 \phi}{\partial t^2} \tag{8.6}$$

governing the vibration of beams.

In order to treat transient field problems of the type of Equations 8.3 to 8.6 by electronic differential analyzer or by digital techniques, it is essential to effect a transformation to reduce the number of independent variables. This is accomplished by means of finite-difference approximations in which a continuous variable such as x, y, z, or t is replaced by an array of discretely spaced points. Solutions are then obtained for these points, and interpolation techniques are used to construct continuous equipotentials or streamlines. When an independent variable is discretized in this manner, the corresponding partial derivative is replaced by an approximate algebraic expression. In computer terms, this means that the operation of integration is replaced by additions and subtractions.

Consider as an example of a very simple transient field problem, the parabolic Equation 8.4. This equation has two independent variables, the space variable x and the time variable t. In applying analog methods, either of these two variables can be kept in continuous form or discretized by the application of finite difference approximations. Therefore, four basic analog approaches exist:

1. Continuous-space-continuous-time (CSCT). Both the time variable and the space variable are kept in continuous form.
2. Discrete-space-continuous-time (DSCT). The left side of Equation 8.4 then becomes an algebraic expression, while the right side remains unchanged.
3. Continuous-space-discrete-time (CSDT). The right side of Equation 8.4 then becomes an algebraic expression, while the left side remains unchanged.
4. Discrete-space-discrete-time (DSDT). Both the x and the t variables are discretized so that all derivatives in Equation 8.4 become algebraic expressions.

If a digital computer is to be employed, only the DSDT method is feasible since no continuous variables can be handled digitally. This classification does not, of course, apply directly to elliptic partial differential equations since these do not contain time as an independent variable. The above classification has been adopted in this chapter, however, since hybrid computer techniques have been useful primarily in transient field problems governed by parabolic, hyperbolic, and biharmonic equations.

ANALOG METHODS

8.2 CONTINUOUS-SPACE-CONTINUOUS-TIME

While continuous-space analog models, such as the electrolytic-tank and resistance-paper models, are widely used to simulate fields governed by Laplace's and Poisson's equations, the application of CSCT techniques to transient field problems has not been practical. The simulation of a field characterized by Equation 8.4, for example, by electrical means would involve the provision for distributed resistance and distributed capacitance. Available dielectrics make possible the modeling of a two-dimensional field with only very, very small distributed capacitivity. This, in turn, implies a very, very rapid time scale and makes very difficult the accurate recording of transient phenomena. The same time scale difficulties exist in the treatment of hyperbolic and biharmonic equations by CSCT techniques. A possible exception to this generalization is the Monte Carlo analog method, briefly discussed in Section 8.7. Here a parabolic equation including one-, two-, or three-space dimensions is solved by using probabilistic techniques without any actual discretization of the space or time domains. However, solutions are provided for only one point in space and time, and no actual modeling or simulation of an entire field is attempted.

8.3 DISCRETE-SPACE-CONTINUOUS-TIME

The most widely used approach to the analog simulation of transient field problems involves the approximation of all derivatives with respect to space variables by finite-difference expressions. Using Equation 8.4 as an example, the x domain is represented by a set of discretely-spaced points, and the potential $\phi_{x,t}$ at a typical point in space and time is expressed in terms of the potentials at neighboring points as

$$\frac{\phi_{x+\Delta x} + \phi_{x-\Delta x} - 2\phi_x}{\Delta x^2} = \frac{1}{\sigma}\frac{\partial \phi_x}{\partial t} \tag{8.7}$$

The second derivative with respect to x has been approximated by the familiar "second central difference." This is obtained by writing

$$\left.\frac{\partial \phi}{\partial x}\right|_{x+\frac{1}{2}\Delta x} \simeq \frac{\phi_{x+\Delta x} - \phi_x}{\Delta x} \tag{8.8a}$$

$$\left.\frac{\partial \phi}{\partial x}\right|_{x-\frac{1}{2}\Delta x} \simeq \frac{\phi_x - \phi_{x-\Delta x}}{\Delta x} \tag{8.8b}$$

$$\frac{\partial^2 \phi}{\partial x^2} \simeq \frac{1}{\Delta x}\left[\left.\frac{\partial \phi}{\partial x}\right|_{x+\frac{1}{2}\Delta x} - \left.\frac{\partial \phi}{\partial x}\right|_{x-\frac{1}{2}\Delta x}\right] \tag{8.8c}$$

where Δx represents the distance between adjacent grid points in the x direction.

A passive analog simulator for Equation 8.4 is constructed by recognizing the formal similarity between Equation 8.7 and Kirchhoff's current-law equation for an electrical node formed by two series resistors and a capacitor to ground, as in Figure 8.1a where ϕ is represented by electrical voltage. The general analog for Equation 8.3 then takes the form of a rectangular network of resistors in a one-, two-, or three-dimensional array with a capacitor linking each node point to ground, as illustrated in Figure 8.1b for a field in one space dimension.

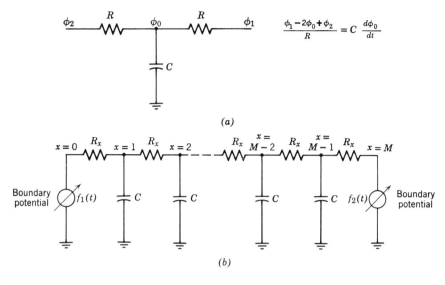

Figure 8.1 Analog circuit for DSCT method: (a) Typical node and (b) complete network.

One application of electronic analog computers to problems of this type involves the simultaneous solution of Equation 8.7 at each node in the space domain. For this purpose an integrator and an adder are required at each grid point. An economy in equipment can be effected if solutions of alternate polarity are acceptable at alternate points in space. The operations of integration and addition can then be combined so that a single integrator performs the computation

$$\phi_x = \frac{\sigma}{\Delta x^2} \int_0^t (\phi_{x+\Delta x} + \phi_{x-\Delta x} - 2\phi_x) \, dt \qquad (8.9)$$

The computer system then takes the form shown in Figure 8.2 and requires n operational amplifiers for n grid points. The initial condition applied to each integrator corresponds to the specified value of the corresponding point in space at time $t = 0$; the transient voltages as recorded at the output of each integrator correspond to the transient field potentials at the corresponding points in the system being simulated. Howe and Haneman[2] have described the application of this technique to a variety of transient field problems.

A major difficulty arises in the utilization of DSCT methods when it is necessary to simulate time-varying or nonlinear fields. Under these conditions, it is necessary to adjust continuously the magnitudes of a

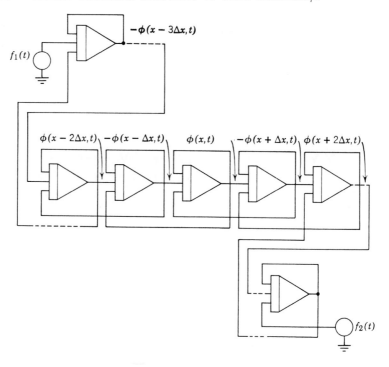

Figure 8.2 Portion of the DSCT analog of diffusion equation.

multitude of circuit elements in the course of a computer run, or to provide a separate nonlinear function generator at each node point. This is always difficult and usually economically impossible, so that this analog approach is essentially limited to linear, constant-parameter fields.

8.4 CONTINUOUS-SPACE-DISCRETE-TIME

Jury[3] has described an application of dynamic memory circuits to the solution of Equation 8.4. In this method, the space variable x is kept in continuous form while the time variable t is approximated by a sequence of discrete steps, $n\, \Delta t$. The finite-difference approximation of Equation 8.4 then becomes

$$\frac{\partial^2 \phi_t}{\partial x^2} = \frac{\phi_t - \phi_{t-\Delta t}}{\alpha\, \Delta t} \tag{8.10}$$

where the function $\phi_{t-\Delta t}$ has been stored from the previous calculation, so that ϕ_t is the only unknown. For the next time increment the equation

8.4/CONTINUOUS-SPACE-DISCRETE-TIME

solved by the computer is

$$\frac{\partial^2 \phi_{t+\Delta t}}{\partial x^2} = \frac{\phi_{t+\Delta t} - \phi_t}{\alpha \Delta t} \tag{8.11}$$

This procedure is repeated for as many increments of time as may be of interest. The computer circuitry used to solve Equation 8.10 is therefore used repeatedly for N computer runs. The larger the number of time increments, N, the longer the time required for solution. In this respect, this approach differs from that presented in the preceding section in which a refinement of the finite-difference net, by reducing the net interval, involves additional node points and, therefore, additional computer equipment in direct proportion to the number of nodes. A general circuit illustrating this approach is shown in Figure 8.3. Note that the computer time variable represents the problem space variable x.

The application of the CSDT technique using conventional electronic analog computer equipment presents three major problems: continuous memory, the determination of the appropriate initial condition for integrator 2, and stability. As indicated by Equation 8.10, the solution for $\phi_{x,t}$, where x is a continuous function in the range $0 \leq x \leq X$ and time is a parameter, requires a knowledge of the function $\phi_{x,t-\Delta t}$ which is the solution for the preceding time step. It is therefore necessary to record the transient solution for each time level and to play back this transient voltage as the solution for the next time level is being generated. The storage of a continuous function by all-analog techniques is always difficult. The method suggested by Jury[3] involves the use of an array of sample-hold units, each storing the solution value for a specific point in space x, the space domain represented by computer time. This introduces truncation errors into the solution, greatly increases the required equipment, and presents considerable switching and timing difficulties.

In order to employ the circuit shown in Figure 8.3, it is necessary to specify appropriate initial conditions for both integrators. The initial

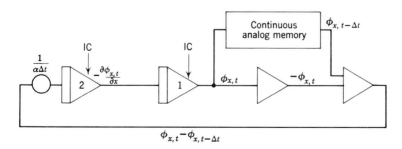

Figure 8.3 CSDT analog of diffusion equation.

condition for integrator 1, corresponding to ϕ at $x = 0$ and $t = 0$, is usually readily available. The initial condition of integrator 2 is rarely specified, however. Most parabolic differential equations in one space dimension take the form of 2-point boundary value problems, so that ϕ at $x = X$ and $t = 0$ is given rather than the first derivative of ϕ at $x = 0$. It then becomes necessary to determine the appropriate initial condition for integrator 2 iteratively by performing a series of trial runs until the solution at $x = X$ corresponds to the specified value.

A final difficulty in the application of the circuit shown in Figure 8.3 lies in the fact that a feedback loop is formed by four operational amplifiers. If the loop gain exceeds unity, the circuit becomes unstable and the solution useless. As indicated, the loop gain is determined by the setting of the potentiometer and is $1/\alpha \Delta t$. It is therefore necessary in applying this technique to make Δt larger than $1/\alpha$ and to accept the consequent truncation errors due to the discretization of the time variable.

The CSDT method has the advantage over the DSCT method that non-linear and time-varying parameters can be accommodated without excessive equipment requirements. For example, if α is a function of the potential ϕ, a function generator is included in the feedback loop, but only one such generator is required. To date, no applications of the CSDT method have been made to problems in two- or three-space dimensions or to hyperbolic and biharmonic equations. A hybrid computer technique using the CSDT approach is described in Section 8.9.

8.5 DISCRETE-SPACE-DISCRETE-TIME

If all independent variables in a partial differential equation are approximated by finite-difference expressions, the partial differential equation is replaced by a system of simultaneous algebraic equations. The second central difference, as derived in Equation 8.8, is generally used to approximate second derivatives with respect to space variables. In the case of the time derivative, such as that in Equation 8.4, there is a choice between so-called forward and backward difference approximations. As explained in more detail in Section 8.6, a backward difference approximation is preferred for computational stability reasons. The finite-difference approximation of Equation 8.4 then takes the form

$$\frac{\phi_{x+\Delta x, t} - 2\phi_{x,t} + \phi_{x-\Delta x, t}}{\Delta x^2} = \frac{1}{\alpha}\left[\frac{\phi_{x,t} - \phi_{x,t-\Delta t}}{\Delta t}\right] \qquad (8.12)$$

or

$$\frac{(\phi_{x+\Delta x, t} - \phi_{x,t})}{\Delta x^2} + \frac{(\phi_{x-\Delta x, t} - \phi_{x,t})}{\Delta x^2} + \frac{(\phi_{x,t-\Delta t} - \phi_{x,t})}{\alpha \Delta t} = 0 \qquad (8.13)$$

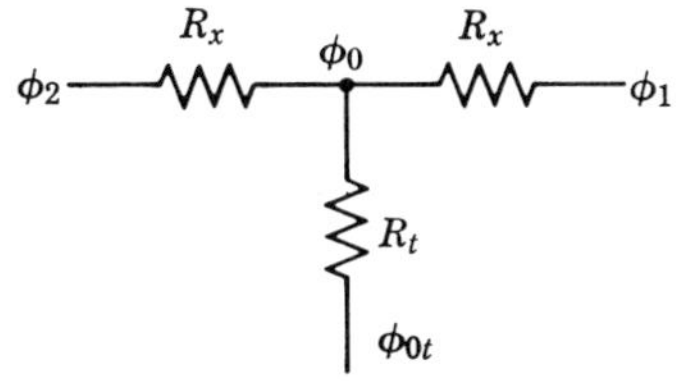

8.5/DISCRETE-SPACE-DISCRETE-TIME

$$\frac{\phi_1 - \phi_0}{R_x} + \frac{\phi_2 - \phi_0}{R_x} + \frac{\phi_{0t} - \phi_0}{R_t} = 0$$

Figure 8.4   Typical node for DSDT method.

For each time level there are as many simultaneous algebraic equations as there are points in the space domain. An analog technique which takes advantage of the ability of electrical networks to solve simultaneous algebraic equations was introduced by Liebmann.[4] The analogy is based on the recognition of the similarity of Equations 8.12 and 8.13 and Kirchhoff's node-law equation of an electrical network formed of three resistors Δx^2, Δx^2, and $\alpha \Delta t$ in magnitude as shown in Figure 8.4. The circuit for the simulation of a one-dimensional field governed by Equation 8.4 is shown in Figure 8.5. The voltages $\phi_{x,t}$ are the unknowns, whereas the voltages $\phi_{x,t-\Delta t}$ are the values obtained in the preceding step in the solution, or are initial conditions for the first solution step.

The solution procedure is:

STEP 1. An electrical network of the type shown in Figure 8.5 is employed to simulate the one-, two-, or three-dimensional field such that an electrical node is provided for each finite-difference grid point in the space domain. At the extremities of the network, voltage supplies are employed to simulate specified boundary conditions.

STEP 2. The voltage supplies representing the potentials at time $t - \Delta t$ are given settings corresponding to the initial conditions specified for the corresponding points in the space domain.

STEP 3. By means of a voltmeter, the potentials existing at the network nodes are measured and recorded. These constitute the solutions for that solution step.

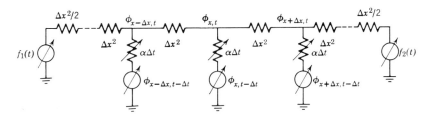

Figure 8.5   Complete analog circuit for DSDT method.

STEP 4. The solution values just recorded are now employed to determine the appropriate new settings for the voltage supplies. Each voltage supply is given a voltage corresponding to the solution recorded at the corresponding node point. The system is now ready to provide the solution for the second step in time.

STEP 5. The potentials existing at the node points are now recorded and transferred to the voltage supplies, and the procedure is repeated for as many time steps as are required to traverse the time domain.

This method employs relatively inexpensive electrical elements and does not suffer from the timing, iteration, and instability problems of the CSDT method. The technique, however, is not directly applicable to transient hyperbolic and biharmonic equations. Nonlinear parameters can be handled by iterative adjustments of all the resistors; this, however, is usually a very tedious, time-consuming procedure. Alternatively, a separate function generator must be supplied at each node point with the attendant expense. A hybrid computer DSDT technique, overcoming many of these disadvantages, is described in Section 8.10.

DIGITAL METHODS

8.6 FINITE-DIFFERENCE TECHNIQUES

When transient field problems are treated on a digital computer, all independent variables must be discretized, since a digital machine is capable only of solving algebraic expressions. In approximating a time derivative by a finite-difference expression, two possibilities exist: "forward difference" and "backward difference" approximations. Equations employing forward differences are generally solved explicitly, a process which is relatively simple computationally but which has the inherent possibility of computational instability. If the ratio of the time increment to the space increment is improperly chosen, round-off errors made in the course of the solution will gradually build up until they overshadow the solution, thus making it worthless. In order to obtain satisfactory solutions by this method, it is necessary to make the time increment relatively small, that is, to take many time-consuming steps. On the other hand, if backward differences are employed, the computational procedure is an implicit one, having no danger of computational instability but requiring the solution of a large number of simultaneous equations at each time level, also leading to long computer runs.

8.6/FINITE-DIFFERENCE TECHNIQUES

Again taking the diffusion equation, Equation 8.4, as an example, the forward difference approximations take the form

$$\frac{\phi_{x+\Delta x,t} + \phi_{x-\Delta x,t} - 2\phi_{x,t}}{\Delta x^2} = \frac{\phi_{x,t+\Delta t} - \phi_{x,t}}{\alpha \Delta t} \qquad (8.14)$$

while a backward difference approximation of $\partial \phi / \partial t$ leads to

$$\frac{\phi_{x+\Delta x,t+\Delta t} + \phi_{x-\Delta x,t+\Delta t} - 2\phi_{x,t+\Delta t}}{\Delta x^2} = \frac{\phi_{x,t+\Delta t} - \phi_{x,t}}{\alpha \Delta t} \qquad (8.15)$$

In Equation 8.14, at any point in the time domain, t, the only unknown term is $\phi_{x,t+\Delta t}$ while all potentials applying to level t are specified initial conditions or are the results obtained in the preceding solution run. Equation 8.14 can therefore be solved explicitly for the unknown potential by simple algebraic steps. However, in Equation 8.15, all terms applying to time level $t + \Delta t$ are unknowns, while the only known term is $\phi_{x,t}$. Equation 8.15 is therefore an equation with one known and three unknowns. It is necessary then to write expressions of the type of Equation 8.15 for every point in the space domain and to solve these equations simultaneously by using appropriate matrix-inversion techniques.

On the surface, Equation 8.14 would appear to be highly preferable. Unfortunately, it can be shown that in the repeated application of Equation 8.14, round-off errors committed in the solution process will grow and make the solution meaningless unless the following inequality is satisfied:

$$\Delta t \leq \frac{\Delta x^2}{2\alpha} \qquad (8.16)$$

To traverse a given time domain, T, condition 8.16 makes it necessary to take many more steps in time, that is, to make Δt much smaller, than in the application of Equation 8.15. Moreover, where the field parameter α is a function of the field potential ϕ, the appropriate Δt can generally not be specified in advance. For these reasons, in most engineering applications the implicit approach (8.15) is preferred.

In implicit integration methods, the unknown potentials applying to a given time step are contained in a system of simultaneous algebraic equations, including all unknown field potentials for that time level. If the space domain has been represented by N finite-difference grid points, it then becomes necessary to invert an $N \times N$ matrix at each time step. The presence of nonlinear parameters makes it further necessary to iterate at each time step, that is, to invert the matrix several times until a convergence condition is satisfied. A digital computer flow chart for such an

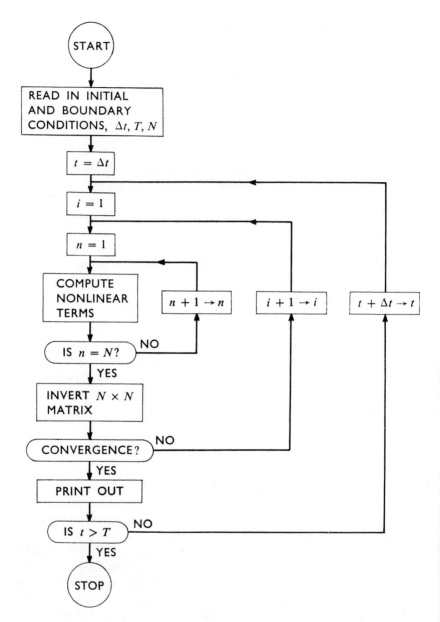

Figure 8.6   Flow chart for digital solution.

implicit technique for the treatment of initial value problems is shown in Figure 8.6, where N is the number of grid points in the space domain, T represents the extent of the time domain, i is an index counting the number of iterative subcycles for each time step, and n is an index counting the number of net points in the space domain. The necessity for the repeated inversion of large matrices (a problem utilizing only 30 grid points in two space dimensions leads to a 900 × 900 matrix) makes detailed nonlinear field simulations prohibitively time consuming, even on the fastest available digital computers.

8.7 MONTE CARLO METHODS

An alternative approach to the solution of partial differential equations by using digital computers is based on the theory of probability. This method serves not primarily to provide the potential distribution everywhere within a one-, two-, or three-dimensional field, but rather is used to determine the potential at one specific point. In principle, the Monte Carlo method involves the selection of some point of interest within the field and the commencing of a series of so-called *random walks* from this point. Each such walk consists of a sequence of small steps, such that each step begins at the end of the preceding step; but the direction of each step is random. Eventually, each random walk will reach some point on the field boundary. The potential at this boundary point is recorded, and a new walk is commenced. Provided enough random walks are taken, and provided that each step in the walk is sufficiently small, it can be shown that the weighted average of the potentials at the boundary intersections converges to the potential existing at the point of interest. So far, the Monte Carlo method has been most useful for the solution of two-dimensional problems governed by Laplace's and Poisson's equations, with Dirichlet boundary conditions (potential specified everywhere along the boundary), but the method is also applicable to other types of field problems.

Consider first the application of the Monte Carlo method to a uniform two-dimensional field governed by Laplace's equation

$$\frac{\partial^2 \phi}{\partial x^2} + \frac{\partial^2 \phi}{\partial y^2} = 0$$

As in the other methods described in this chapter, a finite-difference grid is now specified. If a rectangular grid with $\Delta x = \Delta y = h$ is selected, the finite difference approximation for the potential at a typical grid point is given by

$$\phi_{i,j} = \tfrac{1}{4}(\phi_{i+h,j} + \phi_{i-h,j} + \phi_{i,j+h} + \phi_{i,j-h}) \qquad (8.17)$$

where the subscripts i and j refer to the x and y coordinates, respectively, of the grid points. For nonuniform fields or unequal net spacings, each of the terms in the right side of Equation 8.17 has a different coefficient. The coefficients of the difference equation, which are all $\frac{1}{4}$ in this case, are interpreted as transition probabilities from one point to the neighboring points. Thus Equation 8.17 specifies that if the random walk has progressed to a point i, j, there is an equal probability that any of the four neighboring points will be reached in the next step of the walk.

Table 8.1

Random Number from Sequence A	Random Number from Sequence B	New x Coordinate	New y Coordinate
0, 1, 2, 3, 4	0, 1, 2, 3, 4	i	$j + h$
5, 6, 7, 8, 9	5, 6, 7, 8, 9	i	$j - h$
0, 1, 2, 3, 4	5, 6, 7, 8, 9	$i + h$	j
5, 6, 7, 8, 9	0, 1, 2, 3, 4	$i - h$	j

A point of interest is now selected within the field, and may be assigned the coordinates 0, 0 for convenience. All random walks are commenced at this point, and sequences of random numbers are employed to determine the direction of each step. Consider, for example, the use of two sequences A and B of random numbers, each containing the integers 0, 1, 2, 3, 4, 5, 6, 7, 8, and 9. One integer from each sequence is provided for each step of the calculation. The decision as to the direction of the corresponding step of the random walk can then be made as indicated in Table 8.1. Depending on the combination of the two random numbers, therefore, the random walk will proceed for one step in the positive or negative x or y direction. This stepwise progress is repeated until a grid point lying on the field boundary is reached. The walk is then terminated, and the potential at that point (a specified boundary value) is recorded and placed in an accumulating register. A new random walk is then commenced from the same original point, 0, 0. It can be demonstrated, using certain theorems from the theory of probability, that as the number of such random walks becomes large the average of the boundary potentials at which each walk terminates will approach the true potential at the point 0, 0 of interest. The number of random walks required to produce such convergence to the solution depends on the geometry of the field under study, the uniformity of the boundary conditions, the uniformity of the field parameters, and to a lesser extent on the manner in which the finite difference grid is constructed. The average length of each random walk depends primarily on the distance of the point under consideration from the field boundaries and on the net interval h.

Figure 8.7 is a digital computer flow chart for the basic application of the Monte Carlo method to the solution of a problem governed by Laplace's equation. In this program:

BOX 1. Initial data read into the computer include: the coordinates of all points in the finite-difference grid falling on the field boundaries, using a coordinate system in which the point of interest has the coordinates 0, 0; the potential specified for each of the boundary points; the number M of random walks to be carried out; two sequences A and B of random numbers which may be generated by the computer using special techniques or which may be read from published tables of random numbers, as discussed in Chapter 11.

BOX 2. The index n counting the number of random walks is set to unity.

BOX 3. The indices i and j, denoting the coordinates of the point from which the next step in the random walk is to commence, are set to zero so that each random walk starts from the point of interest.

BOX 4. A new random number is obtained from each of the two random number sequences A and B.

BOX 5. The direction of the next step in the random walk is inferred from the two random numbers. This can be accomplished by reference to a table such as Table 8.1.

BOX 6. The coordinates of the next point in the random walk are computed by adding the changes in coordinates obtained in Box 5 to the coordinates of the present point contained in Box 3.

BOX 7. The coordinates of the next point are examined to determine whether this point lies on the boundary of the field.

BOX 8. If the answer of the question posed in Box 7 is negative, the indices i and j are modified by the coordinate changes Δi and Δj obtained in Box 5, and the operations in Boxes 4, 5, 6, and 7 are repeated.

BOX 9. If the answer to the question posed in Box 7 is affirmative, the boundary potential ϕ_B specified for the boundary point $i + \Delta i$, $j + \Delta j$ is fed into an accumulator register. This register contains the sum of all the boundary potentials thus obtained in the n random walks.

BOX 10. The index n is examined to determine whether a sufficient number of random walks has been performed.

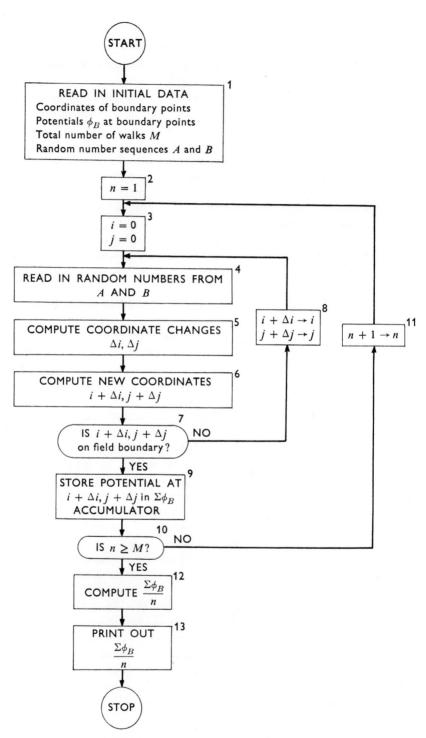

Figure 8.7   Flow chart for Monte Carlo method.

BOX 11. If the answer to the question posed in Box 10 is negative, the index n is advanced by unity, and the steps in Boxes 3 through 9 are repeated.

BOX 12. If the answer to the question in Box 10 is affirmative, the quantity $\Sigma \phi_B$ in the accumulator is divided by the number of random walks which have been carried out to provide an average potential.

BOX 13. The quantity obtained in Box 12 constitutes the solution to the problem, i.e., the potential at the point of interest, and is printed out. The computer then comes to a stop.

A number of modifications and refinements is desirable and may be necessary in the program of Figure 8.7. Usually it is impossible to specify in advance precisely how many random walks M have to be performed in order to achieve a sufficient convergence to the correct solution. For this reason, in place of specifying a fixed number of random walks, the desired statistical variance is specified. Additional computational steps must then be included to determine whether this statistical specification of convergence has been met.

If the coefficients of Equation 8.17 are not all identical, the transition probabilities in the four principal directions are different from each other. Under these conditions Table 8.1 must be replaced by a more complex table which takes the specific transition probabilities into account. If the problem under consideration is governed by Poisson's equation instead of by Laplace's equation, the nonhomogeneous term of this equation must be taken into account at each step in the random walk by adding an appropriate term to the contents of the accumulator register at each step in the random walk. Problems governed by parabolic equations such as Equation 8.4 are treated by considering the time variable in a manner similar to the space variable and by performing the random walks in the time-space system. The Monte Carlo method is particularly suitable for the relatively quick determination of the potential at one or several points of interest in a boundary value problem. The calculation of the potential at enough points to permit the construction of a family of equipotential lines is generally too time consuming by this method

The solution of partial differential equations by the Monte Carlo method can also be effected by analog techniques. Here a mask having the same geometry as the two-dimensional field under study is placed over the face of an oscilloscope The beam of the oscilloscope is placed at coordinates corresponding to the coordinates of the point of interest within the field, and electrical noise generators are employed to deflect the beam in a stepwise manner. When the beam contacts a field boundary, the random walk is terminated and a new one is commenced. Diode-function generators, in conjunction with trigger and gate circuits, are employed to apply to an accumulator register the potentials specified for the points at which the beam intersects the boundary at the end of each walk. For problems of relatively simple geometry, good convergence can be attained after approximately 500 random walks per experimental point. Further details of this method are described by Chuang[5] and reviewed by Tomovic.[6] A hybrid computer Monte Carlo approach is described in Section 8.11.

HYBRID METHODS

8.8 DISCRETE-SPACE-CONTINUOUS-TIME

The chief shortcoming of the analog DSCT analog techniques described in Section 8.3 is the direct proportion between the number of finite-difference grid points in the space domain and the analog hardware required for the simulation. In the case of linear problems, at least one integrator is required for each point; if the problem is nonlinear, at least one function generator must also be supplied at each point. Since the minimization of the truncation error due to finite-difference approximation usually demands a large number of net points, the obtaining of an accurate solution is difficult even for one-dimensional problems.

Accordingly, it has been suggested that hybrid computing techniques be employed to time-share a limited amount of analog equipment. The space domain to be simulated is broken up into two or more sections, and analog elements are provided only for a single section. The analog system is then used repeatedly to provide solutions for each of the sections of the space domain. In the circuit shown in Figure 8.2, the specified boundary conditions are applied to the two extremities of the analog network. If now an analog system is used to represent only a portion of the overall field, the "boundary" conditions at the two ends of this system are generally not known, but are a part of the solution being sought. It is therefore necessary to employ iterative techniques to "match" all the time-shared sections.

Figure 8.8 shows a simple three-integrator system that is useful for solving a linear parabolic equation such as Equation 8.4. A digital computer together with the necessary linkage equipment is employed to simulate adjacent sections in the positive and negative x directions by applying to the analog network the transient potentials which would be generated by integrators immediately to the left and to the right of the section shown, and to record the potentials generated by the analog system.

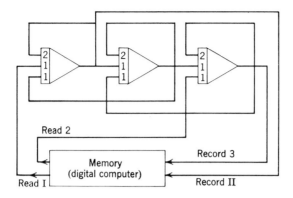

Figure 8.8   Hybrid system for DSCT method.

The latter transients become the "boundary" potentials to be applied to the analog system when it is used subsequently to simulate adjacent sections in the space domain. The analog system is used successively to represent the sections of the space domain, starting with the leftmost section and ending with the rightmost section. Initially, the "boundary" values for the interior sections are arbitrarily assumed. After the first iterative pass through the system, the digital computer will have stored in its memory the first "solution values" generated by the analog system in each section. These, then, serve as the boundary excitations for the succeeding iterative cycle in which the analog system again sequentially simulates each of the sections in the space domain. These iterative cycles are repeated until convergence to the correct solution has been obtained.

The need for these iterations (the number of iterative cycles can become very large if a reasonably high accuracy is desired), as well as the errors introduced in the hybrid loop, greatly limits the applicability of this hybrid approach. Miura[7] has made a detailed analysis of the advantages and limitations of this method and concludes that it has merit particularly in the area of chemical-process simulation where large numbers of nonlinear, one-dimensional diffusion equations must be treated.

8.9   CONTINUOUS-SPACE-DISCRETE-TIME

The two major disadvantages of the CSDT analog method described in Section 8.4 are the requirements for continuous-function storage and for the iterative determination of one of the initial conditions. Both of these tasks are readily handled by hybrid computer methods. The manner in which the analog circuit of Figure 8.3 is modified to this end is illustrated in Figure 8.9. Recall again that in the CSDT method, the problem time-variable is discretized so that solutions are generated at successive time levels. The computer time-variable τ is employed to represent the problem space-variable x so that the solution ϕ_t which is generated represents the potential profile in the interval $0 \leq x \leq X$. In order to provide the necessary input function $\phi_{t-\Delta t}$, the output of integrator 1 is sampled, coded, and stored in the digital computer. During the next computing cycle, the appropriate values are read out of the digital computer memory, reconverted to analog form, and introduced into the analog loop. The digital computer therefore acts as a continuous time-delay element.

The second function of the digital computer in the hybrid system involves the modification of the initial condition applied to integrator 2. In general, when treating one-dimensional parabolic equations, boundary conditions are provided at two separate points in the system, usually denoted as $x = 0$ and $x = X$. Since the computer time-variable is employed to represent x, this implies that the value of ϕ_t is specified for two instants of computer time. The specification for $x = 0$ determines the appropriate initial condition for integrator 1. Since the initial condition of integrator 2 corresponds to $\partial \phi / \partial x$ at $x = 0$, its value cannot, in general, be deduced from the problem specification. Rather, it is necessary to assume this

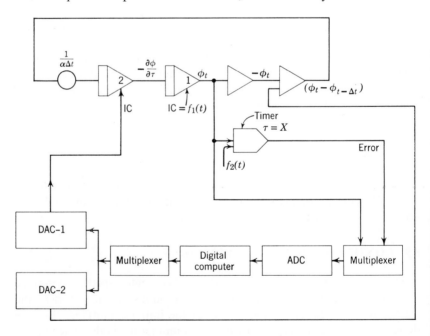

Figure 8.9 Hybrid system for CSDT method.

initial condition arbitrarily and to determine whether it leads to the appropriate value of ϕ at $x = X$. If not, it is necessary to modify the initial condition iteratively until the solution generated by the analog system at the computer time corresponding to $x = X$ differs from the specified boundary condition by no more than ϵ, where ϵ is a specified tolerance. To this end, a comparator is employed to compare the value of ϕ at the computer time corresponding to $x = X$ with the specified boundary value $f_2(t)$. If the difference between these two values exceeds the specified tolerance, ϵ, the digital computer, using a suitable algorithm, calculates and imposes an improved value for the initial condition for integrator 2, and an additional iterative run is made by the analog computer. Finally, when the error magnitude has decreased to below ϵ, the solution generated by the analog computer is accepted as the correct solution, recorded in the digital computer, and the computation of the solution for the next step in the time domain is commenced.

As an example of the application of the CSDT technique to a nonlinear partial differential equation, the following problem was originally treated by Electronic Associates, Inc.[8] Consider the flow of gas in a long transmission line oriented along the x coordinate. Gas with a pressure $P(x, t)$ and a volumetric flow rate $Q(x, t)$ is introduced at $x = 0$. The gas is utilized at $x = X$. The problem is to determine the pressure P and the flow Q throughout the pipe for time $t > 0$. The basic equations governing the flow are derived from the mass-continuity and from the energy-balance equations as

$$\frac{\partial Q}{\partial x} = -AD^2 \frac{\partial P}{\partial t} \qquad (8.18a)$$

$$\frac{\partial P^2}{\partial x} = -\frac{B}{D^5} Q^2 \qquad (8.18b)$$

where D is the diameter of the line and A and B are functions of the temperature; the roughness of the pipe and other environmental conditions are assumed constant. The boundary conditions pertinent to the specific problem are that at $x = 0$ the pressure, P, is maintained constant for all time; and that at $x = X$ the flow rate Q is a sinusoidal function of time. If Equations 8.18a and 8.18b are combined, it is apparent that the problem is one governed by a nonlinear one-dimensional diffusion equation subject to two-point boundary conditions.

In accordance with the CSDT method, a backward difference approximation is made for the time derivative in Equation 8.18a, while the space variable x is left in continuous form and is represented by the computer

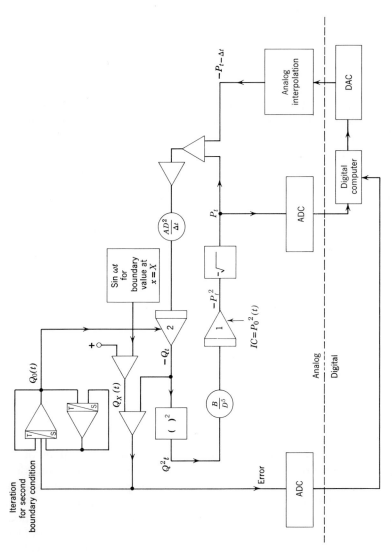

Figure 8.10. Hybrid system for CSDT (*Courtesy of Electronic Associates Inc.*).

time variable. Accordingly, Equations 8.18a and 8.18b become

$$\frac{dQ_t}{dx} = -\frac{AD^2}{\Delta t}(P_t - P_{t-\Delta t}) \tag{8.19a}$$

$$\frac{dP_t^2}{dx} = -\frac{B}{D^5} Q_t^2 \tag{8.19b}$$

where the subscript t indicates the time level at which a solution is sought.

The analog computer flow chart for the treatment of this problem is shown in Figure 8.10. A loop is formed of two integrators and two sign changes just as in Figure 8.3. In place of the block designated as continuous memory, information corresponding to P_t is read into the digital computer for function storage and eventual playback. An analog interpolation circuit is used to smooth the data read out of the digital computer between successive points. The initial condition of the integrator generating $-P_t^2$ is obtained from one of the specified boundary conditions. The initial condition of integrator 2 is not known and must be determined iteratively. It is the function of the two track-and-hold circuits (denoted T/S) at the top of the figure as well as of the digital control circuitry to perform this iteration. An initial guess is made for Q at $x = 0$. An iterative subcycle is then run on the computer and the value of Q at $x = X$ is obtained (note that x actually corresponds to computer time τ). At $x = X$ a comparison is made between the computed and the specified $Q(X, t)$. If the difference between these two values is too large to satisfy a specified error criterion, the initial condition applied to integrator 2 generating Q_t is modified, and a new iterative subcycle is performed. This process is continued until the error criterion is satisfied. The solution is then recorded, and the computer is stepped to the succeeding subcycle to generate P and Q as a function of x for the next time increment.

Considerable care must be employed in the application of the CSDT method in order to avoid excessive errors and computational instability. The following are major sources of error:

1. Analog component tolerance and drift errors.
2. Sampling and reconstruction errors.
3. Truncation errors.
4. Analog instabilities.

The analog errors encountered in the CSDT method are the same that exist in the solution of ordinary differential equations by pure analog methods. Since the solution for one time level is used as a continuous-function input in the next time level, there exists the danger that analog errors, which are within a specified tolerance for a given run, may accumulate as the solution progresses. This problem is compounded by the fact that the function ϕ_t is stored in the digital computer as a set of samples rather than as a continuous function. The function reintroduced into the analog system, therefore, will differ from that originally sampled. To minimize this error, it is necessary to sample as frequently as available digital and conversion equipment permits and to employ a first-order or higher-order reconstruction scheme.

The truncation errors due to the approximation of the time derivative by a finite-difference expression are roughly proportional to Δt. The time step, Δt, cannot be reduced beyond a certain point without causing the analog loop to become electrically unstable so that, in general, a considerable error proportional to $\Delta t(\partial^2\phi/\partial t^2)$ must be accepted. It is possible to compensate for this error by employing a higher-order approximation for the time derivative, but this involves the storage within the digital computer of at least two continuous functions, corresponding to ϕ at $t = t - \Delta t$ and at $t = t - 2\Delta t$. A variety of techniques for obviating analog instabilities for small Δt have been proposed by Witsenhausen[9] and others. However, these have not yet been applied to practical simulations. A unique method for avoiding the instabilities of the analog portion of the simulation has been proposed by Vichnevetsky[14].

8.10 DISCRETE-SPACE-DISCRETE-TIME

The CSDT hybrid method described in Section 8.9 is analog computer oriented. The differential equation is solved within the analog loop, with the digital computer serving primarily for memory and control. In this respect, the CSDT method is similar to most of the other hybrid computer applications described in this book. An alternative approach to hybrid computations, exemplified by the system described by Karplus,[10,11] may be termed *digital computer oriented*. Here the problem to be solved is first programmed for conventional digital computer solution. The flow chart is then carefully analyzed to determine whether the replacement of any of the loops in the digital program by analog techniques would result in an appreciable reduction in overall computation time and present other significant advantages. Analog hardware is therefore considered as comprising subroutines employed in a digital computer solution. Since analog units operate in parallel, the solution time required for the subroutine becomes independent of the complexity of the calculations and is determined entirely by the sampling and conversion rate of the linkage system. The system described below represents but one example of this novel approach to hybrid computations.

8.10/DISCRETE-SPACE-DISCRETE-TIME

The purpose of the computational technique and system to be described is to assist in the solution of engineering field problems governed by equations of the type

$$\frac{\partial}{\partial x}\left(\sigma(\phi)\frac{\partial \phi}{\partial x}\right) = k(\phi)\frac{\partial \phi}{\partial t} + f(\phi) \qquad (8.20)$$

$$\frac{\partial}{\partial x}\left(\sigma(\phi)\frac{\partial \phi}{\partial x}\right) = k(\phi)\frac{\partial^2 \phi}{\partial t^2} + f(\phi) \qquad (8.21)$$

$$\frac{\partial^2}{\partial x^2}\left(\sigma(\phi)\frac{\partial^2 \phi}{\partial x^2}\right) = k(\phi)\frac{\partial^2 \phi}{\partial t^2} + f(\phi) \qquad (8.22)$$

$$\frac{\partial}{\partial x}\left(\sigma(\phi)\frac{\partial \phi}{\partial x}\right) = f(\phi) \qquad (8.23)$$

as well as similar equations in two and three space dimensions, and modified forms of these equations involving terms such as

$$k(\phi)\frac{\partial \phi}{\partial x}, \quad \frac{\partial f(\phi)}{\partial x}, \quad \frac{\partial^2 f(\phi)}{\partial x^2}, \quad \text{and} \quad \frac{\partial^2 f(\phi)}{\partial t^2}$$

For simplicity, the method will be described only for problems in one space dimension, x, but extensions to problems in two and three dimensions are straightforward. To simplify the subscript notation, the scheme illustrated in Figure 8.11 has been adopted. A typical point within the time-space domain is labeled 0, and the adjacent points in the x and t directions are labeled 1 through 10. In general, at any step in a computation, the potentials in the line t_0 and the line $t_0 - \Delta t$ are known information, while the potentials in the line $t_0 + \Delta t$ are to be determined.

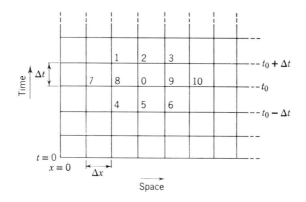

Figure 8.11 Finite difference grid.

In order to permit the utilization of a simple and economic analog system for the treatment of all important nonlinear field problems, equations of the general form of 8.20, 8.21, and 8.23 are first rearranged and transformed to read

$$\frac{\partial^2 \phi}{\partial x^2} + F\left(\phi, \frac{\partial \phi}{\partial x}, \frac{\partial \phi}{\partial t}, \frac{\partial^2 \phi}{\partial t^2}, \ldots\right) = 0 \tag{8.24}$$

Similarly, equations such as 8.22, containing the biharmonic operator, are rewritten as

$$\frac{\partial^4 \phi}{\partial x^4} + F\left(\phi, \frac{\partial \phi}{\partial x}, \frac{\partial \phi}{\partial t}, \frac{\partial^2 \phi}{\partial t^2}, \ldots\right) = 0 \tag{8.25}$$

The finite-difference equation applicable to each net point is then rearranged to take the form

$$\phi_1 - 2\phi_2 + \phi_3 + \phi_i = 0 \tag{8.26}$$

All nonlinear terms in the equations are combined in the term ϕ_i, so that ϕ_i is a highly nonlinear function of the potentials $\phi_0, \phi_1, \ldots, \phi_{10}$. The technique for arranging the nonlinear partial differential equations in this form is described in the references. An analog computer circuit satisfying Equation 8.26 is constructed for each finite-difference grid point in the space domain. These circuits, termed node modules, all operate simultaneously and are interconnected as shown in Figure 8.12. The input of each module corresponds to the "initial condition," ϕ_i, while the output, ϕ_2, constitutes the potential at time $t_0 + \Delta t$; hence, the problem solution for that time step.

The network of the analog-type node modules is connected in a closed loop with a digital computer to form a hybrid computer system. The digital computer is employed to store the solution, the potentials at time $t_0 + \Delta t$ at each node point, in order that these data may be used as initial

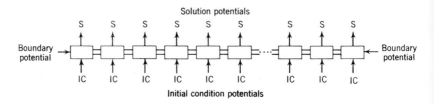

Figure 8.12 Array of node modules.

8.10/DISCRETE-SPACE-DISCRETE-TIME 237

conditions in subsequent time steps. (The digital computer also performs a number of other important functions.) To translate the analog voltages appearing simultaneously at the output terminals of all the node modules into serial digital form, a multiplexer (scanner) is employed to sample in turn the potential ϕ_2 of each node module. These d-c voltages are converted into digital code by an analog-digital converter and stored in the memory of the digital computer. Subsequently, these data, or modified forms thereof, are read out of the digital computer, reconverted to analog form, and applied to the input terminals of the node modules by means of a second scanner (distributor). The type of partial differential equation, any change in Δt, any nonlinearity, and any nonuniformity in the spacing of the net points along the space coordinates are reflected only in ϕ_i. The node modules themselves can therefore be identical to each other and of very simple and economic construction.

The advantage of the hybrid method lies in the fact that all elements of the analog network operate in parallel and relax to the correct solution, i.e., the inverse of the matrix, in less than two microseconds, regardless of the size of the matrix. The total computation time for each iterative subcycle is therefore determined by the length of time needed to translate the serial digital information into parallel analog form and vice versa.

An array of analog node modules is employed to solve the set of difference equations for each step in time. In such an analog system the magnitudes of the dependent variables (the potentials $\phi(x, t)$) appear as d-c voltages proportional in magnitude to the variables represented. Accordingly, the initial conditions for each step in time are applied to the network in the form of d-c voltages, and the solution values are read out of the network also in the form of voltages.

Node modules for the solution of 8.26 can then take the simple form of an analog summing circuit containing one high-gain d-c operational amplifier in combination with a number of precision resistors. It is also feasible to treat Equation 8.26 by utilizing a purely passive analog system. In this case, the analog network becomes a one-, two-, or three-dimensional rectangular grid of precision resistors, and the potential ϕ_i (modified by the addition of the potential ϕ_2) is applied to each network node through another resistor.

It is possible to employ the digital computer for nonlinear function generation and to utilize the analog system only for matrix inversion. Alternatively, an analog function generator can be included in the hybrid loop. This unit is then time-shared so that it generates the desired nonlinear function sequentially for all points.

The computer system, as shown in Figure 8.13, is a closed loop of analog and digital components, utilizing digital function-generation techniques.

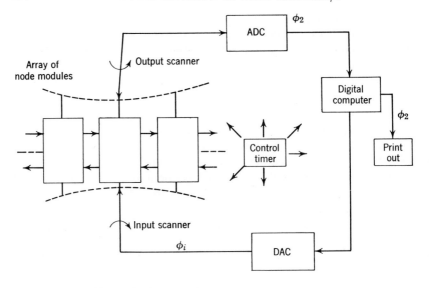

Figure 8.13   Hybrid system for DSDT method.

Of the units shown, only the tasks of the digital computer require further explanation   These can be summarized as follows:

1. The digital computer serves as a memory for the potentials ϕ_2 (the solutions for each computation cycle) and makes these available as initial conditions ($\phi_0, \phi_4, \phi_5, \ldots, \phi_{10}$) in subsequent computation cycles.
2. In the absence of an analog function-generating unit, it serves to generate the potentials ϕ_i.
3. Since ϕ_i is generally a function of the unknown potential ϕ_2, a series of iterative computer runs through the hybrid loop must be made for each time increment. Initial values of ϕ_2 are assumed for each net point, and computations are continued until the potentials ϕ_2 in two successive iterative cycles fail to show any appreciable change. The digital computer serves to test successive values of ϕ_2 for such convergence. Once a specified convergence criterion has been satisfied, the digital computer permits progression to the succeeding time increment.

The operation of the hybrid computing system for the solution of such equations as 8.20 through 8.23 takes the following general steps:

STEP 1.   The one-, two-, or three-dimensional field under study is represented by an array of finite-difference grid points, and a node module is provided for each such point. These modules are interconnected as described earlier, and suitable boundary potentials are applied to those modules adjacent to the field boundaries.

STEP 2. The specified nonlinear functions are introduced into the system either by storing them in the digital computer memory or by suitable adjustments of the analog function generator.

STEP 3. The specified initial conditions and the time increment, Δt, are fed into the digital computer.

STEP 4. The computing system is now set to operate automatically. For each time increment, a series of subcycles is performed until convergence is obtained. The solution for that step of the problem, ϕ_2 for each net point, is then printed out, and the computer automatically steps ahead to the next step in time.

STEP 5. When a specified number of time steps has been completed, the computer comes to a stop.

The advantage of the hybrid technique over a pure digital computation is that the calculation of the nonlinear terms and the inversion of the matrix can be accomplished much more rapidly. Utilizing a pure digital approach, as illustrated in Figure 8.6, the treatment of a two-dimensional nonlinear problem requires at least six multiplications, four divisions, and six subtractions per net point per iterative subcycle By contrast, using commercially available linkage equipment and allowing sufficient settling time for the analog system, the hybrid approach requires less than 15 microseconds per node point per subcycle. In addition, the use of the analog network leads to more rapid convergence, so that considerably fewer subcycles are required for each step in the time domain. If an analog function generator is employed, additional and appreciable savings in computing time are realized. Moreover, the availability of analog function generators permits the convenient adjustment and changing of the nonlinear functions, a feature of particular interest to design engineers.

8.11 MONTE CARLO METHOD

The chief difficulty with the Monte Carlo method described in Section 8.7, when implemented on a digital computer, is the length of time required for the many random walks which must be performed at each point for which a solution is desired. Typically, 500 to 1000 random walks are required to attain solution accuracies better than 1%. The application of hybrid computer techniques permits the random walks to be taken in the analog domain, using analog noise generators, while the digital computer serves for control and for the accumulation and averaging of the potentials

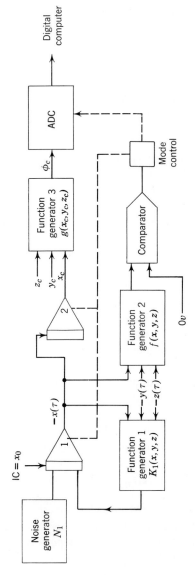

Figure 8.14 Hybrid system for Monte Carlo method.

at the boundary intersections. Little[12] has described the application of this method to elliptical and parabolic partial differential equations when using relatively slow analog equipment. Handler[13] has extended the hybrid Monte Carlo method to a wider variety of partial differential equations and has employed the ASTRAC II ultrahigh-speed hybrid computer.

The application of hybrid techniques is based on the following development. Consider a partial differential equation of the form

$$a_1 \frac{\partial^2 \phi}{\partial x^2} + b_1 \frac{\partial^2 \phi}{\partial y^2} + c_1 \frac{\partial^2 \phi}{\partial z^2} = K_1(x, y, z) \frac{\partial \phi}{\partial x}$$

$$+ K_2(x, y, z) \frac{\partial \phi}{\partial y} + K_3(x, y, z) \frac{\partial \phi}{\partial z} \quad (8.27)$$

where the unknown function $\phi(x, y, z)$ is specified on a simple closed contour C in the x, y, z space. For example, the boundary geometry can be expressed as $f(x, y, z) = 0$ and the potential on this boundary given as $\phi_c = g(x_c, y_c, z_c)$ where the subscript c identifies the boundary coordinates. The diffusion equation, Equation 8.4, can be considered a special case of this equation with $a_1 = 1$, $b_1 = c_1 = K_1 = K_3 = 0$, $K_2 = 1/\alpha$, $y = t$. A series of random walks is now generated by the solution $x(\tau), y(\tau), z(\tau)$, of the stochastic equations of motion

$$\frac{dx}{d\tau} = -K_1(x, y, z) + N_1 \qquad x(0) = x_0$$

$$\frac{dy}{d\tau} = -K_2(x, y, z) + N_2 \qquad y(0) = y_0 \quad (8.28)$$

$$\frac{dz}{d\tau} = -K_3(x, y, z) + N_3 \qquad z(0) = z_0$$

where τ is the computer time variable, and where N_1, N_2, and N_3 are independent white-noise forcing functions with zero mean. As shown in references 5, 12, and 13, these walks will cross the boundary C at random points such that an average of the potential at these boundary intersections will converge to the potential ϕ at x_0, y_0, z_0.

A typical circuit for solving the stochastic equation for one of the independent variables is shown in Figure 8.14. The noise generator, generating N_1, can be either an analog noise generator or a digital pseudorandom noise generator*. The differential equation, Equation 8.28, is

* See Chapter 11.

solved by integrator 1 and function generator 1. The latter unit is not required if K_1 is a constant. Integrator 2 is used as a track-hold unit and tracks the output of integrator 1 until the mode control signal throws it into the hold position. The independent variable $x(\tau)$ along with $y(\tau)$ and $z(\tau)$ is introduced into function generator 2. When the output of this function generator is equal to zero, as detected by the comparator, the random walk has reached a boundary, and a flip-flop unit acts to throw integrator 2 into the hold condition; this causes the output of function generator 3, the potential ϕ_c at the boundary intersection point, to be sent to the digital computer for storage and averaging. Noise generator N_1, function generator 1, and integrators 1 and 2 must be supplied for each independent variable.

As demonstrated by Little and Handler, the treatment of parabolic equations such as Equation 8.4 in one or two dimensions is readily effected using Monte Carlo techniques. In that case, the problem time variable merely replaces one of the independent variables in Equation 8.27, and the specified initial conditions are used as "boundary conditions" along one side of the time-space domain. Poisson's equation and non-Dirichlet boundary conditions, including streamline boundaries, can also be readily handled.

Using a slow analog computer, from 1 to 100 random walks can be taken per second. Using the ASTRAC II, on the other hand, over 1000 random walks in two or three dimensions can be completed in 1 second. Under these conditions, the Monte Carlo method is competitive in its problem-solving capability with other techniques for treating an important class of field problems.

References

1. Karplus, W. J., *Analog Simulation: Solution of Field Problems*, McGraw-Hill, New York, 1958.

2. Howe, R. M., and V. S. Haneman, "The Solution of Partial Differential Equations by Difference Methods Using the Electronic Differential Analyzer," *Proc. IRE*, **41**, 1497–1508, 1958.

3. Jury, S. H., "Solving Partial Differential Equations," *Industrial and Engineering Chem.*, **53**, 177–180, 1961.

4. Liebmann, G., "A New Electrical Method for the Solution of Transient Heat Conduction Problems," *Transactions ASME*, **78**, 655–665, 1956.

5. Chuang, K., L. F. Kazda, and T. Windeknecht, "A Stochastic Method of Solving Partial Differential Equations Using an Electronic Analog Computer," *Project Michigan Report 2900-91-T*, Willow Run Laboratories, Univ. of Mich., June 1960.

6. Tomovic, R., and W. J. Karplus, *High-Speed Analog Computers*, John Wiley, New York, 1962.

7. Miura, T., and J. Iwata, "Time-Sharing Computations Utilizing Analog Memory," *Annals International Assoc. for Analogue Computation (AICA)* **3**, 141–149, 1963.

8. Electronic Associates, Inc., Course Notes, "Hybrid Computation," Princeton, N.J., 1964.

9. Witsenhausen, H. S., "Hybrid Solution of Initial Value Problems for Partial Differential Equations," SM Thesis, Electronic Systems Laboratory, MIT, 1964.

10. Karplus, W. J., "A Hybrid Computer Technique for Treating Nonlinear Partial Differential Equations," *Transactions IEEE*, **EC-13**, 597–605, Oct. 1964.

11. Karplus, W. J., F. C. Rieman, and K. Kanus, "Hybrid Solution of Transient Elastic Beam Problems," *Proc. 4th International Congress, International Assoc. for Analogue Computation (AICA)*, Brighton, England, pp. 516–520, Sept. 1964.

12. Little, W. D., "Hybrid Computer Solution of Partial Differential Equations by Monte Carlo Methods," *Proc. FJCC, AFIPS*, **29**, 181–190, Nov. 1966.

13. Handler, H., "High-Speed Monte Carlo Technique for Hybrid Computer Solution of Partial Differential Equations," Ph.D. Dissertation, Dept. of Elec. Eng., Univ. of Ariz., 1967.

14. Vichnevetsky, R., "A New Stable Computing Method for the Serial Hybrid Computer Integration of Partial Differential Equations," *Proc. AFIPS Spring Joint Computer Conference*, May 1968.

9

Parameter Optimization

9.1 THE NATURE OF THE OPTIMIZATION PROBLEM

Optimization is one of the most important problems in system engineering. One aspect of optimization is the selection of system parameters in such a manner that the performance of the system is as close to optimum as possible, based on a preselected criterion for optimality. Thus it may be desired to minimize cost or energy expenditure or to maximize profit or yield or distance traveled. Many engineers are familiar with the optimization of an analog or digital computer simulation by adjustment of one parameter. The system "payoff function" or "performance criterion function" is evaluated as a function of the parameter while it takes on a number of values in its allowable range, and an optimum or near optimum value is selected. Manual iteration is possible in certain cases even in two-parameter systems. Even here it can be noted that if each parameter is allowed to take on 10 values, a 10 × 10 matrix of parameter values is obtained and one hundred runs are required to determine the optimum. It is clear that automatic methods are required for optimization of systems which depend on more than two parameters.

Consider, for example, the problem of selecting thrust level K and the control system constant τ in order to optimize the response of a satellite vehicle to a particular disturbance. Optimum performance will be defined as that which minimizes both fuel expenditure and the time T_ϵ required to reduce the error resulting from the disturbance to a minimum level ϵ.

9.1/THE NATURE OF THE OPTIMIZATION PROBLEM 245

A performance criterion can be defined as

$$F = k_1 T_\epsilon + k_2 \int_0^{T_\epsilon} [y(\tau, K)]^2 \, dt \tag{9.1}$$

where y represents the attitude deviation and k_1 and k_2 are constants. For each set of parameter values K and τ, there is a corresponding value of the criterion function F. The objective of the system optimization problem is to find computer methods for automatically adjusting K and τ in order to minimize F.

In certain cases it is desired to find a time function [such as a thrust modulation program $K(t)$], rather than a fixed value of the parameter K. In such cases the problem becomes one of *functional optimization* rather than parameter optimization; such problems are examined in Chapter 10.

The objective of this chapter is to discuss hybrid computer techniques for determination of a set of parameter values (α_i) which optimize a performance criterion function $F(\alpha_i)$.

A. Applications

The major applications of the techniques of this chapter lie in the following areas:

1. Process optimization.
2. Model building.
3. Equation solving.

1. The example discussed earlier belongs in the category of process or design optimization. The criterion may be one of minimization of a "cost function," such as Equation 9.1, maximization of a payoff (e.g., the number of pounds of chemical produced per unit time), or the minimization of the system deviation from some prescribed performance. For example, if a desired vehicle trajectory is defined by $y_d(t)$, the criterion function could be defined as

$$F_1 = \int_0^T (y_d - y)^2 \, dt \tag{9.2}$$

or

$$F_2 = \int_0^T |y_d - y| \, dt \tag{9.3}$$

where $y(t)$ is the actual trajectory and T is the duration of the trajectory.

2. Model building refers to the determination of a mathematical model of a system under test, such that the model behavior and the system behavior approximate one another as closely as possible. The model

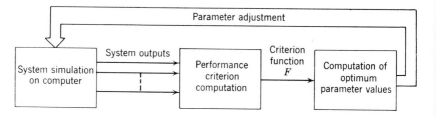

Figure 9.1 Block diagram of design optimization problem.

parameters must be selected to accomplish the optimum match. The block diagrams of Figures 9.1 and 9.2 illustrate these two areas of application.

3. Certain mathematical problems may be formulated in such a way that they reduce to parameter optimization problems. For example, to find the complex roots of a polynomial $P(z)$, where $z = x + jy$, it is necessary to find values of x and y which simultaneously reduce the real and imaginary parts of $P(z)$ to zero. If the polynomial is written as

$$P(z) = u + jv \qquad (9.4)$$

and we formulate a criterion function

$$F = u^2 + v^2 \qquad (9.5)$$

then the minimum of F will correspond to a root of the polynomial. Then, x and y may be considered parameters to be selected such that F is automatically driven to zero. Similarly, two-point boundary value problems, algebraic equations, curve fitting, and other problems may be

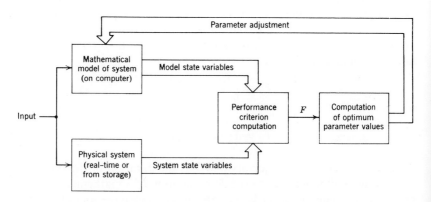

Figure 9.2 Block diagram of modeling of dynamic system.

treated by the methods of this chapter. The emphasis of the chapter, however, will be on the study of applications 1 and 2 as just outlined.

B. Types of Optimization Problems

Two major types of optimization problems may be identified as: (1) static optimization, and (2) dynamic optimization. Static optimization problems are characterized by algebraic equations or inequalities (such as linear or nonlinear programming problems). As the name implies, dynamic optimization problems are concerned with systems described by differential equations, i.e., systems including energy storage elements. Most control and guidance system problems fall into the latter category, as does the modeling of dynamical systems.

C. Portions of the Optimization Problem

It may be seen from the foregoing discussion that the optimization problem consists of the following parts:

1. Decision on a performance criterion function.
2. Isolation of adjustable parameters.
3. Choice and mechanization of a parameter-adjustment method for automatic optimization of the criterion function.

Clearly, portions 1 and 2 belong to the engineering aspects of the particular problem. This chapter is primarily concerned with the description of alternative methods of performing the automatic parameter optimization by means of computers.

In the following sections of this chapter, some of the mathematical background of the parameter optimization problem is reviewed, and various approaches are presented. The emphasis is on so-called "gradient methods" because of their power and generality. The limitations of purely analog and purely digital approaches are discussed and followed by a presentation of hybrid techniques.

9.2 SOME MATHEMATICAL CONSIDERATIONS

In order to present a clear and concise formulation of the parameter optimization problem and the methods of solving this problem by means of computers, it will be necessary to review a few fundamental mathematical concepts. There are three major areas which will be introduced in this section:

1. Finding maxima and minima. This area is important since an optimum solution involves the precise identification of the maximum or minimum

of a cost function and this maximum or minimum must be identified in mathematical terms.
2. Description of dynamic systems. Since the chapter is primarily concerned with the optimization of dynamic systems, it is necessary to state in a concise and clear manner the time domain behavior of these systems. This can be done by means of vectors and matrices and the minimum background for this formulation is introduced here.
3. Criterion functions. The optimization is always referred to a particular criterion function and, consequently, it is important to define carefully what constitutes an acceptable criterion function for the optimization problem.

Finding Maxima or Minima. To find whether a criterion function which describes the system behavior has been maximized, or whether a cost has been minimized, it is important to know how to describe the existence of such an extreme point mathematically. Recall from calculus that if a function of one variable

$$y = f(x) \tag{9.6}$$

has an extremum (i.e., a maximum or a minimum) in an interval, then at that point the derivative of the function must be equal to zero. That is, a necessary condition for the existence of an extremum is

$$\frac{df(x)}{dx} = 0 \tag{9.7}$$

The significance of this statement can be illustrated with reference to Figure 9.3. In this figure there are seven points in the interval (a, b) where condition 9.7 is satisfied. It can be noted from examining this

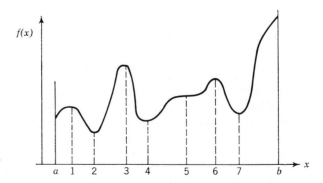

Figure 9.3 A function of one variable.

particular curve that three of those points are maxima, three are minima, and one is a point of inflection. It should be noted first that none of the seven points where condition 9.7 is satisfied is in fact the absolute maximum in this interval, since this maximum occurs on the boundary. Consequently, Equation 9.7 can be used only for the identification of relative maxima or minima in an interval but excluding the boundaries of the interval. Furthermore, a stationary point, such as point 5 in Figure 9.3, may also satisfy the necessary conditions. The significance of these statements is that any search for a maximum or a minimum based on condition 9.7 will only yield a *local* or *relative* maximum or minimum. Considerably more sophisticated techniques are required to find an *absolute* extremum. In the majority of practical cases, it cannot be assumed that the system in fact possesses a single extremum. Consequently, the particular maximum or minimum which the computer finds will depend on the choice of initial condition from which the computer begins its search.

When the optimization problem concerns a function of many variables such as

$$y = f(x_1, x_2, \ldots, x_n) \tag{9.8}$$

then the necessary condition for the existence of a relative extremum is that the partial derivative of the function with respect to each of the n variables be equal to zero simultaneously, that is

$$\frac{\partial f}{\partial x_1} = 0, \quad \frac{\partial f}{\partial x_2} = 0, \quad \ldots, \quad \frac{\partial f}{\partial x_n} = 0 \tag{9.9}$$

In compact form, the set of Equation 9.9 can be expressed by the single equation

$$\mathbf{grad}\, f = \nabla f = \left(\frac{\partial f}{\partial x_1}, \frac{\partial f}{\partial x_2}, \ldots, \frac{\partial f}{\partial x_n}\right) \tag{9.10}$$

since the derivatives in question are in fact the components of the gradient vector.

The remaining discussion in this chapter will be concerned with the finding of a local extremum of a function of n variables. In the parameter optimization problem the variables in question are the adjustable system parameters.

Mathematical Description of Dynamic Systems. Dynamical systems are described by means of differential equations. Thus a system of order n will be described by either a single n-th order differential equation or n first-order equations. Without loss of generality the system can be described by a set of equations of the form

$$\dot{y}_i = g_i(y_1, y_2, \ldots, y_n; t; \alpha_1, \alpha_2, \ldots, \alpha_m) \qquad i = 1, 2, \ldots, n \quad (9.11)$$

where the y_i can be considered to be the corresponding time derivatives of the system output

$$y_{i+1} = \frac{d^i y}{dt^i} \quad (9.12)$$

and the α_j represent the adjustable parameters. To completely characterize a dynamic system, a set of initial conditions is required in addition to the system equations. These would be given by

$$y_i(0) = y_{i0}, \qquad i = 1, 2, \ldots, n \quad (9.13)$$

where y_{i0} represents the value of the $(i-1)$st derivative at the initial time. Since the set of variables y_i constitutes a complete description of the system behavior at any particular time, these variables can be considered as describing the *state* of the system at that time. In more concise mathematical terminology, the variables y_i can be considered as components of a *state vector* **y** and are usually termed the *state variables*. Similarly, the m parameters $\alpha_1, \alpha_2, \ldots, \alpha_m$ can be considered the components of a parameter vector **α**. These two vectors are defined as

$$\mathbf{y} = \begin{pmatrix} y_1 \\ y_2 \\ \cdot \\ \cdot \\ \cdot \\ y_n \end{pmatrix}, \qquad \boldsymbol{\alpha} = \begin{pmatrix} \alpha_1 \\ \alpha_2 \\ \cdot \\ \cdot \\ \cdot \\ \alpha_m \end{pmatrix} \quad (9.14)$$

and the entire set of first-order differential equations described by Equation 9.11 can be stated in the single vector differential equation

$$\dot{\mathbf{y}} = \mathbf{g}(\mathbf{y}; t; \boldsymbol{\alpha}) \quad (9.15)$$

with the initial conditions being contained in the initial state

$$\mathbf{y}(0) = \mathbf{y}_0 \quad (9.16)$$

To summarize, then, the behavior of a dynamical system can be described in terms of a set of quantities called the state variables which contain a sufficient amount of information about the past history of the system so

9.2/SOME MATHEMATICAL CONSIDERATIONS

that, together with knowledge of the inputs to the system and the initial conditions, they completely specify the future behavior of the system. Description of dynamical systems in terms of their state is consistent with modern system theory and provides for an extremely compact interpretation of the behavior of multiparameter systems.

For each set of parameter values the system behavior will be described by means of a solution given by $\mathbf{y}(\boldsymbol{\alpha}^{(1)}, t)$ or $\mathbf{y}(\boldsymbol{\alpha}^{(2)}, t)$. Consider again the example of the preceding section, where it was desired to optimize the response of a satellite vehicle to a particular disturbance. Two parameters were to be specified: the thrust level K and the control system time constant τ. As a result of the selection of these two parameters, an output function $y(t)$ is obtained where y represents the angular oscillation of the vehicle due to the disturbance. The problem of parameter optimization then is that of selecting the two parameters such that some appropriate characteristic dependent on the solutions of the system equations is extremized. Two points can be noted in this connection. First, the process of obtaining a solution of a dynamic system following the selection of parameter values can be viewed as a mapping from a parameter space to a function space. In other words, the solution of a differential equation is equivalent to performing a particular transformation of a point in the function space which represents the particular function $y(t)$. Second, the criterion to be optimized depends on the behavior of the solutions to the differential equation. For example, it may be desired to minimize the deviation of the angular oscillation of the vehicle from some desired response y_D. A criterion function may be formulated as

$$F(y_D, y) = \int_0^T [y_D(t) - y(\boldsymbol{\alpha}^{(1)}, t)]^2 \, dt \qquad (9.17)$$

where $y_D(t)$ represents the desired trajectory and $y(\boldsymbol{\alpha}^{(1)}, t)$ represents the actual response of the system obtained for the particular values of parameters indicated by the vector $\boldsymbol{\alpha}^{(1)}$. The important point to observe in this connection is that the criterion function F is a *functional in the output space* since it depends on the functions y_D and y. However, it is an *ordinary function of the parameters*. Consequently, the formulation of the problem of system optimization in the way indicated here allows the solution to proceed as a problem in ordinary calculus (which is concerned with the maximization or minimization of functions) rather than a problem in the calculus of variations. This is extremely important as far as the computational aspects of the problem are concerned.

Since F is an ordinary function of the parameters $\alpha_1, \alpha_2, \ldots, \alpha_m$, the criterion will be extremized when the gradient of F is equal to 0, as indicated in Equation 9.10. Or, in terms of the notation of this paragraph,

$$\nabla F = \frac{\partial F}{\partial \alpha}\bigg|_{\alpha=\alpha^{(1)}} = \begin{pmatrix} \partial F/\partial \alpha_1 \\ \partial F/\partial \alpha_2 \\ \vdots \\ \partial F/\partial \alpha_m \end{pmatrix}_{\alpha=\alpha^{(1)}} = 0 \qquad (9.18)$$

The problem of parameter optimization then consists in successively adjusting the parameter vector such that on each successive try the criterion function becomes smaller (if minimization is desired) or larger (if maximization is desired). For minimization, the vector $\alpha^{(2)}$ is determined such that

$$F(\alpha^{(2)}, y_D) < F(\alpha^{(1)}, y_D) \qquad (9.19)$$

A Comment on Criterion Functions. Consider the criterion function which is defined above in Equation 9.17. This criterion is based on the square of the deviation of the actual trajectory from the desired trajectory, integrated over the entire duration of the mission. Note that it is a definite integral; consequently, at the end of a particular run it will represent a real number which can be used as a guide on whether the desired maximum or minimum is being approached. Thus F must represent a *distance* in the function space between the desired trajectory $y_D(t)$ and the actual solution $y(t)$. In order to qualify as a *distance function* or *metric* in the solution space, a function F must satisfy the following properties:

(1) $\qquad F(y_1, y_1) = 0$
(2) $\qquad F(y_1, y_2) > 0$
(3) $\qquad F(y_1, y_2) = F(y_2, y_1)$
(4) $\qquad F(y_1, y_2) \leq F(y_1, y_3) + F(y_2, y_3)$

Actually, properties (2) and (3) can be derived from the other two. Properties (1) and (2) can be interpreted very simply from an engineering point of view, since they simply state that any deviation from the desired solution should make the criterion function positive and that when the solution equals the desired solution the criterion should in fact be zero. While these remarks are made with respect to a minimization problem, it is clear that for the maximization problem the criterion function must be made as large as possible.

The important thing to note about these properties is that there is a wide choice of criterion functions available. A particular function may be selected on the basis of mathematical or engineering considerations. Quadratic functions of the kind illustrated in the Equation 9.17 have certain mathematical advantages and, consequently, are in wide use.

In the case of static optimization problems, it is possible to define an instantaneous criterion function rather than one which depends on an integration over a fixed interval such as the one defined in the preceding paragraph. In a static optimization problem, the system behavior is defined by algebraic rather differential equations, that is, the system includes no energy storage elements. Many problems concerned with profit optimization, scheduling, and transportation problems can be cast in the form of static optimization problems. Mathematical programming problems, including linear programming and nonlinear programming, also fit into this general category. In this case, an instantaneous criterion function may be defined such as, for example,

$$F(x_D, x(t)) = [x_D(t) - x(\alpha, t)]^2 \qquad (9.20)$$

where x_D represents the desired value of the solution (assumed to be a constant in this case) and $x(\alpha, t)$ represents the system solution as a function of the parameters and time. The optimization problem in this case can be carried on continuously since the effect of a change in the parameters is reflected immediately in a change in the criterion function. In fact, static optimization problems of this kind are excellently suited to analog computer solution for that reason. A method of solving such problems will be discussed later in the chapter.

9.3 A GEOMETRICAL INTERPRETATION OF THE PARAMETER OPTIMIZATION PROBLEM

A problem of parameter optimization may be visualized conveniently as a problem in hill-climbing. In fact, optimization techniques are sometimes referred to as "hill-climbing methods" in the literature. Consider the situation illustrated in Figure 9.4.*

This figure illustrates the problem faced by a man located somewhere on the side of a mountain who desires to find his house located at the minimum point in elevation in the valley below. His descent presents no problem on a clear day; but if the entire area is shrouded in fog, the path chosen by the traveler to reach the point of minimum elevation corresponds quite closely to the strategy required to implement a minimization problem on the computer. Assume that a rectangular coordinate system is centered

---

* The drawing of Figure 9.4 was inspired by a similar sketch in the book by L. Levine[29].

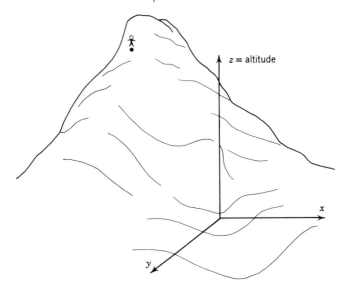

Figure 9.4 Descent from a hillside as a minimization problem.

at the minimum point, with the z-axis of the system representing the altitude or elevation above the minimum and the other two axes representing distances in two perpendicular directions. The traveler's problem, then, if it is to be stated in terms of this coordinate system, is that of choosing the x and y coordinates of his path in such a way that the elevation can represent the desired criterion to be minimized. In terms of the simple illustration in this figure, the parameter optimization problem then consists of determining at each point a direction which will take the traveler further down, i.e., a direction of adjustment of the parameters x and y such that the criterion function will in fact be decreased by traveling along this path. Note that the adjustment of the parameters may be considered continuous or it may be made in discrete steps. At each position along the hillside, there is a direction along which the criterion function changes most rapidly. This is the direction of steepest descent, and it is clear intuitively that following the path of instantaneous steepest descent will eventually get the traveler to his goal. On the other hand, it should also be clear that there are a number of other paths which, while they may not necessarily travel along the steepest descent direction, will still lead the traveler to the point of minimum elevation. The differences among the methods of computer implementation of the optimization problem actually represent nothing more than a choice of path in a descent down the mountainside.

9.3/A GEOMETRICAL INTERPRETATION   255

The situation of Figure 9.4 may also be cast in the form of a contour map as indicated in Figure 9.5. Here the axes have been labeled α_1 and α_2 to indicate that they represent two parameters whose values must be adjusted in order to minimize the criterion function. Contours of constant elevation in the terrain of the previous figure are indicated here simply as contour lines along which the elevation or criterion function has a constant value. Several possible paths starting from an initial condition which will lead to the (local) minimum are indicated. First consider path 1. This path was described in the preceding paragraph as the *path of steepest descent*. Along this path the criterion function changes most rapidly. Consequently, this path coincides with the direction of the gradient vector in the parameter space, and it crosses the contour lines at right angles. An alternate approach to the minimization problem may be seen in path 2. This path represents a series of adjustments in one parameter at a time, beginning with an adjustment in parameter α_2 from the initial condition to a point followed by an adjustment in α_1, etc. In each case the parameter is adjusted until the criterion function reaches a minimum along the particular direction of adjustment. In other words, the traveler heads in the x direction across the hillside until such time as this path leads him to a local minimum in this direction. He stops just before continuing travel would take him to a point of higher elevation again and makes a 90° turn and continues. Still another path is indicated by the number 3 in the figure. This path represents the travels of a man who

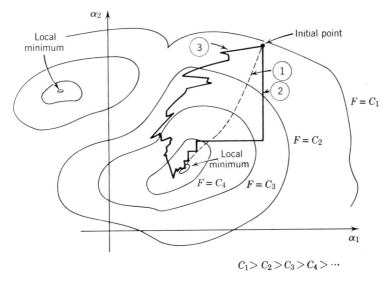

$C_1 > C_2 > C_3 > C_4 > \cdots$

Figure 9.5   Contour lines and paths in parameter space.

takes trial steps from the starting point which are random in length and direction. If the trial step leads to a lower elevation, he moves to the new point. If it does not, he tries another step. While his local path appears random, he does nevertheless move to a point of lower elevation with each step and, given enough time, he arrives at his destination.

The three paths illustrated in this figure represent the three most common strategies utilized in computer techniques of parameter optimization. Path 1 is usually known as the *method of steepest* descent. It may be implemented as a continuous path for static optimization problems, it may be approximated in dynamic optimization problems, or it may be implemented by discrete steps. Path 2 is sometimes known as a "relaxation method" and it refers to the relaxation or adjustment of the parameters one after another in turn. Path 3 is known in the literature as "optimization by random search."

9.4 OPTIMIZATION BY CONTINUOUS STEEPEST DESCENT

Consider first the problem of static optimization. Let the system behavior be described by a set of linear or nonlinear algebraic equations which include a number of adjustable parameters. The criterion function depends on the particular value of those parameters or

$$F = F(\alpha) \tag{9.21}$$

Assume for the moment that the system is simulated on an analog computer so that the criterion function itself represents one of the outputs. It is desired to determine a method of adjusting the parameters such that, starting from an arbitrary initial point α_0, the parameters will move toward the value which optimizes F. It has been indicated above that the path of steepest descent is the path which is normal to the contour lines in the parameter space which represent constant values of the criterion function. Consequently, it can be seen intuitively that the parameters should be adjusted such that their rate of change will be tangential to the gradient vector in this same space. If each component of the changing parameter vector, i.e., each component of $\dot{\alpha}$ is co-linear with the corresponding component of the gradient vector, then the adjustment will in fact be along the path of steepest descent. This can be proved in the following way.

The rate of change of a criterion function with respect to time is given by

$$\frac{dF}{dt} = \frac{\partial F}{\partial \alpha_1}\frac{d\alpha_1}{dt} + \frac{\partial F}{\partial \alpha_2}\frac{d\alpha_2}{dt} + \cdots + \frac{\partial F}{\partial \alpha_m}\frac{d\alpha_m}{dt} \tag{9.22}$$

9.4/OPTIMIZATION BY CONTINUOUS STEEPEST DESCENT

if n parameters are present in the system. This equation can be expressed in vector form as

$$\frac{dF}{dt} = \langle \nabla F, \dot{\boldsymbol{\alpha}} \rangle = \nabla F^T \dot{\boldsymbol{\alpha}} \tag{9.23}$$

where the letter T denotes the transposed vector and

$$\nabla F = \begin{pmatrix} \partial F/\partial \alpha_1 \\ \partial F/\partial \alpha_2 \\ \vdots \\ \partial F/\partial \alpha_m \end{pmatrix}, \quad \dot{\boldsymbol{\alpha}} = \begin{pmatrix} \dot{\alpha}_1 \\ \dot{\alpha}_2 \\ \vdots \\ \dot{\alpha}_m \end{pmatrix} \tag{9.24}$$

That is, the rate of change of F with respect to time is simply the dot or inner product of the two vectors **grad** F and $\dot{\boldsymbol{\alpha}}$. Now, it is desired to maximize the rate of change of the criterion function per unit time, that is, maximize $\dot{F}$. The dot product of two vectors attains its maximum when the two vectors are parallel or, in other words, when corresponding components of the two vectors are proportional to one another. The latter statement can be written as

$$\frac{d\alpha_i/dt}{d\alpha_j/dt} = \frac{\partial F/\partial \alpha_i}{\partial F/\partial \alpha_j} \tag{9.25}$$

or in vector form

$$\frac{d\boldsymbol{\alpha}}{dt} = K \nabla F \tag{9.26}$$

If K is positive, this equation represents an ascent path; if K is negative, it represents a path of descent. The implementation of the strategy indicated in Equation 9.26 is shown in block diagram in Figure 9.6 which indicates schematically the required steps in the process.

The system is simulated on an analog computer and the criterion function is evaluated. The gradient of the criterion function must then be computed. Equation 9.26 then indicates that the implementation of the path of steepest descent requires that each component of the gradient be made proportional to the rate of change of the corresponding parameter with respect to time. Therefore, the output of the gradient computer consists of n components of the gradient, each of which is then proportional to a time rate of change of the n parameters. These are integrated to produce immediately the values of the parameters to be adjusted in the system simulation.

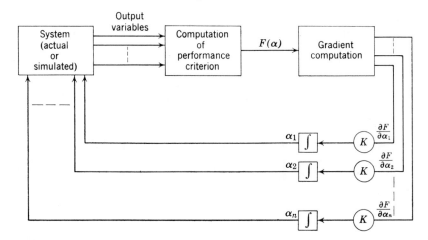

Figure 9.6   Implementation of steepest ascent strategy.

Before discussing methods of computing the components of the gradient, it should be emphasized again that the fundamental step in this development is the statement that the criterion F is an ordinary function of the parameters α_i. If this were not the case, the components of the gradient, which are derivatives of the criterion function with respect to α_1, α_2, etc., would not exist in the strict mathematical sense.

EXAMPLES. The static optimization technique outlined in the preceding paragraph is directly applicable to systems described by algebraic equations or inequalities. The method has been applied to the solution of linear algebraic equations,[19] to the problem of finding roots of polynomials,[10] and to linear and nonlinear programming problems.[18,27] The application of the method to the extraction of real and complex roots of polynomials was discussed briefly in the preceding section. As a better illustration, consider the problem of finding the solution of a set of algebraic equations which may be stated in the form

$$\mathbf{Ax} = \mathbf{b} \tag{9.27}$$

where $\mathbf{x}$ and $\mathbf{b}$ are n-vectors and $\mathbf{A}$ is an $n \times n$ matrix. For analog computer solution of these equations, it is desirable to avoid algebraic loops by solving instead the set of equations

$$\dot{\mathbf{x}} + \mathbf{Ax} = \mathbf{b} \tag{9.28}$$

the steady-state solution of which equals that of Equation 9.27. However, the solution of Equation 9.28 will be unstable unless the matrix $\mathbf{A}$ is

9.4/OPTIMIZATION BY CONTINUOUS STEEPEST DESCENT

positive definite. In order to avoid difficulties which may arise from lack of knowledge of the location of the eigenvalues of this matrix, it is possible to instrument on the computer the equations

$$\dot{\mathbf{x}} + \mathbf{A}^T\mathbf{A}\mathbf{x} = \mathbf{A}^T\mathbf{b} \tag{9.29}$$

where $\mathbf{A}^T$ denotes the transposed matrix. It can be easily verified that the matrix $\mathbf{A}^T\mathbf{A}$ is indeed positive definite and consequently that the solution of system 9.29 is stable. It is also possible to arrive at a stable set of equations leading to the solution of the algebraic system 9.27 by the method of steepest descent as outlined earlier. To accomplish the solution of these equations on the basis of an optimization problem formulation, it is necessary to define an error and an error criterion. We first define an error vector by means of the equation

$$\mathbf{A}\mathbf{x} - \mathbf{b} = \mathbf{e} \tag{9.30}$$

Clearly, $\mathbf{e}$ will be identically equal to zero when $\mathbf{x}$ equals the correct solution of the system. As a criterion function which meets the necessary requirements, it is possible to use the function

$$f = \tfrac{1}{2}\mathbf{e}^T\mathbf{e} = \tfrac{1}{2}\sum_{i=1}^{n} e_i^2 \tag{9.31}$$

which is simply one-half the sum of the squares of the components of the error vector. The gradient of the criterion function is then given by

$$\nabla f = \left(\sum_i e_i \frac{\partial e_i}{\partial x_1}, \sum_i e_i \frac{\partial e_i}{\partial x_2}, \ldots, \sum_i e_i \frac{\partial e_i}{\partial x_n}\right) \tag{9.32}$$

where the derivatives of the components of the error with respect to the parameters x_j can be obtained directly from each algebraic equation in the form

$$\frac{\partial e_i}{\partial x_j} = a_{ij} \tag{9.33}$$

where a_{ij} is the coefficient in the i-th row and j-th column of the original system of algebraic equations. In terms of this substitution, the gradient can be expressed as

$$\nabla f = \left(\sum_i a_{i1}e_i, \sum_i a_{i2}e_i, \ldots, \sum_i a_{in}e_i\right) \tag{9.34}$$

In order to obtain a steepest descent path, it is necessary to instrument the vector equation

$$\dot{\mathbf{x}} = -K\nabla f, \quad K > 0 \tag{9.35}$$

each component of which is then given by

$$\dot{x}_j = -K \sum_{i=1}^{n} a_{ij} e_i \qquad (9.36)$$

Since each component of the error is defined as

$$e_i = \sum_{k=1}^{n} a_{ik} x_k - b_k \qquad (9.37)$$

it is possible to rewrite Equation 9.36 in the form

$$\dot{x}_j = -K \sum_{i=1}^{n} a_{ij} \left(\sum_{k=1}^{n} a_{ik} x_k - b_k \right) \qquad (9.38)$$

This equation is one of the n equations of the same form which must be instrumented on a computer in order to arrive at a stable solution of the algebraic system by means of a gradient method. It can be seen that Equation 9.38 corresponds to one of the scalar equations represented in matrix form in Equation 9.29 above. An analog computer instrumentation of this set of equations is shown in Figure 9.7.

A more common static optimization problem is the so-called *programming problem*. Typically, a linear programming problem can be

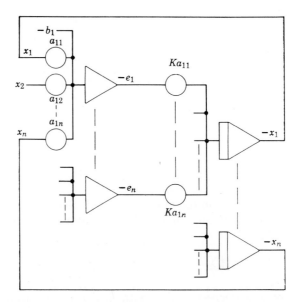

Figure 9.7 Analog solution of algebraic equations using a gradient method.

9.4/OPTIMIZATION BY CONTINUOUS STEEPEST DESCENT

expressed mathematically in this way: It is desired to maximize the "objective function" of n variables

$$F = \sum_{k=1}^{n} \alpha_k x_k, \quad (x_k \geq 0) \tag{9.39}$$

The variables x_k are usually restricted to being positive, as indicated above, since they commonly represent such quantities as prices, distances, and time which cannot be negative. The maximization of the function in Equation 9.39 is subject to a set of restrictions or constraints which can be expressed in the form of normalized inequalities as follows:

$$\sum_{i=1}^{n} \beta_{ij} x_i \leq 1 \tag{9.40}$$

There are generally m such restrictions where $m < n$. In order to take into account the effect of the constraints expressed in these inequalities, the "steepest descent" path must be modified by the fact that there are forbidden regions in the n-dimensional Euclidean space where the optimum is being sought. This can be done by using a slightly more complex expression than the one used above (for example, Equation 9.35) to describe the time variation of the variables x_i. When the optimum being sought is located sufficiently far away from the restricting surface, Equation 9.35 applies. A more general statement, however, can be written as

$$\dot{\mathbf{x}} = K\nabla F - \sum_{j=1}^{m} \delta_j \mathbf{N}_j \tag{9.41}$$

where the δ_j is defined as

$$\delta_j = 0 \quad \text{if} \quad \sum_{i=1}^{n} \beta_{ij} x_i \leq 1$$

$$\delta_j = 1 \quad \text{if} \quad \sum_{i=1}^{n} \beta_{ij} x_i > 1 \tag{9.42}$$

and the vector $\mathbf{N}_j$ is defined as a vector normal to

$$\sum_{i=1}^{n} \beta_{ij} x_i = 1 \tag{9.43}$$

which defines a plane in the n-dimensional parameter space. Consequently, when the restrictions of Equation 9.40 are being satisfied, the equation defining the motion of the point in the parameter space is the steepest descent equation considered previously. However, if the point attempts to cross one of the planes defining the constraint equations, then δ_j changes from 0 to 1 and an increment of velocity normal to and away

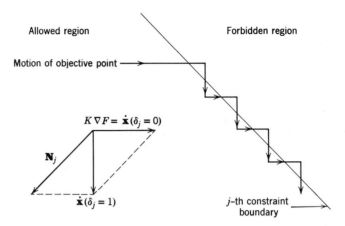

*Figure 9.8 Motion of objective point near **constraint boundary**.*

from the constraint plane is imparted to the traveling point. As a result, the motion of the characteristic point in the parameter space is restricted to the region bounded by the constraint surfaces and seeks the allowable maximum. The motion of the point in the neighborhood of a constraint is illustrated in Figure 9.8.

The instrumentation of these equations on an analog computer requires the use of diodes or similar gating circuits which introduce the additional vectors proportional to the components of $\mathbf{N}_j$ when a constraint is violated. Computers based on the implementation of Equation 9.41 for the solution of linear programming problems have been marketed commercially. Note that the use of analog circuitry for the solution of the equations makes feasible the use of nonlinear algebraic constraints[28] as well as the use of linear constraints of the kind given in Equation 9.40. While the analog method offers a great deal of flexibility for the study of programming problems, the equipment requirements become excessive when hundreds or thousands of variables are involved.

9.5 APPROXIMATIONS TO CONTINUOUS STEEPEST DESCENT FOR DYNAMIC OPTIMIZATION

The previous section has been concerned with the problem of static optimization. Most engineering problems, however, are concerned with the optimization of systems defined by differential equations. The method outlined in the preceding paragraph cannot be applied directly to dynamic systems since the presence of energy storage elements in the system makes impossible the instantaneous computation of the gradient as required for the implementation of the steepest descent path. However, it has been

9.5/APPROXIMATIONS TO CONTINUOUS STEEPEST DESCENT 263

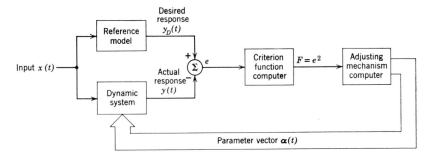

Figure 9.9 Model reference adaptive system.

demonstrated[12] that approximations to the equations of the previous section lead to a feasible optimization technique for cases where the system time constants are sufficiently short compared to the rate of adjustment of the parameters themselves. Before illustrating the development of the approximate method, it is useful to outline the source of the mathematical difficulty. This may be done with reference to Figure 9.9 which shows in block diagram form the so-called model reference adaptive control system. It is desired to adjust the system parameters denoted by the vector α in such a manner that the dynamic system output $y(\alpha, t)$ will approximate as closely as possible the output of the reference model denoted by $y_D(t)$. However, it can be noted that $y(t)$ is not an instantaneous function of the parameters, due to transient effects present in dynamic systems, but rather it depends on the present state *and past history* of both the system and the parameters. Thus, even if it were possible to obtain an instantaneous adjustment of the parameters to their ideal or correct values, the criterion function $F = e^2$ would not instantaneously decrease to its minimum value unless all transients have been dissipated. Consequently, the fundamental assumption made in the mathematical development of Section 9.2, namely, that F is an algebraic function of the parameters, is now violated. Since F depends on the entire time history of the parameters, it is no longer a function in the sense of ordinary calculus but rather a *functional*, and an instantaneous gradient cannot be defined. The only way to circumvent the mathematical difficulty apparent in this process is by allowing the parameters to remain fixed during the computation of the gradient. While the parameters are fixed, say, during an interval of T seconds, it is possible to evaluate the gradient of a criterion function defined, for example, as

$$F = \int_0^T e^2 \, dt \qquad (9.44)$$

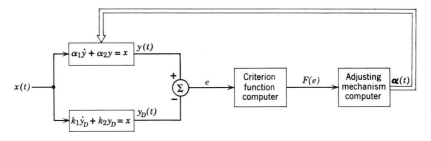

Figure 9.10 Continuous parameter adjustment scheme.

which is indeed an algebraic function of the parameters. Following the evaluation of the gradient vector and the consequent determination of a direction in which the parameters should be adjusted, it is possible to adjust the parameters and again compute the components of the gradient. This process, however, leads to iterative rather than continuous operation. To obtain continuous parameter adjustment with a dynamic system, it is necessary to assume that the rate of adjustment of the parameters will be sufficiently slow compared to the basic time constants of the system itself so that they may be considered approximately constant. The instrumentation of this method on an analog computer leads to serious stability problems when a rapid rate of parameter adjustment is desired. Furthermore, the actual path of adjustment in the parameter space cannot be formulated in the clear and concise fashion in which this is possible for the static optimization case; consequently, a simple geometric interpretation is not possible. The method may be termed an "approximate gradient method," or a "quasi-gradient method."[32] It has been used extensively in adaptive control problems to obtain parameter optimization and to identify system parameters with a "learning model."[12,17,24,30]

To illustrate the analog computer implementation of this method and its difficulties, before proceeding to digital and hybrid techniques, consider the problem of finding the parameters of a mathematical model of a first-order system by means of an approximate continuous gradient method. The block diagram of this method is given in Figure 9.10. In this figure as before, y_D represents the desired and known output of a first-order differential equation with parameters k_1 and k_2; y represents the output of the model whose parameters are denoted by the letters α_1 and α_2. The objective of the problem is to optimize the model by adjustment of the two parameters until an appropriate criterion function which expresses the difference between the two outputs is minimized. It can be seen that in this particular case the correct solution of the problem is possible, provided no disturbances are present, and is given when the

9.5/APPROXIMATIONS TO CONTINUOUS STEEPEST DESCENT

parameters α_1 and α_2 are equal to k_1 and k_2, respectively. To instrument a computer technique for adjustment of the parameters, select the criterion function

$$F = \frac{e^2}{2} = \frac{[y - y_D]^2}{2} \quad (9.45)$$

To find the approximate gradient method equations, ignoring for the moment the mathematical difficulties involved, it is possible to differentiate the criterion function formally in order to obtain the components of the approximate gradient vector. These components are then given by

$$\frac{\partial F}{\partial \alpha_i} = e \frac{\partial e}{\partial \alpha_i} = e \frac{\partial y}{\partial \alpha_i}, \quad i = 1, 2, \ldots n \quad (9.46)$$

and the "steepest descent" equations are defined by

$$\frac{d\alpha_i}{dt} = -Ke \frac{\partial y}{\partial \alpha_i}, \quad i = 1, 2, \ldots n \quad (9.47)$$

The analog computer instrumentation of these two equations is illustrated in Figure 9.11. From the figure it can be seen that the presence of dynamic elements in the system results in nonlinear closed loops with undetermined stability properties. Experience has shown that the adjusting loop is unstable unless the gain K is kept at a value sufficiently low so that the rate of parameter adjustment is slow compared with the time constants of the system itself. The figure does not illustrate the method of obtaining the approximate partial derivatives defined by Equation 9.46. These components, which were introduced in Chapter 5, are terms which reflect the influence of a change in a parameter on the output of the mathematical

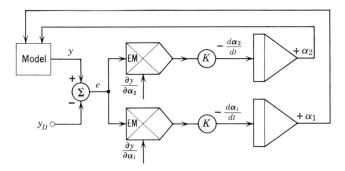

Figure 9.11 Analog implementation of approximate steepest descent.

model and are consequently usually termed "influence coefficients" or "sensitivity coefficients." They may be determined directly on the analog computer by a method due to H. F. Meissinger.[13]

Computation of Influence Coefficients. The influence coefficients such as

$$u_1 = \frac{\partial y}{\partial \alpha_1} \tag{9.48}$$

can be obtained by formally differentiating the differential equation describing the system behavior with respect to the parameter α_1. Consider the equation

$$\alpha_1 \dot{y} + \alpha_2 y = x \tag{9.49}$$
$$\dot{y}(0) = a, \qquad y(0) = b$$

Differentiation of this equation with respect to α_1 leads to

$$\dot{y} + \alpha_1 \frac{\partial^2 y}{\partial \alpha_1 \, \partial t} + \alpha_2 \frac{\partial y}{\partial \alpha_1} = 0 \tag{9.50}$$

Provided that y is continuous in t and α_1, the integration with respect to t and α_1 can be interchanged to yield the new differential equation

$$\alpha_1 \frac{d}{dt}\left(\frac{\partial y}{\partial \alpha_1}\right) + \alpha_2 \left(\frac{\partial y}{\partial \alpha_1}\right) = -\dot{y}$$
$$\frac{d}{dt}\left(\frac{\partial y}{\partial \alpha_1}\right)_{t=0} = 0; \qquad \left(\frac{\partial y}{\partial \alpha_1}\right)_{t=0} = 0 \tag{9.51}$$

It can be noted that this differential equation, whose solution is the desired influence coefficient, is identical in form to the original system differential equation defined by 9.49, differing from it only in the forcing function and in the initial conditions. Consequently, it is possible to instrument on the analog computer a "slave equation" in parallel and identical with the mathematical model equation and provide a coupling term between the two equations. The coupling term will be $-\dot{y}$ or $-y$ depending on whether the influence coefficient with respect to the parameter α_1 or α_2 is being sought. The output of the "slave equation" then is the desired influence coefficient which can be used directly as an input into the multipliers of Figure 9.11. The instrumentation of these equations is illustrated in Figure 9.12.

The slave, or "sensitivity equation" as it is sometimes called, is identical in form with the system differential equation only when the system equation is linear. When the equation is nonlinear the sensitivity equation

9.5/APPROXIMATIONS TO CONTINUOUS STEEPEST DESCENT 267

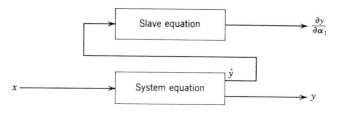

Figure 9.12   Relation of system and slave equations.

generally has a different form and the coupling between the system equation and the sensitivity equation is considerably more complex. Provided that sufficient equipment is available on the computer, it is possible to compute all n of the influence coefficients simultaneously, as required for the instrumentation of Figure 9.11.

Note that the development of the sensitivity equation, as given above, is strictly valid only when the parameters are invariant. This can be seen from another point of view if Equation 9.49 is rewritten as given below

$$\dot{y} = \frac{1}{\alpha_1}(x - \alpha_2 y) \tag{9.52}$$

the solution of the equation can then be written as the integral equation

$$y = \int_0^t \frac{1}{\alpha_1}(x - \alpha_2 y)\, d\tau \tag{9.53}$$

The desired influence coefficient would then be obtained by differentiating this equation with respect to α_1 or α_2. However, it is well known from calculus that differentiation under the integral sign in this equation is permissible only if α_1 and α_2 are independent of the variable of integration, namely time. And yet, as the instrumentation of Figure 9.11 clearly indicates, these variables are in fact dependent on time and are being continuously adjusted. This statement is equivalent to the statement made previously—that the gradient of the criterion function exists only if this function depends in an algebraic sense on the parameters.

In spite of the mathematical limitations outlined above, the method has been applied successfully to both linear and nonlinear problems. As an interesting illustration of the power of the approximate gradient technique, consider the problem of identifying the parameters of a simple mathematical model designed to simulate the behavior of a human operator in a one-dimensional control loop,[9] as illustrated in Figure 9.13.

The human operator indicated in the block diagram detects an error between a desired and actual system output and applies a correcting torque

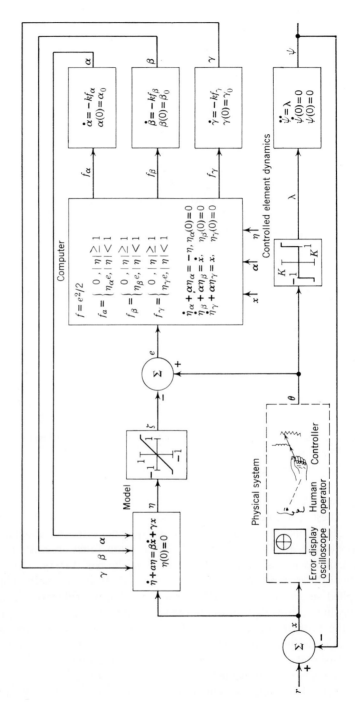

Figure 9.13 Block diagram parameter identification system with human operator.

9.5/APPROXIMATIONS TO CONTINUOUS STEEPEST DESCENT 269

by means of an on-off controller to a vehicle defined simply by its inertia. No torque is exerted on the vehicle until the hand controller reaches the limit of its travel, at which point a fixed level of torque is applied. It is desired to construct a simple mathematical model representing the dynamic behavior which relates the man's visual input to the controller output and, consequently, includes the fact that the controller output is limited. The mathematical model is also indicated in the figure as a first-order linear differential equation followed by a limiter. The objective of the optimization problem is to adjust the parameters, α, β, and γ continuously by means of an approximate gradient method. The criterion function $F = e^2$ is used and the required computations are indicated in Figure 9.13. Following the computation of the approximate influence coefficients, the desired derivatives of the parameters with respect to time are obtained, integrated, and used as inputs to the mathematical model. The typical results of this optimization problem, starting from arbitrary initial values for α, β, and γ, are shown in Figure 9.14.

The first channel in this figure represents the visual input which the operator sees on the oscilloscope. The second trace is his manual controller output; the third trace is the output of the mathematical model. It is desired to make trace 3 as close to trace 2 as possible with the preselected mathematical model and a continuous parameter adjustment technique. The error or difference between the desired (operator's) output and the model output is shown in trace 4. Channel 5 represents the RMS value of the error indicated by a highly damped meter having a 10-second time constant. Channels 6, 7, and 8 show the recorded values of α, β, and γ. With reference to these tracking records, it can be noted that the outputs of the model and of the operator are quite different during the first 100 seconds of the run, while they become quite similar during the last 100 seconds. It can also be seen that approximately 100 seconds are required for the parameters to attain approximately steady values. The presence of high frequencies in the error after 100 seconds indicates the desirability of choosing a higher-order model, if possible, and illustrates the fact that this type of parameter optimization can only select the optimum value of parameters for the particular criterion and for the particular model chosen in the experiment. It is also interesting to note that 100 seconds are approximately equal to three system time constants. Attempts to obtain faster convergence result in instability.

The experimental results are included here to illustrate the fact that approximate gradient methods can be used in a variety of optimization problems when using analog computers. However, since considerable mathematical difficulties are present and the precise optimization strategy is not well defined mathematically, and since stability is an inherent

270 PARAMETER OPTIMIZATION/9

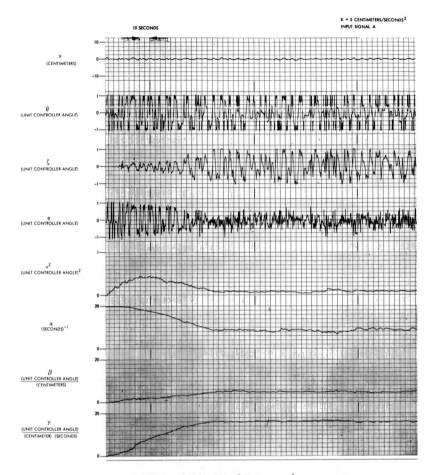

Figure 9.14 Tracking record.

problem in this technique, it is desirable to investigate alternative formulations of a dynamic optimization problem which circumvents these difficulties. A more detailed analysis of continuous parameter optimization methods can be found in the literature.[31,32]

9.6 DYNAMIC OPTIMIZATION BY DISCRETE GRADIENT METHODS

To avoid the serious mathematical problems which are encountered when attempting continuous parameter optimization of dynamical systems,

9.6/DYNAMIC OPTIMIZATION BY DISCRETE GRADIENT METHODS

it is necessary to choose a criterion function which lends itself to iterative optimization techniques. The process begins with the choice of a criterion function defined over finite time such as

$$F = \int_{t_0}^{t_0+T} [y_D(t) - y(\alpha, t)]^2 \, dt \qquad (9.54)$$

where the parameter vector α is assumed to remain constant during the interval from t_0 to $t_0 + T$. Consequently, F is indeed a function of the parameters. Using this criterion function, an iterative gradient method can be instrumented in the following sequence of four steps:

STEP 1. Start at
$$\alpha(0) = \alpha_0; \quad F(\alpha_0) = F_0 \qquad (9.55)$$

STEP 2. At this initial point, given by α_0, evaluate the components of the gradient:
$$\frac{\partial F(\alpha_0)}{\partial \alpha_1}, \quad \frac{\partial F(\alpha_0)}{\partial \alpha_2}, \ldots, \frac{\partial F(\alpha_0)}{\partial \alpha_m} \qquad (9.56)$$

STEP 3. Having available the gradient at point 0, compute a discrete parameter change vector, $\Delta\alpha_0$, by means of the relationship
$$\Delta\alpha_0 = -K\nabla F_0 \qquad (9.57)$$
which corresponds to the continuous steepest descent equations used previously.

STEP 4. Find a new value for the parameter vector α, namely $\alpha^{(1)}$, such that
$$\alpha^{(1)} = \alpha_0 + \Delta\alpha_0 \qquad (9.58)$$

In general, the new parameter vector is given as
$$\alpha^{(i+1)} = \alpha^{(i)} + \Delta\alpha^{(i)} \qquad (9.59)$$

where the change in the parameter vector is defined by
$$\Delta\alpha^{(i)} = -K\nabla F(\alpha^{(i)}) \qquad (9.60)$$

The above series of four steps may be considered a computing algorithm for an iterative method of parameter optimization. A number of unanswered questions remain from this very brief description of the method:

1. How is the gradient computed?
2. How is the computation stopped?
3. How is the step size for one iteration determined?
4. What type of computer is best suited for the implementation of this strategy?

These four points must also be considered in turn.

Gradient Computation. The gradient at each point in the parameter space may be computed either by means of the parameter influence coefficient technique outlined in the preceding section or by means of a discrete approximation to the required derivatives. As noted previously, the parameter influence coefficient technique is particularly applicable when the system to be optimized is linear. In general, regardless of the linearity of the system and providing only that the criterion function possesses continuous partial derivatives with respect to the parameters, the i-th component of the gradient may be approximated at the k-th iteration as

$$\frac{\partial F^{(k)}}{\partial \alpha_i} \cong \frac{F(\alpha_1^{(k)}, \alpha_2^{(k)}, \ldots \alpha_i^{(k)} + \Delta\alpha_i, \ldots \alpha_n^{(k)}) - F(\boldsymbol{\alpha}^{(k)})}{\Delta\alpha_i} \qquad (9.61)$$

where $\Delta\alpha_i$ is an arbitrarily small change in the parameter. Since each evaluation of the criterion function requires T seconds, it is apparent that attempts to obtain the components of the gradient by means of direct analog computation would require n identical simulations in order to obtain n partial derivatives. Conversely, since in general such large amounts of equipment would not be available, it is possible to compute the components of the gradient one at a time, returning the system to its initial condition following each iteration. This alternative requires considerable logic and memory and does not readily lend itself to a conventional analog computer.

Geometrical Interpretation of the Discrete Gradient Method. The discrete approximation to steepest descent may be visualized conveniently in terms of a parameter plane such as that used to illustrate continuous steepest

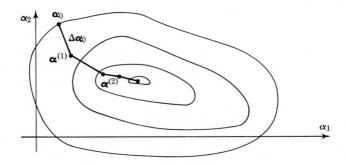

Figure 9.15 *Discrete gradient method.*

9.6/DYNAMIC OPTIMIZATION BY DISCRETE GRADIENT METHODS 273

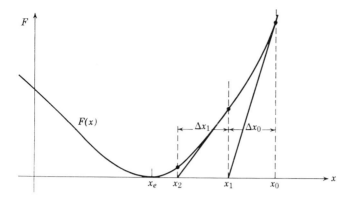

Figure 9.16   Newton-Raphson method.

descent previously. Consider the diagram of Figure 9.15 which again illustrates contours of constant criterion function plotted in the $\alpha_1 - \alpha_2$ plane.

The initial choice of parameter values is designated α_0, and the gradient at α_0 is computed. The vector $\Delta\alpha_0$ then represents the first step in the direction of the gradient from the initial choice of parameter values to point $\alpha^{(1)}$ as illustrated on the figure. Successive applications of this strategy result in a series of straight line steps which lead to the minimum.

Choice of Step Size. The length of each particular step in Figure 9.15 is determined by the value selected for K in Equation 9.57. It can be seen from an inspection of Figure 9.15 that it is possible to select a step size which is too large so that the criterion function will be larger after the step than before. On the other hand, an unnecessarily small step size will result in a wastefully large number of iterations. Consequently, an important problem in discrete gradient methods is the selection of the step size itself.

As a guide to the selection of step size, it is interesting to consider the one-dimensional problem of finding the root of an algebraic equation by the Newton-Raphson method. Consider the diagram of Figure 9.16, where the function $F(x)$ has a root at x_e. The initial assumed value of the variable x is denoted as x_0. If it is assumed that $F(x)$ has a convergent Taylor series in the vicinity of x_0, it is possible to write

$$F(x_0 + \Delta x) = F(x_0) + F'(x_0)\Delta x + 0(\Delta x^2) \qquad (9.62)$$

where the last term on the right-hand side represents second- and higher-order terms. The adjustment of Δx in this method is made along the path tangential to the curve at x_0, i.e., the path along the gradient of F:

$$\Delta x = -KF'(x_0) \tag{9.63}$$

Substituting 9.63 into 9.62 yields the relation

$$F(x_0 + \Delta x) = F(x_0) - K[F'(x_0)]^2 \tag{9.64}$$

It is desired to find a value of K such that the step taken would reduce $F(x)$ to zero or

$$F(x_0) - K[F'(x_0)]^2 = 0 \tag{9.65}$$

Solving for K yields

$$K = \frac{F(x_0)}{[F'(x_0)]^2} \tag{9.66}$$

and the resulting step size is obtained by substituting in 9.63:

$$\Delta x = -\frac{F(x_0)}{F'(x_0)} \tag{9.67}$$

This is the familar Newton-Raphson formula for finding the roots of algebraic equation. Assume for the moment that in the n-dimensional case the minimum of the criterion function is also equal to zero. Then an extension of this strategy to the n-dimensional case may be obtained as follows. Assume again that F possesses a convergent Taylor series in the vicinity of $\boldsymbol{\alpha}_0$ and expand to obtain.

$$F(\boldsymbol{\alpha}_0 + \Delta\boldsymbol{\alpha}_0) = F(\boldsymbol{\alpha}_0) + \nabla F^T \Delta\boldsymbol{\alpha}_0 + O(\Delta\boldsymbol{\alpha}_0^2) \tag{9.68}$$

To adjust along the gradient, the increment $\Delta\boldsymbol{\alpha}_0$ is chosen as follows:

$$\Delta\boldsymbol{\alpha}_0 = -K\nabla F(\boldsymbol{\alpha}_0) \tag{9.69}$$

Substituting this relationship into 9.68 yields

$$F(\boldsymbol{\alpha}_0 + \Delta\boldsymbol{\alpha}_0) = F(\boldsymbol{\alpha}_0) - K[\nabla F^T(\boldsymbol{\alpha}_0) \nabla F(\boldsymbol{\alpha}_0)] \tag{9.70}$$

To find the increment $\Delta\boldsymbol{\alpha}$ which reduces the criterion function to zero, 9.70 is set equal to zero and the coefficient K becomes

$$K = \frac{F(\boldsymbol{\alpha}_0)}{|\nabla F(\boldsymbol{\alpha}_0)|^2} = \frac{F(\boldsymbol{\alpha}_0)}{\nabla F^T(\boldsymbol{\alpha}_0) \nabla F(\boldsymbol{\alpha}_0)} \tag{9.71}$$

If this value of K is substituted into 9.69 the desired step size becomes

$$\Delta\boldsymbol{\alpha}_0 = -\frac{F(\boldsymbol{\alpha}_0) \nabla F(\boldsymbol{\alpha}_0)}{\nabla F^T(\boldsymbol{\alpha}_0) \nabla F(\boldsymbol{\alpha}_0)} \tag{9.72}$$

which is the vector form of the Newton-Raphson iteration process.

9.6/DYNAMIC OPTIMIZATION BY DISCRETE GRADIENT METHODS

In general, the minimum of the criterion function will not be zero and this simple strategy leads to considerable difficulty. This can be seen by examining the absolute magnitude of the step size as the minimum of the criterion function is approached. The magnitude of the step size is given by

$$\|\Delta\alpha^{(i)}\| = \frac{F(\alpha^{(i)})}{\|\nabla F(\alpha^{(i)})\|} \tag{9.73}$$

where the symbol $\|\mathbf{x}\|$ represents the Euclidean norm of a vector $\mathbf{x}$ defined as $\|\mathbf{x}\| = \sqrt{x_1^2 + x_2^2 + \cdots + x_n^2}$. Now, as the minimum is approached, the rate of change of the criterion function becomes smaller and smaller. Consequently, it is possible to write

$$\lim_{i \to \infty} \|\Delta\alpha^{(i)}\| = \lim_{\nabla F \to 0} \frac{F(\alpha^{(i)})}{\|\nabla F(\alpha^{(i)})\|} \tag{9.74}$$

The right-hand portion of this equation tends to infinity unless F is equal to 0 at the minimum. Since this is in general not the case, the value of K obtained from Equation 9.71 can be considered simply an upper bound on the possible step size. A more realistic basis for selection of step size is the so-called "optimum gradient method."

The Optimum Gradient Method.[7] A convenient strategy for implementing the discrete gradient method may be illustrated with reference to Figure 9.17. The process begins as before by finding the gradient direction at point 0. The gain K is selected in such a way that the resulting step leads to the minimum possible value of F along the gradient from point 0. At this point a new gradient is computed, and the largest possible step

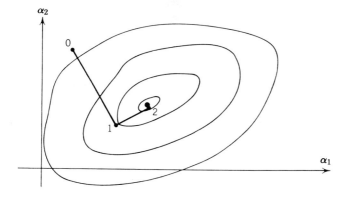

Figure 9.17 Optimum gradient method.

taken along the new direction $\nabla F^{(1)}$. Let the value of K which minimizes F in the gradient direction be denoted by K^*. Then K^* is obtained from

$$\min_{K>0} F(\boldsymbol{\alpha}^{(i)} - K\nabla F^{(i)}) = F(\boldsymbol{\alpha}^{(i)} - K^*\nabla F^{(i)}) \tag{9.75}$$

The new value of the parameter vector becomes

$$\boldsymbol{\alpha}^{(i+1)} = \boldsymbol{\alpha}^{(i)} - K^*\nabla F^{(i)} \tag{9.76}$$

An Algorithm for the Optimum Gradient Method. The value K^* is defined by the minimization of Equation 9.75. In practice, a number of alternate formulations of a method for computing K^* are possible. One convenient way is to select the largest possible step as an initial guess on the basis of the Newton-Raphson method, i.e., let

$$\Delta\boldsymbol{\alpha}^{(i,0)} = -\frac{F(\boldsymbol{\alpha}^{(i)})\,\nabla F(\boldsymbol{\alpha}^{(i)})}{\|\nabla F(\boldsymbol{\alpha}^{(i)})\|^2} \tag{9.77}$$

and then decrease the step by binary steps until the minimum in F is located. If Equation 9.77 is considered Step 1 of the method; one can go on as follows:

STEP 2. For $n = 0$, evaluate:

$$F^{(i,n)} = F\left(\boldsymbol{\alpha}^{(i)} + \frac{1}{2^n}\Delta\boldsymbol{\alpha}^{(i,0)}\right) \tag{9.78}$$

STEP 3. Increase n by integer steps until, at $n = m$:

$$F^{(i,m-1)} > F^{(i,m)} < F^{(i,m+1)} \tag{9.79}$$

STEP 4. If $F^{(i,m)} < F^{(i)}$, the new parameter vector is found from

$$\boldsymbol{\alpha}^{(i+1)} = \boldsymbol{\alpha}^{(i)} + \frac{1}{2^m}\Delta\boldsymbol{\alpha}^{(i,0)} \tag{9.80}$$

STEP 5. If $F^{(i,m)} > F^{(i)}$, i.e., the criterion function is not smaller than at the starting point, m must be increased further until a satisfactory minimum is located.

Stopping the Search. When the minimum is approximately located, it is desirable to find the minimum more accurately than is possible from the selection of integral values of m in Equation 9.80. One possible way of accomplishing the objective is by means of quadratic interpolation using the last three values of the criterion function and finding K^* in that

9.6/DYNAMIC OPTIMIZATION BY DISCRETE GRADIENT METHODS

manner. For example, let the approximate new value of the criterion function be given by

$$F_0 = F(\alpha_0 - K_0^* \nabla F_0) < F(\alpha_0) \quad (9.81)$$

and let two adjacent values, obtained from $2K_0^*$ and $\tfrac{1}{2}K_0^*$ respectively, be given by F_2 and F_1. Then the minimum can be found by quadratic approximation as

$$K_{\min} = \frac{3K^*}{4} \frac{F_2 - 5F_0 + 4F_1}{F_2 - 3F_0 + 2F_1} \quad (9.82)$$

An interpolation formula of this type is convenient in a digital implementation of gradient search problems to avoid the difficulty seen in the continuous case, where the velocity of approach to the minimum tends to zero as the minimum is approached.

Convergence and Stability. The algorithm outlined in the preceding paragraphs requires that the criterion function be tested at each stage before an actual parameter change is implemented to assure that

$$F(\alpha^{(i)} + \Delta\alpha^{(i)}) < F(\alpha^{(i)}) \quad (9.83)$$

Since F is a metric in the parameter space, it is continuous and bounded from below by zero, and the optimum gradient method produces a convergent sequence of values of F.

Digital Computer Implementation. The continuous gradient method of Section 9.4 is ideally suited to analog computer implementation. The iterative gradient method of parameter optimization, on the other hand, logically lends itself to digital computer mechanization. A flow chart based on the "optimum gradient method," as developed by McGhee,[11] is shown in Figure 9.18. This subroutine begins with the computation of the Newton-Raphson step (Equation 9.77) and includes the quadratic approximation to find the minimum.

It is clear that the digital computer lends itself readily to the large number of logical decisions and iterative procedures required to implement this method. Note that the computation of the gradient is not indicated in detail in the subroutine of Figure 9.18 which deals exclusively with determination of the step $\Delta\alpha^{(i)}$ to the minimum along a predetermined direction. While digital techniques are ideally suited to this portion of the optimization program, they are not ideally suited to the determination of the gradient and the evaluation of the criterion when the system to be optimized is highly nonlinear. The problem leads itself very clearly to hybrid computation.

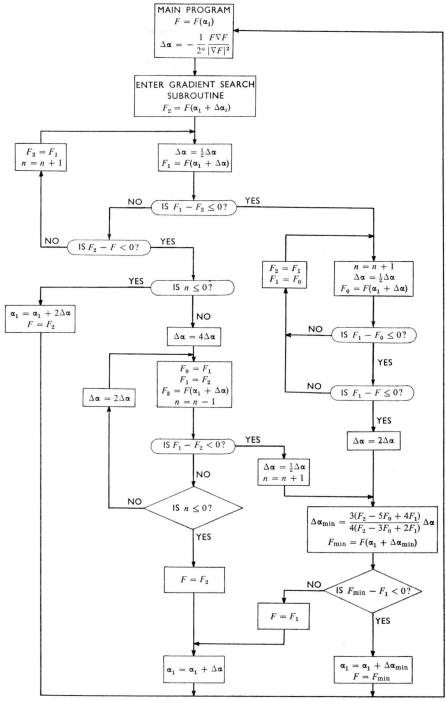

Figure 9.18 Flow chart of scalefactor determination program in optimum gradient method (from McGhee[11]).

9.6/DYNAMIC OPTIMIZATION BY DISCRETE GRADIENT METHODS

Hybrid Computer Implementation. The analog computer is ideally suited for the simulation of nonlinear systems. For the implementation of the discrete steepest descent method, the following procedure offers the advantage of optimum utilization of the capabilities of each computer:

STEP 1. Simulate the system to be optimized on the analog computer and provide for adjustment of the parameters from the digital computer via digital-analog converters.
STEP 2. Program parameter step size determination on the digital computer.
STEP 3. Compute the gradient of the criterion function by discrete approximation (Equation 9.61) using the digital computer as follows:
   (a) Evaluate $F(\boldsymbol{\alpha}^{(0)})$ on analog computer, store in digital computer.
   (b) Change parameters, one at a time, by digital command, and evaluate $F(\alpha_1^{(0)} + \Delta\alpha, \alpha_2^{(0)}, \alpha_3^{(0)} \ldots, \alpha_n^{(0)})$, $F(\alpha_1^{(0)}, \alpha_2^{(0)} + \Delta\alpha, \ldots, \alpha_n^{(0)}) \ldots, F(\alpha_1^{(0)}, \alpha_2^{(0)} \ldots, \alpha_n^{(0)} + \Delta\alpha)$ and store all n values in digital computer.
   (c) Evaluate all n components of the gradient on the digital computer from

$$\frac{\partial F(\boldsymbol{\alpha}^{(0)})}{\partial \alpha_i} = \frac{F(\alpha_1^{(0)}, \alpha_2^{(0)}, \ldots, \alpha_i^{(0)} + \Delta\alpha, \ldots, \alpha_n^{(0)}) - F(\boldsymbol{\alpha}^{(0)})}{\Delta\alpha}$$
$$i = 1, 2, \ldots, n \quad (9.84)$$

   (d) Use resulting gradient $\nabla F(\boldsymbol{\alpha}_0)$ as input to step size subroutine.
STEP 4. Upon completion of one cycle of minimization (which will take $n + 1$ analog computer runs to obtain all n gradient components), adjust parameters on analog side by

$$\boldsymbol{\alpha}^{(1)} = \boldsymbol{\alpha}^{(0)} + \Delta\boldsymbol{\alpha}_{\max}^{(0)} \quad (9.85)$$

as discussed previously.

In block diagram form, this method is shown in Figure 9.19.

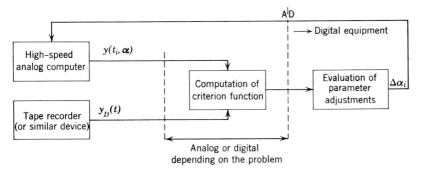

Figure 9.19 Hybrid computer implementation of discrete gradient method.

An alternate procedure, requiring a predetermined step size rather than computation of a $K_{\max}$, has been implemented on a hybrid computer by Witsenhausen.[26] The small step $(\Delta \boldsymbol{\alpha})_s^{(i)}$ is repeated from $\boldsymbol{\alpha}^{(i)}$ along the gradient direction until the criterion function is minimized. Consequently, the logic is very similar to that described here.

9.7 ITERATIVE OPTIMIZATION WITH CYCLICAL PARAMETER ADJUSTMENT

The necessity of storing $n + 1$ values of the criterion function in order to compute the gradient may be circumvented in two ways:

1. $n + 1$ simultaneous systems may be simulated on the analog computer. In general, the equipment requirements preclude this method.
2. An alternate strategy may be formulated based on adjustment of one parameter at a time, rather than all n parameters. This method will be outlined briefly.

Assume that the criterion function at $\boldsymbol{\alpha}^{(0)}$ has been computed and is given by $F(\boldsymbol{\alpha}^{(0)})$. Now compute one component of the gradient by Equation 9.84, say,

$$\frac{\partial F(\boldsymbol{\alpha}^{(0)})}{\partial \alpha_1}$$

Then adjust the parameter $\alpha_1^{(0)}$ only, to yield

$$\alpha_1^{(1)} = \alpha_1^{(0)} - K \frac{\partial F(\boldsymbol{\alpha}^{(0)})}{\partial \alpha_1} \tag{9.86}$$

and define the new parameter vector $\boldsymbol{\alpha}^{(1)}$ as:

$$\begin{aligned}\boldsymbol{\alpha}^{(1)} &= (\alpha_1^{(1)}, \alpha_2^{(0)}, \alpha_3^{(0)}, \ldots \alpha_n^{(0)}) \\ &= (\alpha_1^{(1)}, \alpha_2^{(1)}, \alpha_3^{(1)}, \ldots \alpha_n^{(1)})\end{aligned} \tag{9.87}$$

Parameter α_2 is adjusted to yield $\boldsymbol{\alpha}^{(2)}$, and so forth. The process is continued until $\boldsymbol{\alpha}^{(n)}$ has been obtained, and then the cycle is repeated.

An iterative analog computer implementation of this method has been discussed by Brunner,[1] who considered the criterion function F to be

$$F = \int_0^T \sum_{i=1}^M G_i(t)\epsilon_i^2(t)\, dt \tag{9.88}$$

where the ϵ_j are errors in satisfying the M performance criteria and the $G_i(t)$ are positive weighting functions which make it possible to emphasize

9.7/ITERATIVE OPTIMIZATION WITH CYCLICAL ADJUSTMENT

certain errors more than others in the construction of F. The iteration strategy was based on adjustment of one parameter at a time, with the gain K in Equation 9.86 determined from a modified Newton-Raphson approach, giving the increment $\Delta \alpha_i$ as:

$$\Delta \alpha_i^{(0)} = - \frac{\int_0^T \sum_{k=1}^M G_k(t) \epsilon_k^{(0)}(t) \frac{\partial \epsilon_k^{(0)}}{\partial \alpha_i} dt}{\int_0^T \sum_{k=1}^M G_k(t) \left[\frac{\partial \epsilon_k^{(0)}}{\partial \alpha_i}\right]^2 dt} \quad (9.89)$$

A block diagram showing the implementation of this method is given in Figure 9.20. It can be shown that the step size given by Equation 9.89 always leads to a decrease in the criterion function, i.e., that after the k-th step

$$F(\alpha_1^{(k)}, \alpha_2^{(k)}, \ldots, \alpha_i^{(k)} + \Delta \alpha_i^{(k)}, \ldots, \alpha_n^{(k)}) \leq F(\boldsymbol{\alpha}^{(k)}) \quad (9.90)$$

provided only that the change $\Delta \alpha_i$ is sufficiently small. The equality in Equation 9.90 would hold, to within any desired precision, at the minimum obtainable by adjustment of α_i.

A geometrical interpretation of the optimization strategy discussed here is given in Figure 9.21. Two possible paths, denoted by I and II, are indicated in the figure, depending on the choice of step size. Step sizes larger than those of trajectory II would lead to instability, unless a provision for adjusting the step size is available in the program. The

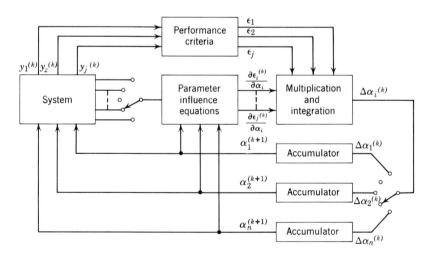

Figure 9.20 Modified Newton-Raphson method.

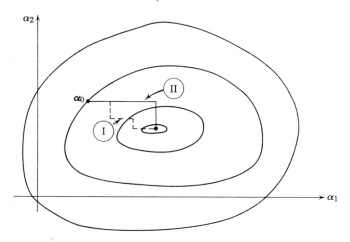

Figure 9.21 Cyclical parameter adjustment.

strategy of one-parameter-at-a-time optimization by using iterative gradient methods can be implemented by using iterative analog computers, digital computers, or hybrid computers.

9.8 OPTIMIZATION BY ONE-PARAMETER SEARCH METHODS

The methods discussed so far have required the computation of the gradient of the criterion function prior to, or simultaneously with, adjustment of parameters. Such a sophisticated adjustment strategy may not be justified by the particular application. Considerably simpler strategies may lead to longer computation time, but may save equipment

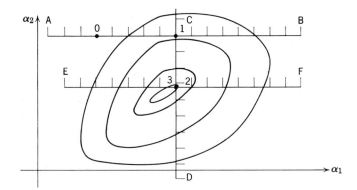

Figure 9.22 One parameter search (relaxation).

9.8/OPTIMIZATION BY ONE-PARAMETER SEARCH METHODS

(for analog implementation) or programming time and storage for complex subroutines (for digital implementation). One such implementation consists of searching for the minimum by adjustment of the parameters, in turn, over their allowable range and choosing the value which minimizes the criterion function in this range. Consider again the two-parameter case, as illustrated in Figure 9.22. Assume that the search is started at point 0 with parameter values $(\alpha_1^{(0)}, \alpha_2^{(0)})$. The allowable range of parameter α_1 (from A to B) is scanned, at increments $\pm \Delta \alpha_1$, as indicated in Figure 9.22 and F is computed at each step. The desired value of α_1 is

$$\alpha_1^{(1)} = \alpha_1^{(0)} \pm k(\Delta \alpha_1) \tag{9.91}$$

where k is a positive integer obtained from

$$\min_k F(\alpha_1^{(0)} \pm k \Delta \alpha_1, \alpha_2^{(0)}) = F(\alpha_1^{(1)}, \alpha_2^{(0)}) = F(\alpha_1^{(1)}, \alpha_2^{(1)}) \tag{9.92}$$

The minimum obtained from 9.92 is denoted point 1 in Figure 9.22. The process is now repeated with respect to parameter α_2, leading to point 2, at which time parameter α_1 is again adjusted. This method does not require computation of the gradient ∇F and requires only a minimum of storage.

An alternate formulation of a search process is to adjust each parameter by $+\Delta \alpha$ and $-\Delta \alpha$, computing and storing F with each change. If there are n parameters, it is necessary to compute and store the numbers

$$F(\alpha_1^{(0)}, \alpha_2^{(0)}, \ldots, \alpha_i^{(0)} \pm \Delta \alpha, \ldots, \alpha_n^{(0)}) \qquad i = 1, 2, \ldots, n \tag{9.93}$$

The maximum is stored and the corresponding parameters denoted $\alpha_1^{(1)}$, $\alpha_2^{(1)} \ldots, \alpha_n^{(1)}$ and the process is repeated. This strategy is illustrated for the two-parameter case in Figure 9.23. Five values of F must be stored at each stage of iteration, but less total computation is required than for the previous method since it is not required to scan the whole range of the parameter.

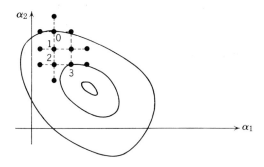

Figure 9.23 Neighboring grid method.

9.9 OPTIMIZATION BY RANDOM SEARCH

The parameter optimization methods discussed in previous sections of this chapter are primarily suited to optimization of criterion functions with unique minima or maxima. Furthermore, they may fail to converge or may converge only very slowly if the criterion function-parameter space exhibits "ridges"[33] or if the criterion function is only piecewise differentiable or piecewise continuous. Both of these difficulties are likely to arise in connection with nonlinear systems. This section presents an approach to finding a global optimum by means of a modified sequential random perturbation technique implemented on a hybrid computer.

Random search techniques for parameter optimization were originally proposed by Brooks.[34] They were successfully implemented on analog computers by Munson and Rubin[14] and Favreau and Franks.[35] A hybrid computer implementation using up to four parameters, a variable step size, and probability distribution changes keyed to past successes with an algebraic criterion function was studied by Mitchell.[36,37] The method has also been applied to the optimization of a nonlinear dynamic system with nine parameters.[38,39] Furthermore, the effects of initial conditions and analog computer errors can be included in the optimization program in a systematic way. The method is illustrated by application to the optimization of a satellite acquisition system.[38]

A. Problem Formulation

The dynamical system to be optimized is described by the differential equation:

$$\dot{\mathbf{x}} = \mathbf{f}(\mathbf{x}; t; \boldsymbol{\alpha}) \qquad (9.94)$$

where $\mathbf{x}$ is the $n \times 1$ state vector, $\mathbf{f}$ is an $n \times 1$ vector function, $\boldsymbol{\alpha}$ is an $m \times 1$ parameter vector, and t represents running time. It is desired to study this dynamic system over a large class of initial conditions. Define $\mathbf{X}_0$ as the set of all $n \times 1$ initial state vectors which are of interest to be studied and an element of $\mathbf{X}_0$ as $\mathbf{x}_0$. A unique solution of 9.94 is solely dependent on $\mathbf{x}_0$, $\boldsymbol{\alpha}$, and t and, therefore, can be represented as:

$$\mathbf{x} = \mathbf{x}(\mathbf{x}_0; t; \boldsymbol{\alpha}) \qquad (9.95)$$

A cost or criterion function can be written ordering the desirability of the particular choice of $\boldsymbol{\alpha}$ for a given $\mathbf{x}_0$ as:

$$J(\mathbf{x}_0; \boldsymbol{\alpha}) = \int_0^t g(\mathbf{x}; t; \boldsymbol{\alpha})\, dt \qquad (9.96)$$

where g is a scalar function of $\mathbf{x}$, $\boldsymbol{\alpha}$, and t. Examples of J may be fuel consumed or time required for satellite acquisition for a given parameter and initial condition set.

For a given initial condition and parameter setting, Equation 9.96 provides a scalar value describing the "quality" of the dynamic system in a quantitative fashion. As it is desired to study the effect of the parameter settings over the entire space of initial conditions, a new overall criterion function must be defined, which synthesizes the various values of $J(\mathbf{x}_0; \boldsymbol{\alpha})$ obtained for different initial conditions. This provides a single measure of the "quality" of a selection $\boldsymbol{\alpha}$ over the entire space of initial conditions. Examples of F are:

$$F_1 = \max_q J(\mathbf{x}_0, \boldsymbol{\alpha}) \qquad \mathbf{x}_0 \in \mathbf{X}_0$$

$$F_2 = \sum_{j=1}^{q} J(\mathbf{x}_{0j}, \boldsymbol{\alpha}) \qquad \mathbf{x}_0 \in \mathbf{X}_0 \qquad (9.97)$$

$$F_3 = \sum_{j=1}^{q} \max J(\mathbf{x}_{0j}, \boldsymbol{\alpha}) \qquad \mathbf{x}_{0j} \in \mathbf{X}_j \subset \mathbf{X}_0$$

For a given criterion function F, a computer algorithm is desired which finds the optimum $\boldsymbol{\alpha}$ which will be denoted by $\boldsymbol{\alpha}^*$:

$$F^*(\boldsymbol{\alpha}^*) = \min_{\boldsymbol{\alpha}} F(\boldsymbol{\alpha}) \qquad (9.98)$$

B. An Algorithm for Random Search Optimization

Strictly speaking, pure random search refers to a computation of the criterion function at a number of randomly chosen points in the parameter space and selection of the particular parameter values yielding the smallest value of $F(\boldsymbol{\alpha})$. However, such a sequence of randomly selected parameter vectors does not take advantage of the local continuity properties of most criterion function surfaces. Consequently, the strategy to be discussed below should more properly be referred to as "sequential random scanning" or "random creep."[37] Assume that the initial parameter vector is designated by $\boldsymbol{\alpha}^{(0)}$. Now, choose an increment $\Delta\boldsymbol{\alpha}^{(0)}$ by selecting the individual parameters $\Delta\alpha_i^{(0)}$, $i = 1, 2, \ldots, m$ from m Gaussian sequences of random numbers with mean zero and variance C_i. Then, if the m random sequences are independent, the orientation and length of the parameter increment $\Delta\boldsymbol{\alpha}^{(0)}$ will be random, and a trial value

$$\boldsymbol{\alpha}' = \boldsymbol{\alpha}^{(0)} + \Delta\boldsymbol{\alpha}^{(0)} \qquad (9.99)$$

is obtained. The criterion function $F' = F'(\boldsymbol{\alpha}')$ is computed and compared to $F_0 = F(\boldsymbol{\alpha}^{(0)})$. If there is an improvement, the parameters are updated by letting $\alpha' = \alpha^{(1)}$. If there is no improvement, the trial step is abandoned and a new trial step is chosen. This basic strategy is illustrated in the flow chart of Figure 9.24.

Now, from the standpoint of computer implementation, the differential equations can be solved on the analog computer for each trial value α'. The random increments $\Delta\alpha_i$ can be obtained by sampling analog noise generators, by generating pseudo-random sequences in the digital computer, or by construction of special devices such as shift register noise generators with several independent outputs (see Chapter 11).

C. Modification of the Basic Algorithm

In order to take maximum advantage of the properties of the criterion surface, the basic strategy can be modified in a number of ways. Successive steps can be made correlated in such a way as to favor a successful

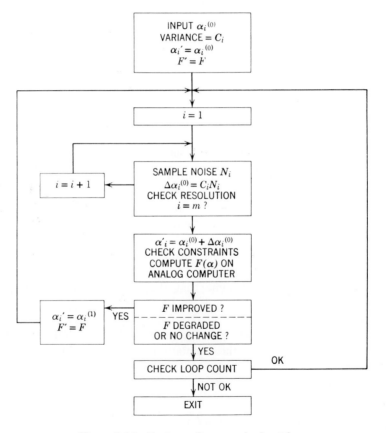

Figure 9.24 Basic random search algorithm.

direction. The mean of the distribution of steps can be biased in the direction of a successful step or after a specified number of successive successes.[36] Thus the j-th trial step could be computed from

$$\Delta \boldsymbol{\alpha}^{(j)} = C^{(j)} \mathbf{N}^{(j)} + \mathbf{b}^{(j)} \tag{9.100}$$

where $\mathbf{N}^{(j)}$ is a column vector of Gaussian random samples, $C^{(j)}$ is a diagonal matrix of variances, and $\mathbf{b}^{(j)}$ is a bias vector which may be altered after one or more successes (or failures).

The term "absolute biasing" has been used to denote the repeated use of a successful random step as long as continued success is attained. That is, if $\Delta \boldsymbol{\alpha}^{(j)}$ is successful, we choose

$$\Delta \boldsymbol{\alpha}^{(j+1)'} = \Delta \boldsymbol{\alpha}^{(j)} \tag{9.101}$$

and test for success. If $\boldsymbol{\alpha} \Delta^{(j)}$ was a failure, we choose

$$\Delta \boldsymbol{\alpha}^{(j+1)'} = -\Delta \boldsymbol{\alpha}^{(j)} \tag{9.102}$$

Such a technique will be referred to as "absolute positive and negative directional biasing."

It is also possible to adjust the variance of the distribution of step sizes. For example, as a local minimum is approached, the variance can be decreased in order to decrease the probability of overshooting the optimum.

The basic algorithm logically divides into two parts, concerned with the search for a local minimum and the global minimum respectively.

D. Search for a Local Minimum

This strategy, using absolute positive and negative biasing, is illustrated in Figure 9.25. It consists of the following steps for the computation of the j-th trial step:

STEP 1. A Gaussian random vector $\mathbf{N}^{(j)}$ is obtained. In the present study the components $N_i^{(j)}$, $i = 1, 2, \ldots, m$, were obtained from successive trial samples of an analog high-frequency noise generator. The sampling frequency was sufficiently low compared to the noise bandwidth to insure that successive samples are essentially uncorrelated.

STEP 2. A trial step is computed with magnitude constraints imposed on all parameters.

STEP 3. The analog computer is used to compute a new value of $F = F'$.

STEP 4. F' is compared with $F^{(j)}$. If $F' < F^{(j)}$, we let $F' = F^{(j+1)}$ and $\alpha' = \alpha^{(j+1)}$. If $F' > F^{(j)}$, the increment $-\Delta \boldsymbol{\alpha}^{(j)}$ is tried.

STEP 5. The number of trials which leads to either no change or an increase in F are counted and used as a stopping criterion.

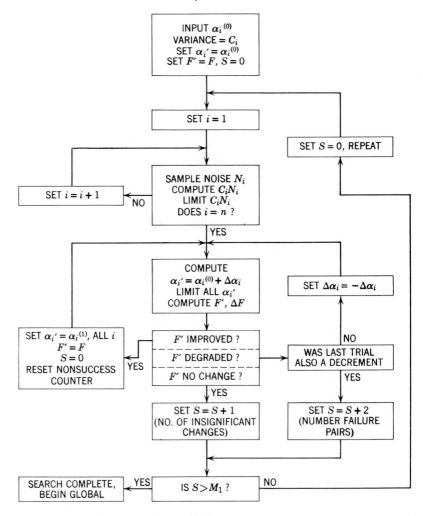

Figure 9.25 Absolutely biased local random search flow diagram.

E. Search for the Global Optimum

Once a local optimum has been found, it must be tested to determine whether it is indeed the global optimum. A reasonable approach is to again use a random search. This allows much flexibility in that the statistics of the search may be adjusted to correspond to estimates of the location of other likely optima. In many cases, the space near the local optimum is the most likely location for an even better optimization criterion. The strategy then consists of randomly sampling numbers corresponding to the parameters (as in the local random search), starting with a very small variance and expanding this variance slowly if no better points are found. If an improvement is found, a local search strategy is once again initiated. Figure 9.26 gives the flow detail of the global random search algorithm.

F. A Note on Convergence

It has been stated by Korn[37] that the random search technique will converge whenever the gradient technique does. Clearly, for the local optimization algorithm, convergence can be assured since the strategy results in a sequence of criterion function values which is monotonically decreasing and bounded from below by zero. However, the rate of

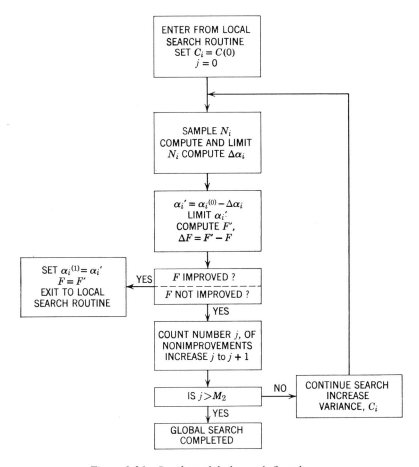

Figure 9.26   Random global search flow diagram.

convergence is another matter. Rastrigin, who published one of the early papers on random search optimization,[40] has also investigated its convergence properties.[41,42] In the case of unimodal criterion functions, where constant F contours are hyperspheres in the parameter space, he shows that the mean rate of progress of the random search method in the gradient direction exceeds that of the steepest descent method when more than three parameters are involved. However, no such proof is available for the nonlinear case.

Global search, which in the limit samples the parameter space everywhere, will converge to the global optimum with probability one. However, any computer implementation is finite and therefore cannot insure the location of the global optimum.

9.10 A DETAILED EXAMPLE: A SATELLITE ACQUISITION OPTIMIZATION

The general acquisition problem considered is that of aligning a single axis of a satellite parallel to a desired vector and driving the angular rotation about this axis to zero (one axis acquisition problem). The reference coordinate frame is a three-dimensional, cartesian set $(\mathbf{i}_1, \mathbf{i}_2, \mathbf{i}_3)$ with the $\mathbf{i}_3$ axis defining the desired pointing direction. The kinematic representation consists of the three direction cosines of the satellite axis to be aligned with the reference, or 3 axis. The control system can consist of any collection of sensors (that can be described by the six variables) whose outputs are processed by a compensation network, which can be described by the parameters to be optimized. The outputs of the networks are then used to drive angular acceleration devices (torquers). In order to provide a more concrete control equation, the sensors are assumed to be three rate sensors and two direction cosine sensors. The control laws were chosen to be proportional but saturable control. The equations expressing the acquisition dynamics, kinematics, control law, several potential optimization criteria and the end-of-run criterion are given in Table 9.1. Six initial conditions (three attitudes and three rates) were specified for each trial optimization.

A. Analog Simulation

The above-mentioned equations were simulated on a Beckman 2132 analog computer. The end-of-run criterion indicated in Table 9.1 is necessary to determine when acquisition is complete. This is especially critical when the time of acquisition is the optimization criterion. The criterion selected is the absolute value of a weighted sum of the variables biased by some fixed voltage (see C below). The criterion is assumed

9.10/A SATELLITE ACQUISITION OPTIMIZATION

Table 9.1 Simulated Acquisition Equations

Item Description	Equations	Discussion
Kinematics	$\dot{a}_{13} = \omega_2 a_{33} + \omega_3 a_{23}$ $\dot{a}_{23} = \omega_1 a_{33} - \omega_3 a_{13}$ $\dot{a}_{33} = \omega_2 a_{13} - \omega_1 a_{23}$	a_{i3}—Direction cosine from i body axis to 3 inertial ω_i—Body rate about i body axis
Dynamics	$\dot{\omega}_1 = [(I_2 - I_3)/I_1]\omega_2\omega_3 + \gamma_1$ $\dot{\omega}_2 = [(I_3 - I_1)/I_2]\omega_1\omega_3 + \gamma_2$ $\dot{\omega}_3 = [(I_1 - I_2)/I_3]\omega_1\omega_2 + \gamma_3$	I_i—Inertia about i principal inertia axis γ_i—Control angular acceleration about i axis
Control laws	$\gamma_1 = -\alpha_1(a_{23} + \alpha_3\omega_1)$ $\gamma_2 = -\alpha_2(-a_{13} + \alpha_4\omega_2)$ $\gamma_3 = -\alpha_5\omega_3$	α_i ($i = 1, 2, 3, 4, 5$)—control law gain constants
End-of-run criterion	$\sum_{i=1}^{3} b_i \lvert\omega_i\rvert + \sum_{i=1}^{2} c_i \lvert a_{i3}\rvert - K = 0$	b_i, c_i—Weighting and scaling constants K—a constant
Minimum fuel criterion	$F = \int_0^T \sum_i \lvert\gamma_i\rvert \, dt$	
Minimum time criterion	$F = \int_0^T dt$	
Constraints	$\alpha_7, \alpha_9, \alpha_6$ α_8, α_{10} $\alpha_{11}, \alpha_{12}, \alpha_{13}$	Limits on magnitude of $\gamma_1, \gamma_2, \gamma_3$, respectively Saturation values of a_{13} and a_{23}, respectively Saturation values of $\omega_1, \omega_2, \omega_3$, respectively

satisfied when this sum becomes zero. The bias is necessary as noises and drifts by the analog would prevent an unbiased zero from ever occurring. The bias level selected is more a function of actual voltage levels than their equivalent variable units.

B. Digital-Analog Interface

The digital computer provides those functions it is best suited for. It provides the optimization algorithm, initialization search procedure, controls the analog modes (initial condition, operate), provides voltages for the initial conditions of the variables and parameter settings, and reads the optimization criterion from the analog.

To better understand the operations of the various elements of the optimization process, a logical flow diagram of the entire computer operations in the optimization phase is shown in Figure 9.27. An optimization criterion is selected (for example, minimum fuel or time). A set of initial conditions for the variables is selected, based on the space search that has already been run. A best initial guess for the parameters is made. The error criterion (F) for this initial setting will be assumed already calculated. A new point in parameter space is selected by the optimization algorithm. For each set of variable initial conditions, three analog computer runs are made, giving a J for each run. The three J's are compared to check parity (the reasons will be explained later). If their standard deviation is within tolerance, the mean J for the three is selected as the criterion for this set of initial conditions. This is the

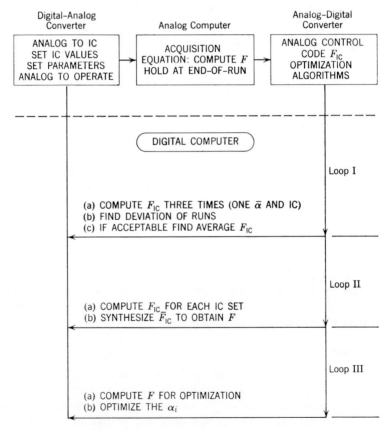

Figure 9.27 Optimization master logic flow diagram.

innermost loop (loop I). This process is repeated for each set of initial conditions (loop II). The J's for each initial condition set are averaged to give F. This is the F used to determine (loop III) whether the explored point is better than the present point. Loop III minimizes time or fuel as the error criterion.

C. Computer Error Detection

The speed and accuracy of convergence of the optimization procedure are directly related to the accuracy and repeatability of the analog simulation. Most malfunctions were of a momentary nature so that for over 99 % of the time no more than two successive runs were adversely affected. In light of this knowledge, the computer was programmed to prevent computer malfunctions from negating an algorithm or consuming excessive time in trouble shooting.

D. Overload Detection

An overload indicated either an unstable differential equation or an equipment malfunction. The computer was programmed to halt on overload.

E. Time-Limit Detection

Certain parameter combinations may result in extremely long optimization times. Occasionally, however, a computer malfunction could have the same effect. The computer was programmed to stop on a maximum time and try the problem again. If time is exceeded again, it is printed out, a large value is assigned to the optimization criterion, and the optimization continues automatically.

F. Standard Deviation Tests

The use of confidence tests provides a powerful tool for improving the accuracy of analog computer studies. Assume that errors in analog computer results are normally distributed with zero mean and a known standard deviation σ. The assumption of zero mean can be justified if the computer is balanced frequently and σ can be estimated from the sample variance V_s. Then confidence tests can be used to determine the number of runs needed to satisfy a particular accuracy criterion. For example, suppose it is desired to be confident with .95 probability that the computer has no more noise than when σ was estimated. For $N = 3$ runs, we require that $\sqrt{V_s} \leq 3.0\sigma$. When the allowable variance is exceeded, it is concluded that the computer is not operating properly. The computer makes three runs with the same inputs. If the standard

Table 9.2 Numerical Values for Acquisition Optimization Study

Parameter	Description	Assumed Value of Range of Values	Units
ω_{im}	Maximum initial rate	± 70	Milliradians/sec
ω_{ia}	Average initial rate	± 26	Milliradians/sec
a_{ijn}	Nominal direction cosines	$a_{13} = a_{23} = 0$ $a_{33} = +1$	—
$a_{ij\,\min}$	Minimum initial attitudes	± 1†	—
$a_{ij\,\max}$	Maximum initial attitudes	$+1, \pm 1/\sqrt{3}$†	—
$(I_2 - I_3)/I_1$	Inertia ratios	$+0.311$	—
$(I_3 - I_1)/I_2$	Inertia ratios	-0.802	—
$(I_1 - I_2)/I_3$	Inertia ratios	$+0.643$	—
α_1	Roll position gain	0.001 to 0.100	Sec^{-2}
α_2	Pitch position gain	0.001 to 0.100	Sec^{-2}
α_3	Roll rate to position gain	0.001 to 0.100	Sec^{-2}
α_4	Pitch rate to position gain	0.100 to 10.0	Sec
α_5	Yaw rate gain	0.100 to 10.0	Sec
α_6	Yaw torque limit	0.001 to 0.100	Rad/sec^2
α_7	Roll torque limit	0.001 to 0.100	Rad/sec^2
α_8	Pitch direction cosine limit	0.01 to 1.00	—
α_9	Pitch torque limit	0.001 to 0.100	Rad/sec^2
α_{10}	Roll direction cosine limit	0.01 to 1.00	—

† All physically realizable combinations.

deviation is less than 1 volt,‡ operation is assumed normal and the optimization continues. If the standard deviation is greater than 1 volt,‡ malfunction is assumed. In this case, the large standard deviation is printed out and another set of three runs attempted. If this set is accepted, the first set is discarded and the optimization is continued. The operator may stop the computation if he desires.

Note that if a set of runs is accepted, the three values of F are averaged, thus increasing the accuracy of the optimization criterion by $\sqrt{3}$.

G. Results of the Optimization Study

The specific numerical values used in the optimization are given in Table 9.2. A maximum of nine independent parameters were studied at one time. Figure 9.28 shows a typical optimization using acquisition time as the criterion for the absolutely biased local random search. The nine parameter settings are plotted versus real time. The vertical lines indicate when the computer is in reset. Three trials are simulated and compared before the case is accepted as just discussed. The optimization

‡ One volt is an arbitrary number which depends on the quality of the particular analog computer. In many cases, 0.1 volt may be a reasonable value.

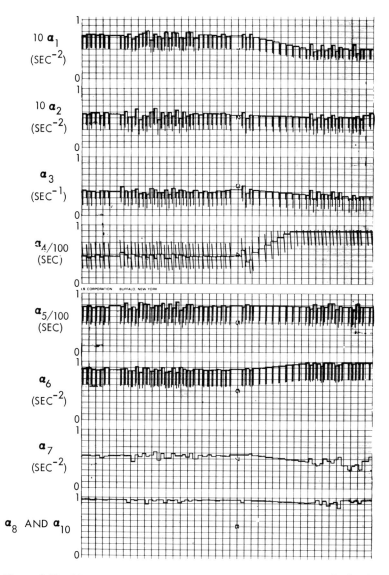

Figure 9.28 Nine parameter minimum-time optimization using absolutely biased local random search.

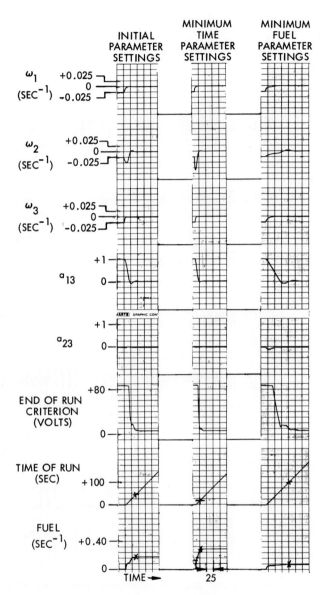

Figure 9.29 Comparison of minimum-time and minimum-fuel optimal solution with preoptimization solution.

runs shown considered just one initial condition set. To indicate the improvements in system performance during the optimization for one initial condition set, Figure 9.29 gives the response of five independent variables (ω_x, ω_y, ω_z, a_{13}, a_{23}) as defined in Table 9.1, the end-of-run criterion, time of run, and fuel consumption for: (1) the initial parameter settings, (2) the optimized minimum time settings, and (3) the optimized minimum fuel settings.

The major results of the study were the following:

(a) Absolute biasing is an efficient way of improving the convergence of a random search process.
(b) A successful strategy for changing the variance of the distribution of steps during the local search was not found. A uniform variance yielded convergence to a local optimum as rapidly as any variance adjustment strategy attempted. In this study, a variance equal to 4% of the range between maximum and minimum limits on each parameter was used.
(c) The random global search technique proved to be very useful. The strategy that appears most useful is that of using the local optimum as the origin and initiating the purely random (no biases) search with a very small variance. After each search, the variance is widened until any point in the parameter space has some chance of being inspected. For the most successful strategy found in the study, the variance was initially set equal to .5% of the span (upper limit to lower limit) of each parameter which was incremented by a factor of 1.02 every trial until the variance reached 50% of the span. This strategy assumes that the probability of a further improvement is highest near the local optimum and decreases linearly as the distance from the local optimum. In nearly every search for which the global optimization strategy was used, improvements were found.
(d) With nine free parameters three hundred to five hundred runs were made, with a limit of one hundred or two hundred runs in the global search subroutine.

An application of the algorithm outlined above to the optimization of a three-mode controller has recently been published by Ung.[43]

9.11 DYNAMIC OPTIMIZATION IN THE PRESENCE OF NOISE

The discussion in this chapter to this point has been based on the assumption that measurements of the system to be optimized or matched could be made without measurement noise, and that the system itself was deterministic in nature. In practice, one or both of these assumptions may not be justifiable, and stochastic or random effects must be taken into account. A detailed treatment of the statistical optimization problem is beyond the scope of this chapter. Some techniques are discussed in Chapter 11 in connection with the simulation of random processes. However, two general comments can be made here.

The criterion function itself becomes a random variable, since its value, even with a fixed set of parameters, will change with each measurement. Consequently, it may be necessary to make many measurements for each set of parameter values, compute the mean of the measurements, and formulate the problem so that the function to be optimized is the mean or expected value of the criterion function. Theoretically, an infinite number of measurements would be required. In practice, a finite number is taken and, consequently, a nonzero probability exists that the estimated mean of the criterion function is wrong and the parameters are adjusted incorrectly.

Steepest descent methods may be applied to the optimization problem in the presence of noise by the following technique. A number of measurements are taken at various points in the parameter space and a criterion function surface is fitted to the set of measurements by linear or nonlinear regression analysis.[25] The gradient method is then applied to this fitted surface as in the deterministic case. This method has been applied to identification of nonlinear dynamic systems.[11] Alternatively, methods such as quasilinearization[44] can be employed for parameter optimization.

References

1. Brunner, W., "An Iteration Procedure for Parametric Model Building and Boundary Value Problems," *Proc. Western Joint Computer Conference*, 519–533, 1961.
2. Chernoff, H., and J. B. Crockett, "Gradient Methods of Maximization," *Pacific J. Math.*, **5**, 33–50, 1955.
3. Chestnut, H., R. R. Duersch, and W. M. Gaines, "Automatic Optimizing of a Poorly Defined Process," *IEEE Transactions on Applications and Industry*, 32–41, March 1963.
4. Curry, H. B., "The Method of Steepest Descent for Nonlinear Minimization Problems," *Quart. Applied Math.*, **2**, 258–261, Oct. 1944.
5. Elkind, J. I., and D. M. Green, "Measurement of Time-Varying and Nonlinear Dynamic Characteristics of Human Pilots," *ASD Tech. Rpt.* 61-225, U.S. Air Force, Dec. 1961.

6. Eykhoff, P., "Process Parameter Estimation," in *Progress in Control Engineering*, ed. by O. Elgerd, Macmillan, London, 1964.
7. Feldbaum, A. A., "Automatic Optimalizer," *Automation and Remote Control*, **19**, 718–740, Aug. 1958.
8. Graupe, K. K., "The Analog Solution of Some Functional Analysis Problems," *AIEE Transactions Communications and Electronics*, pp. 793–799, Jan. 1961.
9. Humphrey, R. E., and G. A. Bekey, "A Technique for Determining the Parameters in a Nonlinear Model of a Human Operator," *Report No. 9865-6003-MU000*, Space Technology Laboratories, Mar. 1, 1963.
10. Levine, L., and H. F. Meissinger, "An Automatic Analog Computer Method for Solving Polynomials and Finding Root Loci," *1957 IRE Convention Record*, **5**, Part 4, 164–172, Mar. 1957.
11. McGhee, R. B., "Identification of Nonlinear Dynamic Systems by Regression Analysis Methods," Ph.D. Dissertation, Univ. Southern Calif., Depart. of Elec. Eng., June 1963.
12. Margolis, M., "On the Theory of Process Adaptive Control Systems: The Learning Model Approach," Ph.D. Dissertation and Report No. 60-32, Univ. of Calif. at Los Angeles, Department of Eng., May 1960.
13. Meissinger, H. F., "The Use of Parameter Influence Coefficients in Computer Analysis of Dynamic Systems," *Proc. Western Joint Computer Conference*, pp. 181–192, May 1960.
14. Munson, J. K., and A. Rubin, "Optimization by Random Search on the Analog Computer," *IRE Transactions on Electronic Computers*, EC-8, 200–209, June 1959.
15. Norkin, K. B., "On One Method of Automatic Search for the Extremum of a Function of Many Variables," *Automation and Remote Control*, **22**, 534–538, 1961.
16. Ornstein, G. N., "Applications of a Technique for the Automatic Determination of Human Response Equation Parameters," Ph.D. Dissertation, Ohio State Univ., Dept. of Psychology, 1961.
17. Potts, T. F., G. N. Ornstein, and A. B. Clymer, "The Automatic Determination of Human and other System Parameters," *Proc. Western Joint Computer Conference*, **19**, 645–660, May 1961.
18. Pyne, I. B., "Linear Programming on an Electronic Analog Computer," *AIEE Transactions Communications and Electronics*, pp. 139–143, May 1956.
19. Rogers, A. E., and T. Connolly, *Analog Computation in Engineering Design*, McGraw-Hill, New York, 1960.
20. Stakhovskii, R. I., "Twin Channel Automatic Optimalizer," *Automation and Remote Control*, **19**, 729–740, July 1958.
21. Stakhovskii, R. I., L. N. Fitsner, and A. B. Shubin, "Automatic Optimizers and Their Use for Solving Variational Problems and for Automatic Synthesis," *Proc. First IFAC Congress on Automatic Control*, Butterworths, London, 1962.
22. Tompkins, C. B., "Methods of Steep Descent," in *Modern Mathematics for the Engineer* (E. F. Beckenbach, ed.), McGraw-Hill, pp. 448–479, 1956.
23. Vichnevetsky, R., "One Dimension Iteration on the Analog Computer," Electronic Associates, Inc., European Computation Center, E.C.C. Report No. 43, March 1961.
24. Wertz, H. J., "Adaptive Control of Systems Containing the Human Operator," Ph.D. Dissertation, Univ. of Wis., Dept. of Elec. Eng., 1962.
25. Williams, E. J., *Regression Analysis*, John Wiley, New York, 1959.
26. Witsenhausen, H. S., "Hybrid Techniques Applied to Optimization Problems," *Proc. Joint Computer Conference*, 377–392, 1962.

27. Neustadt, L. W., "Applications of Linear and Nonlinear Programming Techniques," *Proc. 3rd International Analog Computer Conference*, Opatija, Yugoslavia, pp. 197–200, Sept. 1961.

28. Moser, J. H., C. O. Reed, and H. L. Sellars, "Nonlinear Programming Technique for Analog Computation," *Chem. Engineering Progress*, **51**, 59–62, June 1961.

29. Levine, L., *Methods for Solving Engineering Problems Using Analog Computers*, McGraw-Hill, New York, 1964.

30. Bekey, G. A., H. F. Meissinger, and R. E. Rose, "A Study of Model Matching Techniques for the Determination of Parameters in Human Pilot Models," *NASA Contractor Report*, **CR-143**, Jan. 1965.

31. Bekey, G. A., and R. B. McGhee, "Gradient Methods for the Optimization of Dynamic System Parameters by Hybrid Computation," in *Computing Methods in Optimization Problems* (A. V. Balakrishnan and L. W. Neustadt, eds.), Academic Press, New York, 305–327, 1964.

32. Meissinger, H. F., and G. A. Bekey, "An Analysis of Continuous Parameter Identification Methods," *Simulation*, **6**, 95–102, Feb. 1966.

33. Wilde, D. J., *Optimum Seeking Methods*, Prentice-Hall, Englewood Cliffs, N.J., 1964.

34. Brooks, S. H., "A Discussion of Random Methods for Seeking Maxima," *Operations Research*, **6**, 244–251, Mar. 1958.

35. Favreau, R. R., and R. G. Franks, "Statistical Optimization," *Proc. 2nd International Analog Computer Conference*, 1958.

36. Mitchell, B. A., "A Hybrid Analog-Digital Parameter Optimizer for ASTRAC-II," *Proc. AFIPS Spring Joint Computer Conference*, **25**, 271–285, 1964.

37. Korn, G. A., *Random Process Simulation and Measurements*, McGraw-Hill, New York, 1966.

38. Sabroff, A. E., R. Farrenkopf, A. Frew, and M. Gran, "Investigation of the Acquisition Problem in Satellite Attitude Control," Air Force Tech. Rpt., AF FDL-TR-65-115, Dec. 1965.

39. Bekey, G. A., A. Sabroff, M. H. Gran, and A. Wong, "Parameter Optimization by Random Search Using Hybrid Computer Techniques," *Proc. AFIPS Computer Conference*, **29**, 191–200, Nov. 1966.

40. Rastrigin, L. A., "Extremal Control by the Method of Random Scanning," *Automation and Remote Control*, **21**, 891–896, 1960.

41. Rastrigin, L. A., "The Convergence of the Random Search Method in the Extremal Control of a Many-Parameter System," *Automation and Remote Control*, **24**, 1337–1342, 1963.

42. Gurin, L. S., and L. A. Rastrigin, "Convergence of the Random Search Method in the Presence of Noise," *Automation and Remote Control*, **26**, 1505–1511, 1965.

43. Ung, M. T., "Sample Optimization Hybrid Program," *Simulation*, **10**, 21–28, Jan. 1968.

44. Kumar, K. S. P., and R. Sridhar, "On the Identification of Control Systems by the Quasi-Linearization Method," *IEEE Trans. on Automatic Control*, **AC-9**, 151–154, Apr. 1964.

10

Optimal Control Problems

10.1 INTRODUCTION

The optimization problems considered in the previous chapter were concerned with the selection of a set of *parameter values* which led to the best performance of a given system (as defined by an appropriate criterion function). In contrast, the problems considered in this chapter require the determination of a set of *input* or *control functions* which will lead to optimum performance. The system equations are assumed known in advance. Thus we are concerned with selecting input functions from some class of functions, rather than selecting parameter values from some admissible set. It is evident that the optimum selection of functions is a more difficult mathematical problem than the parameter optimization problem. The latter can often be solved by using the tools of ordinary calculus, while the former requires the calculus of variations. For example, we may want to transfer a space vehicle from some orbit to another (specified) orbit while minimizing fuel expenditure, or be assured that concentrations of certain chemicals in a process reach specified values in minimum time.

Consider the first-order system of Figure 10.1. Optimum performance of this system may require the minimization of both control effort and output amplitude; a typical criterion could be

$$J(u) = \int_{t_0}^{t_f} [u^2(t) + x^2(t)] \, dt \tag{10.1}$$

where $x(t)$ is the system output, $u(t)$ is the control input to be found, and $T_0 = (t_f - t_0)$ is the time interval over which the optimization is

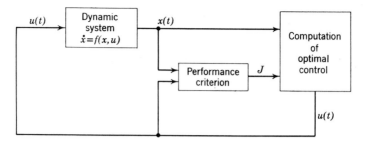

Figure 10.1 Block diagram of optimal control problem.

performed. The interval T_0 may be fixed in advance or it may be unknown (for example, in minimum-time problems). The objective of the problem is to find the particular control $u^*(t)$ (the optimal control) which minimizes $J(u)$; this can be written as

$$J^*(u^*) = \min_u \int_{t_0}^{t_f} [u^2(t) + x^2(t)]\, dt \qquad (10.2)$$

where J^* represents the optimum value of the performance criterion. Before proceeding to a more careful formulation of the problem, consider the following observations:

1. The cost J is a *functional* rather than a function, since the entire time history of $u(t)$ must be specified to obtain a numerical value of J. It is often called the "cost functional" or "criterion functional."
2. In general, $u(t)$ cannot be completely arbitrary but must be subject to certain constraints. For example, the magnitude of $u(t)$ in Figure 10.1 may be limited to some maximum value, so that

$$|u(t)| \leq C_1$$

or a limited amount of control energy may be available, in which case

$$\int_{t_0}^{t_f} u^2(t) \leq C_2$$

where C_1 and C_2 are constants. The set of functions $u(t)$ which satisfy the constraints is usually called the "admissible set" of control functions. If $\mathbf{u}(t)$ represents a control vector, then each component $u_i(t), i = 1, 2, \ldots, r$ of $\mathbf{u}(t)$ may be appropriately constrained.
3. Similarly, the system state variables may be restricted in some way. For example, the objective of the problem may be to find the control vector $\mathbf{u}(t)$ which transfers an aerospace vehicle from some starting

position to a given final coordinate in minimum time (or with minimum energy expenditure), without exceeding a given maximum velocity due to structural weakness.

4. From a computer point of view, the intuitively obvious approach to the problem would be to attempt a method similar to that of searching for optimum parameter values. Thus, in principle, we could simulate the system on a computer and try all possible admissible functions $u(t)$ as inputs. Only certain of these will satisfy the boundary conditions on $x(t)$ at $t = t_0$ and $t = t_f$. For all of these the cost functional could be recorded, and the particular $u(t)$ which gives the smallest cost selected as $u^*(t)$. Unfortunately, this brute force approach is completely unfeasible. If one visualizes $u(t)$ as represented by some kind of expansion in orthogonal functions $\varphi_i(t)$, say,

$$u(t) = \sum_{i=1}^{\infty} a_i \varphi_i(t)$$

then the problem is equivalent to a parameter optimization problem with the unknowns being the infinite number of parameters a_i, $i = 1, 2, \ldots, \infty$. Clearly, a more systematic approach is needed.

5. In certain cases, particularly with stochastic systems, it may be desirable to find an optimum control u^* which depends on the system output $x(t)$; in that case the resulting optimum system will be a *closed-loop* (feedback) system.

Hybrid computers have been applied to the solution of a limited number of optimal control problems. However, the vast majority of such problems are treated by purely digital methods for a variety of reasons which will emerge in the following discussion. The purposes of this chapter are to present the mathematical formulation of optimal control problems and to review the hybrid computer techniques which have been employed for their solution.

10.2 MATHEMATICAL FORMULATION

The purposes of this section are to formulate the optimal control problem mathematically and to review the major results which lead to computational solutions. The formulation is heuristic and no proofs of the statements are included; for more detailed and rigorous developments the reader is referred to the works of Athans and Falb,[22] Feldbaum,[23] Tou,[24] Pontryagin et al.,[19] or Leitmann.[25]

To state the problem precisely, we must: (a) describe the dynamic system under consideration and its initial state, (b) define the desired or goal state, (c) define the admissible set of controls, and (d) select a cost of performance criterion. Once the problem is stated, the methods of solution can be investigated.

Consider a dynamic system described by the nonlinear vector differential equation

$$\dot{\mathbf{x}}(t) = \mathbf{f}[\mathbf{x}(t), \mathbf{u}(t)] \qquad (10.3)$$

where $\mathbf{x}(t) = [x_1(t), x_2(t), \ldots, x_n(t)]$ is the *state vector* of the system at time t and $x_i(t)$, $i = 1, 2, \ldots, n$ are the *state variables*, $\mathbf{u}(t)$ is an r-dimensional *control vector* with $r \leq n$, and $\mathbf{f}$ is a vector valued function of $\mathbf{x}(t)$ and $\mathbf{u}(t)$. The components of $\mathbf{f}$ are assumed differentiable with respect to all arguments. The initial state of the system at time t_0 is given by

$$\mathbf{x}(t_0) = \mathbf{x}_0 \qquad (10.4)$$

In this chapter, t_0 will always be taken as the time origin $t_0 = 0$. The control vector is subject to certain constraints; all controls which satisfy these constraints will belong to a set U, so that for all *admissible* controls

$$\mathbf{u}(t) \subset U \quad \text{for all } t\dagger \qquad (10.5)$$

The objective of the problem will be to find an admissible control $\mathbf{u}(t)$ which transfers the system from its initial state $\mathbf{x}(t_0)$ to a desired state at the terminal time t_f while extremizing the cost functional. In some cases, $\mathbf{x}(t_f)$ is unspecified and this is called a "free right-end problem." To insure that the final state $\mathbf{x}(t_f)$ is in fact the desired state, we select a *goal* or *target set* S and require that

$$\mathbf{x}(t_f) \in S \qquad (10.6)$$

Finally, depending on the optimization task at hand, we select a *cost functional*,

$$J(\mathbf{u}) = \int_{t_0}^{t_f} L[\mathbf{x}(t), \mathbf{u}(t)] \, dt \qquad (10.7)$$

In this equation L is a scalar valued function of the state vector $\mathbf{x}(t)$ and control vector $\mathbf{u}(t)$. Consequently, $J(\mathbf{u})$ is also a scalar.

It is now possible to state the optimal control problem precisely as follows:

Given the system described by Equation 10.3, it is desired to find an admissible control vector $\mathbf{u}(t)$ which transfers the system from the initial state $\mathbf{x}_0$ to the final state $\mathbf{x}(t_f)$ such that $\mathbf{x}(t_f) \in S$ and the cost functional of Equation 10.7 is minimized.

† The symbol $\in$ may be read as "belongs to the set" or "is a member of the set."

To accomplish the optimization, it is possible to use a number of approaches such as the classical calculus of variations, dynamic programming, or the Pontryagin Maximum Principle.

We begin by considering the basic results of the Pontryagin Maximum Principle. To state these results, certain additional quantities must be defined. Consider first the n components of the system Equation 10.3

$$\dot{x}_i = f_i(x_1, x_2, \ldots, x_n; u_1, u_2, \ldots, u_r), \quad i = 1, 2, \ldots, n \quad (10.8)$$

We now introduce a vector $\mathbf{p}(t)$, each component of which is related to the system equations:

$$\dot{p}_i = -\sum_{j=1}^{n} \frac{\partial f_j}{\partial x_i} p_j, \quad i = 1, 2, \ldots, n \quad (10.9)$$

The vector $\mathbf{p}(t)$ is known as the *adjoint* or *costate* vector because of its relationship to the state vector $\mathbf{x}(t)$.

When the system is linear, it can be described by the vector-matrix equation

$$\dot{\mathbf{x}} = A(t)\mathbf{x} + B(t)\mathbf{u} \quad (10.10)$$

where $A(t)$ and $B(t)$ are $n \times n$ and $n \times r$ matrices of coefficients, respectively. It is easy to see that in this case application of the definition 10.9 leads to the relation

$$\dot{\mathbf{p}} = -A^T(t)\mathbf{p} \quad (10.11)$$

for the adjoint vector, where the superscript T refers to matrix transposition. Note that the adjoint system is also linear.

In addition to $\mathbf{p}(t)$, it is also necessary to introduce a scalar quantity known as the Hamiltonian, denoted by H and defined as follows:

$$\begin{aligned} H(\mathbf{x}, \mathbf{p}, \mathbf{u}) &= \mathbf{p}^T \mathbf{f}(\mathbf{x}, \mathbf{u}) + L(\mathbf{x}, \mathbf{u}) \\ &= \sum_{i=1}^{n} p_i f_i(\mathbf{x}, \mathbf{u}) + L(\mathbf{x}, \mathbf{u}) \end{aligned} \quad (10.12)$$

where $L(\mathbf{x}, \mathbf{u})$ is the integrand of the cost functional of Equation 10.7. To simplify the notation, it is customary to define a new state variable $x_0(t)$ such that

$$\dot{x}_0 = f_0(\mathbf{x}, \mathbf{u}) = L(\mathbf{x}, \mathbf{u}) \quad (10.13)$$

so that if $t_0 = 0$, we have

$$x_0(t_f) = J, \quad x_0(t_0) = 0 \quad (10.14)$$

If relation 10.13 is added to the system equations previously defined, the dimensionality of the state vector is increased to $(n + 1)$. Using the augmented state and adjoint vectors, the Hamiltonian becomes

$$H(\mathbf{x}, \mathbf{p}, \mathbf{u}) = \sum_{i=0}^{n} p_i f_i(\mathbf{x}, \mathbf{u}) \qquad (10.15)$$

From Equations 10.9 and 10.13, we see that the relation between the state and adjoint variables can be expressed in terms of the Hamiltonian as

$$\dot{x}_i = \frac{\partial H}{\partial p_i}, \qquad \dot{p}_i = -\frac{\partial H}{\partial x_i}, \qquad i = 0, 1, 2, \ldots, n \qquad (10.16)$$

Expression 10.16 is known as the "Hamiltonian system" or as "the canonical equations."

Consider now a restricted optimal control problem: Given the dynamic system of Equation 10.3, we wish to transfer the system from an initial state $\mathbf{x}_0$ to a final state $\mathbf{x}_f$ by using an admissible control vector $\mathbf{u}(t)$ such that the cost function $J(\mathbf{u})$ is minimized. Let the $\mathbf{u}(t)$ which accomplishes this objective be denoted by $\mathbf{u}^*(t)$, the corresponding system trajectory by $\mathbf{x}^*(t)$, and the corresponding cost by $J(\mathbf{u}^*) = J^*$. Then, the Pontryagin Maximum Principle states that:

In order that the admissible control $\mathbf{u}(t)$ and the corresponding trajectory $\mathbf{x}(t)$ be optimal, so that

$$J^*(\mathbf{u}^*) = \min_{\mathbf{u}(t) \in U} J(\mathbf{u}) \qquad (10.17)$$

it is necessary that there exist a nonzero, continuous vector $\mathbf{p}^*(t)$ such that:

(a) the canonical equations are satisfied for $\mathbf{u}^*(t)$, $\mathbf{x}^*(t)$, and $\mathbf{p}^*(t)$,
(b) the Hamiltonian is maximized for all time in the interval under consideration and for all admissible controls, that is:

$$H[\mathbf{x}^*(t), \mathbf{p}^*(t), \mathbf{u}^*(t)] \geq H[\mathbf{x}^*(t), \mathbf{p}^*(t), \mathbf{u}(t)] \qquad (10.18)$$

for $t_0 \leq t \leq t_f$ and $\mathbf{u}(t) \in U$ and,
(c) *at the terminal time*

$$p_0(t_f) \leq 0 \qquad (10.19)$$

If the control $\mathbf{u}(t)$ is discontinuous, then Equation 10.18 need only be satisfied at points of continuity of $\mathbf{u}(t)$. The proof of the Maximum Principle is found in the literature.[19,22]

On first inspection of the Maximum Principle, it may appear that nothing has been gained; namely, that we have simply traded the minimization of J for the maximization of H. However, the Maximum Principle is in fact a statement of a *necessary condition* for a control to be optimal; in this sense it is analogous to the necessary condition that is encountered in ordinary calculus minimization problems (that the first partial derivative of a function be zero at a minimum).† Thus, while we have not completely solved the problem, the Maximum Principle enables us to test possible controls for optimality, to select all those which meet this necessary condition, and to compare the cost $J(\mathbf{u})$ associated with each of these candidate functions to find the global optimum.

Note that the above statement of the Maximum Principle applies to only the specific problems considered. When final state $\mathbf{x}_f$ is not fixed but constrained to some set S of the state space, we must also take into account the so-called "transversality conditions." The Principle can also be broadened to apply to nonautonomous systems, i.e., systems where time dependence appears explicitly in the right-hand side of Equation 10.3.

10.3 COMPUTATIONAL CONSIDERATIONS

From a computer point of view, the optimization may now be viewed as: (1) simultaneous solution of the system and adjoint differential equations, and (2) computation and optimization of the Hamiltonian over the admissible controls $\mathbf{u}(t)$. Since there are $n + 1$ first-order system differential equations and also $n + 1$ adjoint first-order differential equations, it is evident that $2n + 2$ independent boundary conditions are needed to obtain a unique solution. The boundary conditions are obtained as follows:

(a) n initial conditions for the system at $t = t_0$ are given.
(b) If the target state is fixed, n additional final conditions at $t = t_f$ are given. If the target is not a fixed point but a region, then a combination of target specifications and transversality conditions is used to obtain n conditions on the system and adjoint variables at $t = t_f$.
(c) By definition, the cost function begins at zero so that

$$x_0(t_0) = 0$$

(d) From Equation 10.9, $p_0(t_f) = $ constant (here taken as -1) provides the last boundary condition.

Thus there are indeed enough boundary conditions, but unfortunately some conditions apply at $t = t_0$ and others at $t = t_f$. No initial conditions for the adjoint equation are known. Consequently the optimization

† The Maximum Principle is also a sufficient condition when the system and the cost function are linear in both the state variables and the control.

problem has been transformed into the solution of a nonlinear two-point boundary value problem. Since computers only solve initial value problems, some type of iteration scheme is required. Furthermore, this iteration is troubled by problems of sensitivity and scaling.

The scaling problem arises from the fact that if the system equations are stable, the adjoint equations are unstable, or vice versa. This difficulty is easily seen if we consider a stable first-order system, say

$$\dot{x} = -ax(t) + bu(t)$$

where $a > 0$. The solution of the homogeneous equation is

$$x(t) = e^{-at}x(0)$$

However, from Equation 10.11, the adjoint equation is

$$\dot{p} = ap$$

the solution of which is

$$p = e^{+at}p(0)$$

Thus, if the terminal time is unknown, the magnitude of the exponential terms may be very difficult to estimate and a separate scaling subroutine (or iteration loop) may be required if these equations are solved on analog computers. The large range of floating-point variables in digital computers generally eliminates this problem in digital solution.

The sensitivity problem makes analog solution of system and adjoint equations very difficult. This can be seen from the expression in the preceding paragraph. If the unknown initial conditions of the adjoint are denoted by $p(0)$, then the sensitivity of the solution of the adjoint equation to perturbation of these initial conditions is given by

$$\mu_0 \triangleq \frac{\partial p(t)}{\partial p(0)}$$

The sensitivity function μ_0 can be obtained as the solution of the sensitivity equation

$$\dot{\mu}_0 = a\mu_0, \; \mu_0(0) = 1$$

which is

$$\mu_0(t) = e^{at}$$

Consequently, at the terminal time,

$$\mu_0(t_f) \triangleq \frac{\partial p}{\partial p(0)}(t_f) = e^{at_f}$$

Thus, the adjoint variables at t_f may be extremely sensitive to slight changes in the initial conditions. Highly sensitive problems are not well suited to the limited precision of analog computer solutions. And yet, as will be seen below, hybrid solution of optimization problems normally requires that the dynamic equations be solved on the analog computer and that the iteration logic and criterion functions be evaluated digitally. Consequently, hybrid solution of optimal control problems always requires considerable care in the areas of scaling and accuracy.

To reduce the problems outlined above, it is sometimes possible to integrate the system equations *forward* in time (with assumed initial conditions for the adjoint equations) and then integrate the adjoint equations *backward* in time to obtain better initial conditions.

A computational solution of the optimal control problem may involve a double set of iterations:

1. An iteration procedure is needed to find the unknown *initial* conditions for the adjoint equations.
2. An iteration procedure is needed to ensure that out of the possible trajectories which satisfy the boundary conditions, only the one which maximizes the Hamiltonian is selected. When the Hamiltonian can be maximized with respect to $\mathbf{u}(t)$ analytically (e.g., by differentiating), then only a single set of iterations is required.

In block diagram form, the optimal control problem may now be viewed, as in Figure 10.2. The analog integrators are shown for convenience only; in a specific case, the integrations may be done digitally.

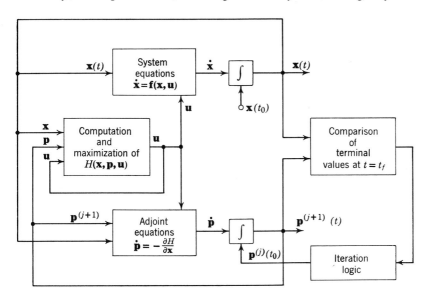

Figure 10.2 Block diagram of optimization process.

In the following pages, the basic method shown will be discussed in detail in connection with the hybrid computer solution of several specific problems.

10.4 LINEAR TIME-OPTIMAL CONTROL PROBLEM

Consider now a system described by the linear equation

$$\dot{\mathbf{x}}(t) = A(t)\mathbf{x}(t) + B(t)\mathbf{u}(t); \quad \mathbf{x}(t_0) = \mathbf{x}_0 \quad (10.20)$$

It is desired to transfer this system from $\mathbf{x}_0$ to the origin of the state space (i.e., $\mathbf{x}(t_f) = \mathbf{0}$) in minimum time by using an amplitude constrained control $\mathbf{u}(t)$ such that

$$|u_i(t)| \leq M$$

Let us first examine the analytical foundation of the problem. The cost function is

$$J = \int_{t_0}^{t_f} dt \quad (10.21)$$

since the time interval $(t_f - t_0)$ is to be minimized. Note that t_f is unknown. As a result of our previous definition

$$L(\mathbf{x}, \mathbf{u}) = f_0(\mathbf{x}, \mathbf{u}) = 1 \quad (10.22)$$

The Hamiltonian for the system is

$$H = \sum_{i=0}^{n} p_i f_i \quad (10.23)$$

or

$$H = p_0 + \mathbf{p}^T A \mathbf{x} + \mathbf{p}^T B \mathbf{u} \quad (10.24)$$

where $\mathbf{p}$ has only n components and $f_0 = 1$ from Equation 10.22. From an inspection of this equation, it is clear that in order to maximize H by choice of $\mathbf{u}$, we want the product $\mathbf{p}^T B \mathbf{u}$ to be always as positive as possible (no matter what the other terms are since we have no direct control over them). This can be accomplished by always choosing each component of $\mathbf{u}$ to be at its maximum value (constrained to $\pm M$ by the statement of the problem) and of the same sign as the corresponding component of $\mathbf{p}^T B$, that is:

$$u_k^* = M \operatorname{sgn} \sum_{i=1}^{r} p_i B_{ik}, \quad k = 1, 2, \ldots, r \quad (10.25)$$

In other words, the optimal control is bang-bang, being equal to either $+M$ or $-M$, with only switching times being unknown. To obtain them, the adjoint variables must be calculated. Since the adjoint variables satisfy the canonical Equations 10.16, we have

$$\dot{p}_i = -\frac{\partial H}{\partial x_i}, \quad i = 0, 1, \ldots, n \qquad (10.26)$$

Now the computer problem is clear. The Maximum Principle gave us a form for the optimal control, but its numerical evaluation leads to a two-point boundary value problem. We have $(n + 1)$ system equations and $(n + 1)$ adjoint equations, with $(n + 1)$ given initial conditions

$$x_0(t_0) = 0, \quad x_i(t_0) = x_{i0}, \quad i = 1, 2, \ldots, n \qquad (10.27a)$$

and n final conditions

$$x_i(t_f) = 0, \quad i = 1, 2, \ldots, n \qquad (10.27b)$$

The remaining boundary condition is obtained from 10.19:

$$p_0(t_f) = \text{constant} \leq 0$$

and can be arbitrarily chosen as $p_0(t_f) = -1$.[22]

Alternative computer techniques for solution of the above system of equations differ primarily in the choice of iterative methods for finding the unknown *initial* conditions for the adjoint equations.

A. Solution by Trial and Error

For small problems the unknown initial conditions can be found by trial and error. For example, using a repetitive analog computer and a multichannel oscilloscope to display the state variables, the $p_i(t_0)$, $i = 1, 2, \ldots, n$ can be adjusted manually until the desired final conditions $x_i(t_f)$, $i = 1, 2, \ldots, n$ are achieved. Such a technique is described by Anderson and Gupta[1] for a linear third-order system with a minimum time performance index. For systems higher than third order, a trial-and-error method is doomed to failure.

B. Solution by Random Search

Maybach[15] has investigated the use of a modified random walk technique for finding the initial conditions of the adjoint equations. Random perturbations are applied to the assumed values of the $p_i(0)$, $i = 1, 2, \ldots, n$ until an appropriately selected criterion function is maximized. For the minimum time problem where $\mathbf{x}(t_f) = 0$, the function

$$F(t) = -\sum_{i=1}^{n} \alpha_i x_i(t)$$

where the α_i, $i = 1, 2, \ldots, n$ are weighting factors, was used. The computation is stopped when any of the x_i reach zero, and the value of $F^j(t)$ and the final time $t_f{}^j$ (where j represents the trial number) are recorded. If all the x_i go to zero simultaneously, then $F^j(t) = 0$ and $t_f{}^j = t^*$, the optimal time. Otherwise, the initial conditions are perturbed and a new run made. Using the extremely high computation rates of the University of Arizona ASTRAC II computer, convergence times of a few seconds were obtained for a second order system with several thousand runs. The approach was troubled by the high sensitivity of the solution to the initial conditions of the adjoint variables and solution of a third-order example led to severe scaling problems.

C. Solution by Neustadt's Method

A particularly elegant iteration technique for computing the initial conditions of the adjoint system is due to Neustadt.[27] The solution of the linear differential Equation 10.20 is given by

$$\mathbf{x}(t) = \Phi(t)\left[\mathbf{x}_0 + \int_0^t \Phi^{-1}(\tau)B(\tau)\mathbf{u}(\tau)\, d\tau\right] \tag{10.28}$$

where $\Phi(t)$ is the $n \times n$ nonsingular fundamental (or state transition) matrix of the system.

If we let $t = t_f$ and let $\mathbf{x}(t_f) = \mathbf{0}$, then Equation 10.23 can be solved for $\mathbf{x}_0$ to yield

$$\mathbf{x}_0 = -\int_0^{t_f} \Phi^{-1}(\tau)B(\tau)\mathbf{u}(\tau)\, d\tau \tag{10.29}$$

In the time-optimal problem the initial state $\mathbf{x}_0$ is given, but the minimum time t_f is unknown. Equation 10.29 represents all those initial states from which the solution can be brought to the origin in time t_f. The set of such initial conditions (for admissible controls) (known as the *recoverable set*) has certain important mathematical properties on which the iteration is based. The time-optimal control problem may now be stated as follows: Find an admissible control $\mathbf{u}(t)$ which satisfies 10.29 for a given $\mathbf{x}_0$ and for the smallest possible t_f.

From the previous discussion, it may be recalled that the form of the optimal control for this problem was given by Equation 10.25. To emphasize the dependence of the optimal control vector on the choice of initial conditions, we can write Equation 10.25 as

$$u_k(t, \boldsymbol{\eta}) = M \operatorname{sgn}\left(\mathbf{p}^T(t, \boldsymbol{\eta})B(t)\right)_k, \qquad k = 1, 2, \ldots, r \tag{10.30}$$

where $\boldsymbol{\eta} = \mathbf{p}(t_0)$. Substituting the control obtained with a trial initial condition into the right-hand side of Equation 10.29, we obtain the quantity

$$\mathbf{z}(t, \boldsymbol{\eta}) = -\int_0^t \Phi^{-1}(\tau) B(\tau) \mathbf{u}(\tau, \boldsymbol{\eta}) \, d\tau \qquad (10.31)$$

Evidently, if the correct $\mathbf{p}(t_0)$ is used, $\mathbf{z}(t, \boldsymbol{\eta})$ will equal $\mathbf{x}(t_0)$. Based on the properties of the recoverable set, Neustadt has proposed the following iteration scheme:

1. Choose an initial condition vector, say $\boldsymbol{\eta}^j$ for the (first) j-th trial, ($j = 1$).
2. Integrate the equation to find $\mathbf{z}(t, \boldsymbol{\eta}^j)$ and stop the integration when

$$\langle \boldsymbol{\eta}^j, (\mathbf{z}(t, \boldsymbol{\eta}^j) - \mathbf{x}_0) \rangle = 0 \qquad (10.32)$$

where the symbol $\langle , \rangle$ indicates the inner product. This time is denoted by $T(\boldsymbol{\eta}^j)$
3. The $(j + 1)$-st initial condition is obtained from

$$\boldsymbol{\eta}^{j+1} = \boldsymbol{\eta}^j - k[\mathbf{z}(T(\boldsymbol{\eta}^j), \boldsymbol{\eta}^j) - \mathbf{x}_0] \qquad (10.33)$$

where k is a (scalar) scalefactor.

The objective of this calculation is to find the $\boldsymbol{\eta}$ which maximizes the $T(\boldsymbol{\eta})$. The initial condition vector which does this will also cause the boundary conditions to be satisfied.

The basic technique alone, proposed by Neustadt[27,28] has been implemented on hybrid computers by Gilbert,[8,9] and Paiewonsky et al.[18] In block diagram form, the solution technique is shown in Figure 10.3; a flow chart of the digital program is shown in Figure 10.4. The computation of the optimum scalefactor k in Equation 10.33 has been shown to be a difficult problem.[18] The references also indicate problems encountered in timing and control, as well as accuracy requirements in the determination of the zeros of Equation 10.32.

To illustrate the nature of the solution, an iterative analog simulation of the above method has been used for solution of a simple second-order problem with a well-known solution[4,18]:

EXAMPLE. Consider the second-order system

$$\ddot{x} + \omega^2 x = u \qquad (10.34)$$

or, equivalently

$$\begin{aligned} \dot{x}_1 &= x_2, & x_1(0) &= c_1 \\ \dot{x}_2 &= -\omega^2 x_1 + u, & x_2(0) &= c_2 \end{aligned} \qquad (10.35)$$

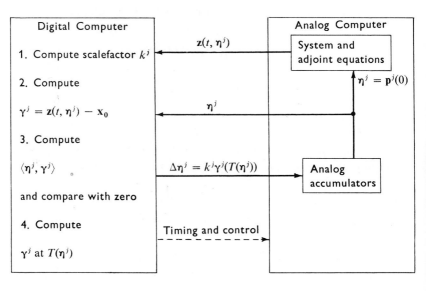

Figure 10.3 Hybrid solution of time-optimal problem by Neustadt's method.

where the magnitude of the control is constrained:

$$|u| \leq 1 \tag{10.36}$$

It is desired to drive the solution to the point $x_1(t_f) = 0$, $x_2(t_f) = 0$ in minimum time. The optimal control for this problem is given by

$$u^* = \text{sgn}\left[-\eta_1 \frac{\sin \omega t}{\omega} + \eta_2 \cos \omega t\right] \tag{10.37}$$

It is easy to show that the Equations 10.31 for this problem become:

$$\left.\begin{aligned} z_1(t, \boldsymbol{\eta}) &= \int_0^t \left(\frac{\sin \omega \tau}{\omega}\right) \mathbf{u}(\tau, \boldsymbol{\eta}) \, d\tau \\ z_2(t, \boldsymbol{\eta}) &= -\int_0^t (\cos \omega \tau) \mathbf{u}(\tau, \boldsymbol{\eta}) \, d\tau \end{aligned}\right\} \tag{10.38}$$

The first set of initial conditions was chosen as

$$\boldsymbol{\eta}^1 = \frac{-\mathbf{x}(0)}{\|\mathbf{x}(0)\|} \tag{10.39}$$

where the symbol $\|.\|$ represents the Euclidean norm of the vector defined as

$$\|x\| = (x_1^2 + x_2^2)^{1/2}$$

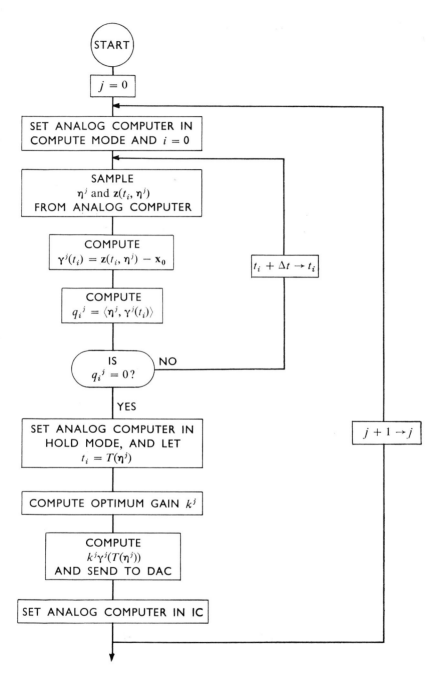

Figure 10.4 Flow chart for digital portion of time-optimal control problem.

The quantity defined in Equation 10.32 becomes

$$\eta_1^j[z_1(t, \boldsymbol{\eta}^j) - x_1(0)] + \eta_2^j[z_2(t, \boldsymbol{\eta}^j) - x_2(0)] \leq 0 \qquad (10.40)$$

This quantity is monitored; when it equals zero, the analog computer is placed into hold, the time is sampled, and the next step in the iteration computed according to the rule of 10.33.

Computational results on this problem are discussed by Paiewonsky.[18]

10.5 SOLUTION BY MEANS OF SENSITIVITY FUNCTIONS

An interesting approach to the evaluation of unknown initial conditions for the adjoint equations based on the use of sensitivity equations has been used by Halbert, Landauer, and Witsenhausen.[10,21] The basic idea was used by Brunner[3] for analog solution of boundary-value problems, using the "influence coefficient" method of Meissinger.[29]

To illustrate the idea, consider a second-order system given by

$$\dot{x}_1 = f_1(x_1, x_2), \qquad \dot{x}_2 = f_2(x_1, x_2) \qquad (10.41)$$

with the boundary conditions

$$x_2(0) = x_{20}, \qquad x_1(t_f) = x_{1f}$$

Thus, we have split boundary conditions with x_{10} being unknown. The dependence of the desired final values x_f on the unknown initial condition can be expressed by the relationship

$$x_{1f} = x_1(x_{10}, x_1(t_f), t_f) \qquad (10.42)$$

If we perturb the unknown initial condition by δx_{10}, the resulting solution at $t = t_f$ will be perturbed by δx_{1f}. The perturbed equation can be written as

$$x_{1f} = x_1(x_{10} + \delta x_{10}, x_1(t_f) + \delta x_{1f}, t_f) \qquad (10.43)$$

Expanding in a Taylor series and keeping only the first two terms yields:

$$x_{1f} = x_1(x_{10}, x_1(t_f), t_f) + \frac{\partial x_1}{\partial x_{10}}\bigg|_{t=t_f} \delta x_{10} + \delta x_{1f} \qquad (10.44)$$

Subtracting the nominal solution 10.43 from the perturbed solution 10.44 yields the variational equation

$$\frac{\partial x_1}{\partial x_{10}}\bigg|_{t=t_f} \delta x_{10} + \delta x_{1f} = 0 \qquad (10.45)$$

10.5/SOLUTION BY MEANS OF SENSITIVITY FUNCTIONS

If the sensitivity coefficient is denoted by α_{11}, Equation 10.45 can be written as

$$\delta x_{10} = -\alpha_{11}^{-1}\delta x_{1f} \tag{10.46}$$

That is, if in response to an arbitrary initial condition $x_1(0)$, we obtain a final value which differs from the known final value x_{1f} by δx_{1f}, Equation 10.46 tells how much the initial condition was in error, within the limits of the approximation incurred by keeping only the first two terms of the Taylor's series. For linear systems, the solution is exact; for nonlinear systems, iteration is required, with the iteration algorithm being given by:

$$x_1^{(j+1)}(0) = x_1^{(j)}(0) - \alpha_{11}^{-1}\delta x_{1f} \tag{10.47}$$

where j denotes the iteration step and α_{11} is the sensitivity coefficient which must be computed.

Applying this technique to the optimal control problem only requires the assumption that perturbations of the unknown initial condition of the adjoint variables can be related to perturbations in the final values of the state variables by the relation

$$A_f \delta \mathbf{p}_0 = -\delta \mathbf{x}_f \tag{10.48}$$

where $\mathbf{p}_0$ and $\mathbf{x}_f$ are the n-dimensional adjoint vector at $t = t_0$ and state vector at $t = t_f$, respectively. The $n \times n$ matrix A_f (the sensitivity matrix) has coefficients α_{ij} defined by

$$\alpha_{ij} = \left.\frac{\partial x_i}{\partial p_j(0)}\right|_{t=t_f} \tag{10.49}$$

Solving for $\delta \mathbf{x}_0$ and assuming that A_f is nonsingular

$$\delta \mathbf{p}_0 = -A_f^{-1}\delta \mathbf{x}_f \tag{10.50}$$

which is a generalization of 10.46. The iteration algorithm now becomes

$$\mathbf{p}_0^{j+1} = \mathbf{p}_0^{j} - A_f^{-1}\delta \mathbf{x}_f^{j} \tag{10.51}$$

It is evident that the sensitivity problems cited in Section 10.3 may still cause computational difficulties. However, the method has been used successfully to study orbital transfer problems of space vehicles.[10]

In block diagram form, the computer mechanization of this system is shown in Figure 10.5. All the basic operations shown can be implemented on the analog computer, with the digital computer providing the timing, control, and storage of the sensitivity matrix A_f. This matrix must be recomputed periodically if the initial condition vector moves far from its assumed value. This can be done by disconnecting the iteration loops,

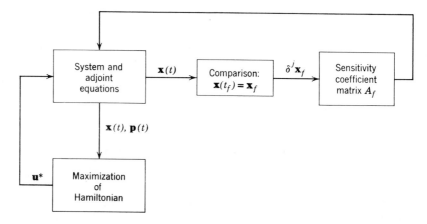

Figure 10.5 Block diagram of computer solution of optimal control problem by sensitivity method.

perturbing the initial conditions by $\Delta p_i(0)$, $i = 1, 2, \ldots, n$, and recording the resulting changes in $\mathbf{x}(t_f)$, namely $\Delta x_i(t_f)$, $i = 1, 2, \ldots, n$. Evidently, inversion of A_f is most conveniently done digitally.

10.6 SOLUTION BY CONTINUOUS STEEPEST DESCENT

In the time-optimal control problems considered in Section 10.4, it has been possible to solve explicitly for the control $\mathbf{u}$ which extremizes the Hamiltonian, thus reducing the problem to one of searching for boundary conditions. In general, this is not possible, and a dual iteration process may be required:

(a) an iteration to satisfy the far-end boundary conditions, and
(b) an iteration to satisfy the requirement that $H(\mathbf{x}, \mathbf{p}, \mathbf{u})$ be maximized for all admissible $\mathbf{u}$.

Analog steepest descent techniques (see Chapter 9) can be utilized to perform (b) continuously by implementing the relationships

$$\frac{du_i}{dt} = K \frac{\partial H}{\partial u_i}, \quad i = 1, 2, \ldots, n \quad (10.52)$$

where K is a suitable large scalar constant. The integrations are performed with nonresetting integrators so that the optimization of H with respect to $\mathbf{u}$ can be performed for all time, including the initial conditions. This approach has been proposed by Korn[13] and de Backer[6] and implemented

10.6/SOLUTION BY CONTINUOUS STEEPEST DESCENT

by Maybach[15,16] on a hybrid computer. The technique is illustrated in block diagram form in Figure 10.6. A one-dimensional problem was solved by Steinmetz on an analog computer.[20] Maybach[26] has shown that continuous steepest descent has certain computational advantages, even in certain situations where an explicit form for $\mathbf{u}(\mathbf{x}, \mathbf{p})$ can be obtained. We shall examine the application of the method to two specific examples.

A. Linear Time-Optimal Control[15,16]

Given the system

$$\ddot{x} = u; \quad |u| \leq 1; \quad \dot{x}(t_f) = x(t_f) = 0; \quad x(0) = c_1; \quad \dot{x}(0) = c_2 \tag{10.53}$$

and

$$J(u) = \int_0^{t_f} dt \tag{10.54}$$

In state variable notation, the system equation becomes

$$\dot{x}_0 = 1, \quad \dot{x}_1 = x_2, \quad \dot{x}_2 = u \tag{10.55}$$

and the adjoint equations

$$\dot{p}_0 = 0, \quad \dot{p}_1 = 0, \quad \dot{p}_2 = -p_1 \tag{10.56}$$

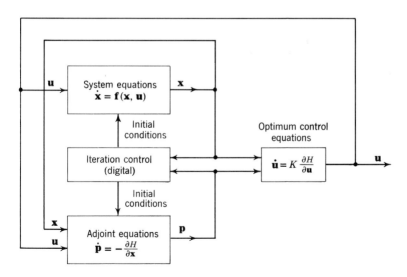

Figure 10.6 Solution by continuous steepest descent.

The Hamiltonian for this problem is

$$H = \sum_{i=0}^{2} p_i f_i = p_0 + p_1 x_2 + p_2 u \qquad (10.57)$$

Applying the optimization rule of Equation 10.52, we obtain the adjustment equation

$$\frac{du}{dt} = K \frac{\partial H}{\partial u} = K p_2, \qquad |u| \leq 1 \qquad (10.58)$$

Note that as K becomes large, the performance of the circuit will in fact approach the known solution

$$u = \text{sgn } p_2 \qquad (10.59)$$

The usual problems of sensitivity are encountered in searching for the unknown initial conditions of the adjoint system. Maybach[16] has shown that convergence can be obtained in 1 to 70 seconds, depending on the initial conditions. (Using the extremely high-speed hybrid system at the University of Arizona, as many as 35,000 iterations were required.) The unknown initial conditions on the adjoint system and the optimal time were found by a random search routine.[26]

B. Optimization of Product Yield in a Chemical Reaction System[20]

Consider a tubular chemical reactor of length L in which a product A decomposes into a product B which in turn decomposes into a product C, with reaction rates k_1 and k_2, respectively.

$$A \xrightarrow{k_1} B \xrightarrow{k_2} C$$

The reactions are assumed irreversible. The reaction rates are temperature dependent, being given by

$$k_i = G_i e^{-E_i/RT}, \qquad i = 1, 2 \qquad (10.60)$$

where the G_i are constants, the E_i are activation energies for the components in question, and R is the gas constant. E_1, E_2, G_1, G_2, and R are known in advance. The optimization problem is to find the temperature profile $T(y)$, where y is distance along the reactor, which maximizes the yield of product B at $y = L$.

To use the Maximum Principle on this problem, we define the following variables:

$x_1(y)$ = concentration of product A at point y

$x_2(y)$ = concentration of product B at point y

The concentration of C is obtained simply by performing a material balance knowing the total inlet flow since no material is destroyed in the reactor.

The system equations then become

$$\frac{dx_1}{dy} = -k_1(T)x_1$$

$$\frac{dx_2}{dy} = k_1(T)x_1 - k_2(T)x_2 \tag{10.61}$$

with initial conditions (at the inlet) being given as

$$x_1(0) = x_{10} \quad \text{and} \quad x_2(0) = x_{20}$$

The performance criterion is

$$J = x_2(y_f) \tag{10.62}$$

where $y_f = L$ refers to the final reactor length (the outlet). The adjoint equations are

$$\frac{dp_1}{dy} = k_1(T)(p_1 - p_2)$$

$$\frac{dp_2}{dy} = k_2(T)p_2 \tag{10.63}$$

where the final values of the adjoint variables are known since this is a free-end point problem,[22] to be

$$p_1(y_f) = 0, \quad p_2(y_f) = -1$$

The Hamiltonian for the problem is

$$H = \sum_{i=1}^{2} p_i f_i = -p_1 x_1 k_1(T) + p_2 x_1 k_1(T) - p_2 x_2 k_2(T) \tag{10.63a}$$

For computer simulation, we let computer time t take the place of the distance variable y. The generation of the optimal temperature profile, $T(t)$ is then obtained from 10.52:

$$\frac{dT}{dy} = -K\frac{\partial H}{\partial T} = -K\left[\frac{\partial H}{\partial k_1}\frac{\partial k_1}{\partial T} + \frac{\partial H}{\partial k_2}\frac{\partial k_2}{\partial T}\right] \tag{10.64}$$

$$\dot{T} = K\left[(p_1 x_1 - p_2 x_1)\frac{\partial k_1}{\partial T} + p_2 x_2 \frac{\partial k_2}{\partial T}\right] \tag{10.65}$$

where the dependence of k_1 and k_2 on T is given by Equation 10.60.

For computer solution, the equations to be solved include: (1) the system Equations 10.61, (2) the adjoint Equations 10.63, and (3) the control Equation 10.65. Iteration techniques are used to generate the unknown initial conditions of the adjoint system. Results are given in the paper by Steinmetz.[20]

10.7 OTHER ANALOG AND HYBRID TECHNIQUES

A number of other techniques for using analog and hybrid computers for the solution of optimal control problems have been investigated. Several such methods will be outlined briefly in this section.

Rather than using steepest descent to extremize the Hamiltonian at each point along the trajectory (as done in the preceding section), Lee[14] discretizes time along the trajectory and, at each point, extremizes the Hamiltonian by a random search technique. This technique involves a local iteration loop at each point along the trajectory, which is nested within the major iteration loop for finding unknown initial conditions of the adjoint equations.

An interesting iterative computation procedure for the solution of linear control problems with criterion function

$$J = \mathbf{x}^T(t_f)\mathbf{x}(t_f) = \sum_{i=1}^{n} x_i^2(t_f) \qquad (10.66)$$

has been proposed by Fancher. The application of the technique was studied by the use of an analog computer for a third-order plant. A variation of the technique for the solution of time-optimal problems is also discussed.[7]

Miura, Tsuda, and Iwata[34] have proposed a method of implementing the Maximum Principle on a hybrid computer which guarantees convergence of the iteration process, even in certain cases when the initial values chosen for the adjoint equations are far from the correct values.

A paper by Anderson and Gupta[35] claims that iteration can be avoided completely by using "penalty function"[22] techniques and solving the problem backwards in time. Unfortunately, this is not true in general and the conclusions of this paper are erroneous.

10.8 GRADIENT METHODS IN FUNCTIONAL OPTIMIZATION

The use of gradient methods for the optimization of ordinary functions was discussed in some detail in Chapter 9. It will be recalled that given a function $F(\mathbf{a})$, where $\mathbf{a}$ is an n-parameter vector,

$$\mathbf{a} = (a_1, a_2, \ldots, a_n) \qquad (10.67)$$

the gradient was defined by

$$\nabla F(\mathbf{a}) = \left[\frac{\partial F}{\partial a_1} \frac{\partial F}{\partial a_2} \cdots \frac{\partial F}{\partial a_n}\right]^T \qquad (10.68)$$

Consequently, the *direction* of steepest descent can be defined by a unit

10.8/GRADIENT METHODS IN FUNCTIONAL OPTIMIZATION

vector **l** normal to the contour surfaces of F, where

$$l_i = \frac{\partial F/\partial a_i}{\sqrt{\sum_{i=1}^{n}(\partial F/\partial a_i)^2}}, \quad i = 1, 2, \ldots, n \quad (10.69)$$

The discrete gradient methods were based on substituting for the i-th trial value of the parameter vector, $\mathbf{a}^{(i)}$, the new value

$$\mathbf{a}^{(i+1)} = \mathbf{a}^{(i)} - K_1 \nabla F(\mathbf{a}^{(i)}) \quad (10.70)$$

where K_1 is a positive scalar constant.

It is now reasonable to ask whether such an approach may be extended to the optimization of functions, i.e., to the determination of a sequence of control vectors $\mathbf{u}^{(i)}(t)$ which optimize a functional $J(\mathbf{u})$. The resulting gradient would define a direction in *function space* rather than in *parameter space* as before. Such an approach was originally suggested by Courant[1] and successfully implemented by Kelley,[30] Bryson and Denham,[31] and others.

The major problem with the development of the gradient in function space is that $J[\mathbf{u}(t)]$ is defined indirectly by the differential equations rather than directly as in parameter optimization problems. To overcome this difficulty, the adjoint equations are introduced and the optimization is based on the Hamiltonian.

To illustrate the method in somewhat more detail, let the system be described by

$$\dot{\mathbf{x}}(t) = \mathbf{f}[\mathbf{x}(t), u(t)], \quad \mathbf{x}(t_0) = \mathbf{x}_0 \quad (10.71)$$

where $\mathbf{x}(t)$ is an $(n \times 1)$ vector and, for simplicity, we use a scalar control $u(t)$. The extension to the vector control case is found in the references.[30,31] We again introduce the adjoint variables:

$$\dot{p}_i = -\sum_{j=1}^{n} \frac{\partial f_j}{\partial x_i} p_j, \quad i = 1, 2, \ldots, n \quad (10.72)$$

The objective of the optimization will be to find the minimum of a function of the final values of the state variables, i.e., the cost function

$$J = J[\mathbf{x}(t_f)] \quad (10.73)$$

Let us assume that t_f (and t_0) are given. The problem then could represent a fixed-time rendezvous or landing maneuver with the objective of finding an angle-of-attack program which minimizes some function of the position, velocity, and acceleration at the terminal time.

Assume now that we have an initial estimate of a reasonable control, say $u^0(t)$. Substituting this into Equation 10.71 will yield a *nominal trajectory* $\mathbf{x}^0(t)$. The resulting cost will be $J^0 = J[\mathbf{x}^0(t_f)]$. If we modify the control by adding a perturbation $\delta u(t)$, we have

$$u^{(1)}(t) = u^0(t) + \delta u(t) \tag{10.74}$$

and the resulting trajectory will be

$$\mathbf{x}^{(1)}(t) = \mathbf{x}^0(t) + \delta \mathbf{x}(t) \tag{10.75}$$

where $\delta \mathbf{x}(t)$, the trajectory perturbation, can be obtained as the solution of the perturbation equation

$$\delta \dot{\mathbf{x}}(t) = F(t)\delta \mathbf{x}(t) + G(t)\delta u(t) \tag{10.76}$$

where $F(t)$ is the $(n \times n)$ matrix of partials $\partial f_i/\partial x_j$ and $G(t)$ is the $(n \times 1)$ matrix of partials $\partial f_i/\partial u$, all evaluated along the nominal trajectory. Evidently we seek a control perturbation $\delta u(t)$ such that the resulting change in the cost function,

$$\delta J = J[\mathbf{x}^{(1)}(t_f)] - J[\mathbf{x}^0(t_f)] \tag{10.77}$$

is negative and as large as possible.

It is easy to show that by solving the perturbation equation, we get

$$\delta J = \int_{t_0}^{t_f} \left(\sum_{i=1}^{n} p_i \frac{\partial f_i}{\partial u} \right) \delta u \, dt + \sum_{i=1}^{n} p_i \delta x_i \bigg|_{t=t_0} \tag{10.78}$$

The adjoint variables $p_i(t)$ are determined by solving Equations 10.72 backwards in time with the "initial" conditions

$$p_i(t_f) = \frac{\partial J}{\partial x_i}\bigg|_{t=t_f} \tag{10.79}$$

Now, if there are no constraints on $\mathbf{x}(t_f)$, we can see by inspection of Equation 10.78 that the greatest change in δJ will be produced by choosing

$$\delta u(t) = K \sum_{i=1}^{n} p_i \frac{\partial f_i}{\partial u} \tag{10.80}$$

where K is a constant. It can be noted, by reference to the definition of the Hamiltonian (10.15), that 10.80 can be written as

$$\delta u(t) = K \frac{\partial H}{\partial u} \tag{10.81}$$

The change $\delta u(t)$ leads in the "steepest descent" direction to the minimum J. Typically, we look for a small change of J, say $|\delta J|^{(i)} = .05$ for the i-th iteration. This restriction enables us to find the constant K in 10.80

10.8/GRADIENT METHODS IN FUNCTIONAL OPTIMIZATION

by substituting in 10.78:

$$K = \frac{\delta J}{\int_{t_0}^{t_f}\left[\dfrac{\partial H}{\partial u}\right]^2 dt} \qquad (10.82)$$

where $\delta J = -.05 J^0$. Substituting this scalefactor into 10.81 yields

$$\delta u(t) = \frac{\partial H/\partial u}{\int_{t_0}^{t_f}\left[\dfrac{\partial H}{\partial u}\right]^2 dt}\, \delta J \qquad (10.83)$$

It is interesting to compare this "direction" with the steepest descent direction in parameter space defined by 10.69.

In summary, the computational solution of the problem by functional steepest descent consists of the following steps:

STEP 1. Assume an initial control $u^0(t)$.
STEP 2. Integrate the system Equation 10.71 forward from t_0 to t_f in order to evaluate the partial derivatives $\partial f_i/\partial u$ and $\partial f_j/\partial x_i$ along the nominal trajectory.

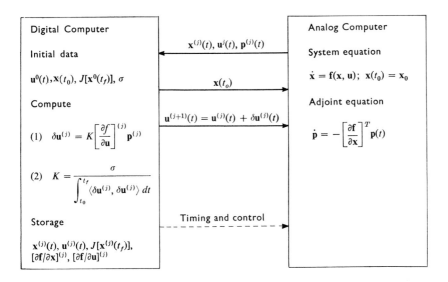

Figure 10.7 Hybrid implementation of discrete iterative gradient method of trajectory optimization.

STEP 3. Solve the adjoint Equation 10.72 backwards in time using the boundary conditions 10.79.

STEP 4. Using the partial derivatives of Step 2 and the adjoint variables from Step 3, evaluate the increment in the control $\delta u(t)$ using 10.80. (The scalefactor K in 10.80 may have to be evaluated first. An additional check may be required to insure that the perturbation $\delta u(t)$ is sufficiently small to justify the use of first-order perturbation equations.)

STEP 5. Substitute $u^0(t) + \delta u(t)$ for $u^0(t)$ and repeat the process.

Techniques of the type outlined above have been used very successfully in the optimization of reentry trajectories and similar problems. However, due to the storage requirements involved in the iteration from $u^{(i)}(t)$ to $u^{(i+1)}(t)$, the solution of such problems by the method outlined above has been done almost exclusively by digital computers. A block diagram of a possible hybrid solution is shown in Figure 10.7.

10.9 APPROXIMATE STEEPEST DESCENT BY FAST-TIME REPETITIVE COMPUTATION

A technique reported by Wingrove and Raby[32,33] is well suited to the hybrid computer implementation of the steepest descent method in function space, because it does not require the solution of the adjoint equations. Rather than find a change in the cost function, δJ, analytically, as in the preceding section, this method is based on introducing small perturbation "impulses" into the control $u^0(t)$, as illustrated in Figure 10.8. The system equations are solved with a control impulse of amplitude Δu and width

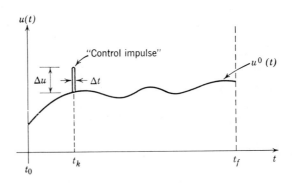

Figure 10.8 Control "impulse" at time t_j during repetitive computation.

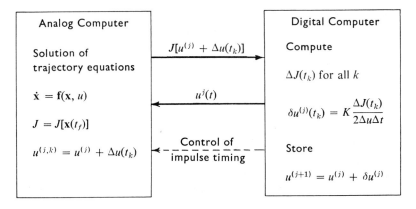

Figure 10.9 "*Control impulse*" *method of implementing steepest descent trajectory optimization.*

Δt at time t_k where $t_0 \leq t_k \leq t_f$. The solution is then repeated with a negative impulse at time t_k. The resulting changes in the cost function can yield an estimate of the perturbation δJ. Calling the perturbation ΔJ, we have

$$2 \, \Delta J(t_k) = J(u^0 + \Delta u) - J(u^0 - \Delta u) \qquad (10.84)$$

From Equation 10.78, for a sufficiently small Δu and Δt, it can be seen that

$$\sum_{i=1}^{n} p_i \frac{\partial f_i}{\partial u} \bigg|_{t=t_k} = \frac{\partial H}{\partial u} \bigg|_{t=t_k} \cong \frac{\Delta J(t_k)}{2 \, \Delta u \, \Delta t} \qquad (10.85)$$

Thus, repeated solution of the system equations with "control impulses" introduced at a sufficient number of points in the interval $[t_0, t_f]$ can be used to obtain the increment δu in the control function. The technique can be readily extended to cases where there are final value constraints and where the control vector has more than one component. A block diagram of the mechanization is shown in Figure 10.9.

While this technique has been used successfully with hybrid computers, its accuracy, convergence, and solution time require further study.

10.10 SUMMARY AND EVALUATION

A critical evaluation of the analog and hybrid computer techniques presented above indicates that all of them suffer from the serious limitations inherent in the limited precision of analog computers. The Maximum

Principle makes it possible to reduce the search for an optimum control to a search for the optimum values of a finite set of parameters, namely a set of initial or final conditions needed to solve the two-point boundary value problem. Such iteration appears to be well suited to analog computers since it requires the repetitive high-speed solution of a set of differential equations. However, the nature of the system equations is such that either the forward or reverse equations suffer from sensitivity problems, as indicated in Section 10.3, and the high speed of the repetitive analog computer needs to be balanced with its limited precision.

On the other hand, perhaps the most successful digital computation techniques are based on steepest descent in function space which requires the storage of large amounts of information from iteration to iteration. While the hybrid implementation of these methods shows considerable promise, further study is required before the application of hybrid techniques to optimum control problems can gain widespread acceptance.

References

1. Courant, R., "Variational Methods for the Solution of Problems of Equilibrium and Vibrations," *Bull. Amer. Math. Soc.*, **49**, 1–23, Jan. 1943.
2. Athans, M., "The Status of Optimal Control Theory and Applications for Deterministic Systems," *IEEE Transactions on Automatic Control*, **AC-11**, 580–595, July 1966.
3. Brunner, W., "An Iteration Procedure for Model Building and Solution of Boundary Value Problems," *Proc. Fall Joint Computer Conference*, pp. 519–533, 1961.
4. Balakrishnan, A. V., and L. W. Neustadt, eds., *Computing Methods in Optimization Problems*, Academic Press, New York, 1964.
5. Darcy, V. J., and R. A. Hannen, "An Application of an Analog Computer to Solve the Two-Point Boundary Value Problem for a Fourth-Order Optimal Control Problem," *IEEE Transactions on Automatic Control*, **AC-12**, 67–74, Feb. 1967.
6. de Backer, W., "The Maximum Principle, its Computational Aspects and its Relation to Other Optimization Techniques," *EURATOM Report* 590e, European Atomic Energy Commision, Ispra, Italy, 1964.
7. Fancher, P. S., "Iterative Computation Procedures for an Optimum Control Problem," *IEEE Transactions on Automatic Control*, **AC-10**, 346–348, July 1965.
8. Gilbert, E. G., "Hybrid Computer Solution of Time-Optimal Control Problems," *Proc. AFIPS Conference*, **23**, 197–204, 1963.
9. Gilbert, E. G., "The Application of Hybrid Computers to the Iterative Solution of Optimal Control Problems," in *Computing Methods in Optimization Problems* (A. V. Balakrishnan and L. W. Neustadt, eds.), Academic Press, New York, pp. 261–284, 1964.

REFERENCES

10. Halbert, P. W., J. P. Landauer, and H. S. Witsenhausen, "Hybrid Solution of Adaptive Path Control," *Proc. AIAA Simulation for Aerospace Flight Conference*, pp. 335-349, 1963 (reprinted in *Simulation*, **2**, R24-R42, June 1964).
11. Howard, D. R. and Z. V. Rekasius, "Error Analysis with the Maximum Principle," *IEEE Transactions on Automatic Control*, **AC-9**, 223-229, July 1964.
12. Knapp, C. H., and P. A. Frost, "Determination of Optimal Control and Trajectories Using the Maximum Principle in Association with a Gradient Technique," *IEEE Transactions on Automatic Control*, **AC-10**, 189-193, April 1965.
13. Korn, G. A., "Enforcing Pontryagin's Maximum Principle by Continuous Steepest Descent," *IEEE Transactions on Electronic Computers*, **EC-13**, 475-476, Aug. 1964.
14. Lee, E. S., "Optimizing Pontryagin's Maximum Principle on the Analog Computer," *Proc. Joint Automatic Control Conference*, pp. 524-531, 1963.
15. Maybach, R. L., "Solution of Optimal Control Problems on a High Speed Hybrid Computer," *Simulation*, **7**, 238-245, Nov. 1966.
16. Maybach, R. L., "Optimum Control by Pontryagin on the Hybrid Computer," *IEEE Region VI Conf. Record*, **1**, 410-419, 1966.
17. Mitchell, B. A., "A Hybrid Analog-Digital Parameter Optimizer for ASTRAC II," *Proc. Spring Joint Computer Conference* (reprinted in *Simulation*, **4**, 179-187, Mar. 1965).
18. Paiewonsky, B. et al., "Synthesis of Optimal Controllers Using Hybrid Analog-Digital Computers," in *Computing Methods in Optimization Problems* (A. V. Balakrishnan and L. W. Neustadt, eds.), Academic Press, New York, pp. 285-303, 1964.
19. Pontryagin, L. S., V. C. Boltyanskii, R. V. Gamkrelidze, and E. F. Mischenko, *The Mathematical Theory of Optimal Processes*, John Wiley, New York, 1962.
20. Steinmetz, H. L., "Using Pontryagin's Maximum Principle to Solve One-Dimensional Optimization Problems, With and Without Constraints, on an Iterative Analog Computer," *Simulation*, **4**, 382-389, June 1965.
21. Witsenhausen, H., "A Heuristic Approach to the Maximum Principle," *Simulation*, **2**, 25-30, June 1964.
22. Athans, M. and P. Falb, *Optimal Control*, McGraw-Hill, New York, 1966.
23. Feldbaum, A. A., *Optimal Control Systems*, Academic Press, New York, 1965.
24. Tou, J. T., *Modern Control Theory*, McGraw-Hill, New York, 1964.
25. Leitmann, G., *Optimization Techniques*, Academic Press, New York, 1962.
26. Maybach, R. L., "The Solution of Optimal Control Problems on an Iterative Hybrid Computer," Ph.D. Dissertation, Univ. of Ariz., Dept. of Elec. Eng., 1967.
27. Neustadt, L. W., "Synthesizing Time-Optimal Controls," *J. Math. Analysis and Applications*, **1**, 484-493, 1960.
28. Neustadt, L. W., "Time Optimal Control Systems with Position and Integral Limits," *J. Math. Analysis and Applications*, **3**, 406-427, 1963.
29. Meissinger, H. F., "The Use of Parameter Influence Coefficients in Computer Analysis of Dynamic Systems," *Proc. Western Joint Computer Conference*, **17**, 181-192, 1960.
30. Kelley, H. J., "Gradient Theory of Optimal Flight Paths," *J. Amer. Rocket Soc.*, **30**, 947-953, 1960.
31. Bryson, A. E., Jr., and W. F. Denham, "A Steepest Ascent Method of Solving Optimum Programming Problems," *J. Applied Mechanics*, **29**, Ser. E, 247-257, 1962.
32. Wingrove, R. C., J. S. Raby, and F. D. Crane, "A Method of Trajectory Optimization by Fast-Time Repetitive Computations," *NASA Technical Note TN-D-3404*, April 1966.

33. Wingrove, R. C., and J. S. Raby, "Trajectory Optimization Using Fast-Time Repetitive Computation," *Proc. AFIPS Computer Conference*, **29**, 799–808, 1966.

34. Miura, T., J. Tsuda, and J. Iwata, "Hybrid Computer Solution of Optimal Control Problems by the Maximum Principle," *IEEE Trans. Electronic Computers*, **EC-16**, 666–670, Oct. 1967.

35. Anderson, M. D., and S. C. Gupta, "Backward-time Analog Computer Solutions of Optimal Control Problems," *Proc. AFIPS Spring Computer Conference*, **30**, 133–139, 1967.

11

Random Processes

11.1 INTRODUCTION AND BASIC DEFINITIONS

The study of random processes offers a fertile field for the application of hybrid computer techniques. In a typical computer study, it is necessary to simulate a dynamic system with random initial conditions, random parameters, or random forcing functions; to sample one or more of the system state variables at certain times; and to compute functions of these sampled values. To insure that the probability of obtaining an erroneous result by chance is sufficiently small, very large numbers of runs may be required (ten thousand runs are not unusual). If the system is nonlinear, digital computer solution times will in general be excessive, and the best way to solve such problems is by hybrid techniques: the system is simulated on a repetitive analog computer, sampled values are transmitted to the digital computer, and statistical functions are computed digitally. A number of examples of such applications will be found in the following pages.

While random process studies may be performed on general-purpose hybrid computers, a number of special-purpose systems have been designed and constructed. These systems feature analog computers with extremely high bandwidth (1–10 MHz), integrator switching timed to a few nanoseconds, digital control, and a relatively small digital computer. Probably the best such computer is the University of Arizona Statistical Repetitive Analog/Hybrid Computer (ASTRAC II) constructed under the direction of Professor G. A. Korn. Many examples in this chapter are based on his work, as published both in the authoritative reference book on the subject of random process simulation[1] and a series of papers by him and his students.[2]

This chapter describes computer techniques for the estimation of certain statistical functions. The emphasis is on applications and not on theory. Consequently, the definitions and concepts are used heuristically and with no attempt at rigor. For a more rigorous treatment of the basic ideas of probability theory and random processes, the reader is urged to consult the basic reference books in the field.[3,4,5] It should also be noted that statistical functions of random variables evaluated on computers are, in general, also random variables. Consequently, the *measurements* made on computers to determine probability density functions or spectral density functions are in fact only *estimates* of these functions. Consequently, we need to study the statistical properties of the estimators to determine in what sense they approach the quantities we desire to measure. Detailed analysis of the questions of measurement and estimation are found in a number of references.[6,7]

11.2 MEASUREMENT OF PROBABILITY DISTRIBUTIONS

Consider the sample functions of a real, continuous random process $x(t)$ as shown in Figure 11.1. We assume that there is an infinite collection (ensemble) of these functions, denoted by $\{^{j}x(t)\}$, $j = 1, 2, 3, \ldots$, and

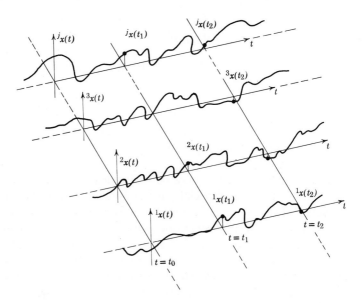

Figure 11.1 Sample functions of a continuous random process.

11.2/MEASUREMENT OF PROBABILITY DISTRIBUTIONS

$-\infty < t < \infty$. The basic definitions of statistical functions are obtained from such an ensemble. Evidently, for measurement purposes, we must use a limited number of records of finite duration.

Consider first the statistical properties of the ensemble at a particular sampling time t_1. The *first-order probability distribution function* (or cumulative distribution function) is defined as

$$F(X, t_1) = \text{prob}\,[x(t_1) \leq X] \tag{11.1}$$

i.e., the distribution function at a time t_1 is defined as the probability that the random variable $x(t_1)$ takes on any value less than or equal to the value X. If the probability distribution function is continuous and differentiable (almost everywhere), we can define the *probability density function* as

$$p(X, t_1) = \frac{\partial}{\partial X} F(X, t_1) \tag{11.2}$$

so that

$$F(X, t_1) = \int_{-\infty}^{X} p(x, t_1)\, dx \tag{11.3}$$

A. Measurements Across the Ensemble

In order to use computer techniques to estimate these functions, we recall the definition of the derivative as a limit:

$$p(X, t_1) = \lim_{\Delta X \to 0} \frac{F(X + \Delta X, t_1) - F(X, t_1)}{\Delta X}$$

$$= \lim_{\Delta X \to 0} \frac{\text{prob}\,(X < x(t_1) < X + \Delta X)}{\Delta X} \tag{11.4}$$

Provided that ΔX is sufficiently small, this expression can be approximated as

$$p(X, t_1) \cong \frac{1}{\Delta X} \text{prob}\left[X - \frac{\Delta X}{2} < x(t_1) \leq X + \frac{\Delta X}{2}\right] \tag{11.5}$$

In terms of the sketches of Figure 11.1, this means that one has to sample the ensemble of random functions $^j x(t)$ at a time t_1 and estimate the probability that these samples fall within a window of height ΔX centered on a particular value X. In block diagram form, such a probability distribution analyzer is shown in Figure 11.2. The sampled functions $^j x(t_1), j = 1, 2, \ldots, N$, appear at the output of the sample-hold circuit. In order to make possible narrow ΔX windows, within the resolution of analog equipment, a limited amplifier is used to obtain the quantity $K[^j x(t_1) - X]$ where K is a suitably large number (say 100) so that 100 mv

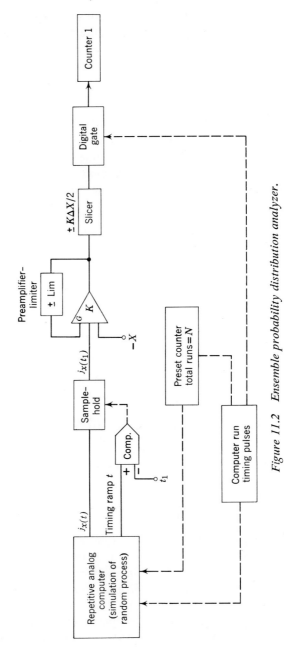

Figure 11.2 Ensemble probability distribution analyzer.

windows are possible. The slicer and gate form the quantity $f[x(t_1)]$, such that

$$f[x(t_1)] = 1 \quad \text{for} \quad X - \frac{\Delta X}{2} < {}^j x(t_1) \leq X + \frac{\Delta X}{2}$$

$$f[x(t_1)] = 0 \quad \text{otherwise} \tag{11.6}$$

If after N computer runs, the counter reading is n, then

$$p_N(X, t_1) \cong \frac{n}{N} \tag{11.7}$$

is an estimate of the desired ensemble probability density. It is evident that $p_N(X, t_1)$ is also a random variable. It can be shown[1] that Np_N is a binomial random variable and, consequently, the errors of estimation can be determined. For practical purposes, it is sufficient to note that in many cases

$$E[Np_N(X, t_1)] = Np(X, t_1) \tag{11.8}$$

i.e., a reliable estimate of the desired density function can be obtained by averaging (finding the expected value of) a sufficiently large number of samples. Practical circuits for the operations indicated above are given by Korn[1,9] and Cameron.[8] (Expected values are discussed further in Section 11.3 below.)

If the sequence $\{{}^j x(t_1) - X\}$ is used as an input to a simple analog comparator, then averaging similar to that above yields estimates of the cumulative distribution function $F(X, t_1)$.

B. Time Averages: Sampled Record

If $x(t)$ can be assumed to be a stationary process, the circuit of Figure 11.2 can also be used to estimate the probability density from time samples of a single record. Let ${}^j x(t)$ be sampled periodically. Then the output of the sample-hold circuit will be the sequence $\{{}^j x(t_k)\}$, $k = 0, 1, 2, \ldots, N$, where $(t_k - t_{k-1}) = \Delta t$ is the sampling interval. The slicer and gate will then form the quantity $f[{}^j x(t_k)]$, such that

$$f[{}^j x(t_k)] = 1 \quad \text{for} \quad X - \frac{\Delta X}{2} < {}^j x(t_k) \leq X + \frac{\Delta X}{2}$$

$$f[{}^j x(t_k)] = 0 \quad \text{otherwise} \tag{11.9}$$

If after N time samples the counter reading is n, then

$$\frac{1}{N} \sum_{k=1}^{N} f[{}^j x(t_k)] = \frac{n}{N} \cong p_N(X) \tag{11.10}$$

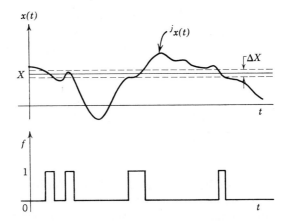

Figure 11.3   Probability density measurement from continuous record.

is an estimate of the desired probability density function. Note that Equation 11.10 yields the time average of the sequence $\{f[^{j}x(t_k)]\}$.

C. Time Averages: Continuous Record

It is also possible to work directly with an analog record $^{j}x(t)$ and generate the function $f[^{j}x(t)]$ defined by Equation 11.6 on a continuous basis as shown in Figure 11.3. Then, an appropriate averaging technique can be used to estimate the probability density, since

$$E[f(^{j}x(t)] = p(X) \qquad (11.11)$$

D. Computational Considerations

From the standpoint of reliability and simplicity, the sampled data estimates given above are more desirable. It is evident that the operations of sampling, holding, amplification, and slicing can be done by analog techniques, while counting and gating are best done digitally. Thus, the implementation of a probability distribution analyzer lends itself well to the use of analog computers with patchable logic elements. On the other hand, it can also be seen that the sampled values $^{j}x(t_1)$ or $x(t_k)$ can be converted to digital form, with all the rest of the operations being performed digitally.

E. Joint Probability Estimates

The joint probability distribution function of two random variables $x(t)$ and $y(t)$ is defined as

$$F[X, t_1; Y, t_2] = \text{prob }[x(t_1) \leq X \text{ and } y(t_2) \leq Y] \qquad (11.12)$$

11.3/ESTIMATION OF AVERAGES

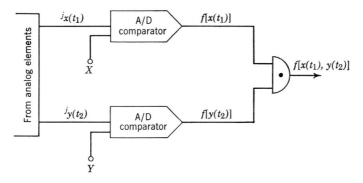

Figure 11.4 Circuit for estimating joint probability distribution of two random variables.

The combination of analog and digital techniques is evident in Figure 11.4, where the analog-digital comparators have outputs of 1 or 0. The output of the AND gate is 1 only when the outputs of both comparators are 1. As before, the average or expected value of f is obtained by counting, since,

$$E\{f[x(t_1), y(t_2)]\} = F[X, t_1; Y, t_2] \qquad (11.13)$$

11.3 ESTIMATION OF AVERAGES

One of the most common and most useful descriptions of a random process is an estimate of the mean or average of a large number of experiments (or, for time averages, of a long time of observation). Similarly, mean squared values are extremely useful in many situations (e.g., as a measure of the effective output of a linear system).

A. Ensemble Averages

An average of measurements taken across the ensemble is known as the *expected value*, mathematical expectation, ensemble average, or simply as the *mean*. It is defined in terms of the *moments* of the probability density function. Thus, the *mean* of the process $x(t)$ at $t = t_1$ is defined as the first moment:

$$E[x(t_1)] = \int_{-\infty}^{+\infty} x p(x, t_1) \, dx \qquad (11.14)$$

and the *mean squared value* is defined as the second moment of $p(x, t_1)$:

$$E[x^2(t_1)] = \int_{-\infty}^{+\infty} x^2 p(x, t_1) \, dx \qquad (11.15)$$

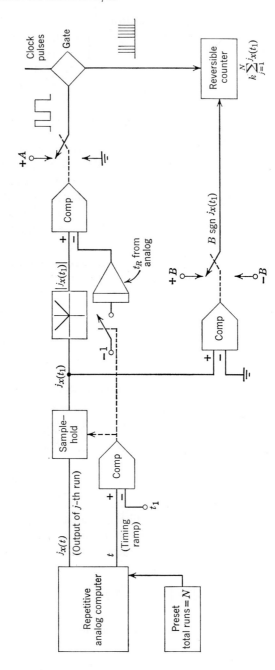

Figure 11.5 Block diagram of sample average measuring circuit.

11.3/ESTIMATION OF AVERAGES 339

From these definitions, it may be assumed that we could estimate the ensemble mean and ensemble mean square by first estimating the probability density function by the techniques of the previous section and then substituting the estimate into the integrals of Equations 11.14 and 11.15. In practice, this procedure is neither practical nor reliable. Instead, one commonly uses the approximations

$$\widetilde{x_N(t_1)} = \frac{1}{N}\sum_{j=1}^{N} {}^j x(t_1) \qquad (11.16)$$

$$\widetilde{x_N^2(t_1)} = \frac{1}{N}\sum_{j=1}^{N} {}^j x^2(t_1) \qquad (11.17)$$

which converge to the expected values (under appropriate conditions) as $N \to \infty$.† It should be noted that $E[x(t_1)]$ and $E[x^2(t_1)]$ are *not* random variables but functions of t_1. On the other hand, their estimates, 11.16 and 11.17, are random variables which depend on the particular set of N measurements taken. As before, if the process is known to be stationary, the time of observation t_1 is immaterial and can be chosen arbitrarily.

A convenient hybrid method for measuring $\widetilde{x(t_1)}$ is to sample each record at time $t = t_1$, convert the sampled value to an equivalent pulse duration, and then count clock pulses during the pulse-on time, as illustrated in Figure 11.5. Note that since the sampled values ${}^j x(t_1)$, $j = 1, 2, \ldots, N$ can be either positive or negative while time durations are always positive, a separate polarity sensor is needed to reverse the direction of counting during negative samples. Other practical circuits are given by Korn.⁽¹⁾ If the analog computer output in Figure 11.5 is $x^2(t)$ rather than $x(t)$, the circuit yields the sample mean square, from Equation 11.17.

B. Time Averages From Continuous Data

If the process can be assumed stationary and ergodic, time averages and ensemble averages equal each other with probability one. In such a case, the mean value may be estimated from time averages. The time average of $x(t)$ is defined as

$$\langle x(t) \rangle = \lim_{T \to \infty} \frac{1}{T} \int_{-T/2}^{+T/2} {}^j x(t)\, dt \qquad (11.18)$$

where ${}^j x(t)$ is a particular sample function of the process. The definition is mostly of theoretical interest. In practice, finite time averages are used

---

† In this chapter, superscript bars will denote time average measurements and superscript wavy lines will denote ensemble average measurements.

and we define

$$\overline{x(t)}_T = \frac{1}{T} \int_0^T {}^j x(t)\, dt \tag{11.19}$$

which is a random variable. It can be shown[1] that $\overline{x(t)}_T$ is an unbiased estimate of the desired ensemble average, so that

$$E[\overline{x(t)}_T] = E[x(t)] \tag{11.20}$$

However, in practice, the measurement of Equation 11.19 requires open-loop analog integration (as shown in Figure 11.6a) which is nearly always undesirable due to error accumulation. Furthermore, it can be shown that in order to obtain accurate estimates (i.e., estimates with a low variance), the required integration time T is inversely proportional to the frequency content of $x(t)$. In other words, low-frequency signals

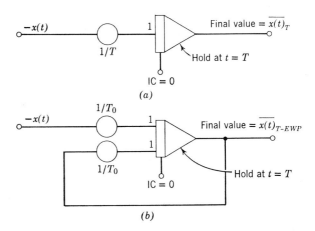

Figure 11.6 *Circuits for estimating time-averages from continuous data.*

may require long integration times. Estimates of the mean output of low-frequency analog noise generators may require several hours of integration if we desire to insure that the probability of obtaining a particular result by chance is sufficiently low (say less than .05). A further discussion of this point is included in the section on noise generation (Section 11.6 below).

A more useful analog measurement is that of the *exponentially weighted-past* (*EWP*) average.

$$\overline{x(t)}_{T-EWP} = \frac{1}{T_0} \int_0^T (e^{-(T-t)/T_0})^j x(t)\, dt \tag{11.21}$$

11.3/ESTIMATION OF AVERAGES   341

Because of the exponential weighting, values of $^jx(t)$ more than a few time constants in the past will have a negligible effect on the final answer. Furthermore, it can be shown that, for sufficiently large integration times T,

$$E[\overline{x(t)}_{T-EWP}] = E[\overline{x(t)}_T] = E[x(t)] \qquad (11.22)$$

so that reliable unbiased estimates may be obtained from the simple filter of Figure 11.6b. Note that the output of circuit (b) can be read continuously.

The variance of the estimates 11.19 or 11.21 needs to be considered carefully in the choice of integration time T.[1,2,6,7]

If an estimate of $E[x^2(t)]$ is desired, a squaring device is used to obtain $x^2(t)$ as the input to either of the averaging circuits of Figures 11.6a or b.

C. Time Averages From Sampled Data

Estimates of the mean and mean square can also be computed from time samples of a single record $^jx(t)$ by the expressions

$$\overline{x(t_n)}_N = \frac{1}{N} \sum_{n=1}^{N} {}^jx(t_n) \qquad (11.23)$$

$$\overline{x^2(t_n)}_N = \frac{1}{N} \sum_{n=1}^{N} {}^jx^2(t_n) \qquad (11.24)$$

It is evident that the circuit of Figure 11.5 can be used for this type of measurement as well, by sampling a single record at the instants t_n, $n = 1, 2, \ldots, N$ rather than successive records at a particular time.

Exponentially weighted-past sampled data averages are obtained from solution of the difference equation

$$y(t_n + 1) = \alpha y(t_n) + x(t_n), \quad y(t_0) = 0, \quad n = 1, 2, \ldots, N \quad (11.25)$$

where

$$\lim_{n \to \infty} E[y_n] = \frac{1}{1 - \alpha} E[x] \qquad (11.26)$$

An analog implementation of this circuit is given in Figure 11.7. Alternatively, the difference Equation 11.25 can be solved digitally.

D. Confidence Limits

As indicated above, measured quantities such as $\overline{x(t)}_T$ are random variables. Consequently, considerable care is needed in the interpretation of a single measurement. The practical design of any experiment requires some measure of the variability of our estimate, so that we may choose

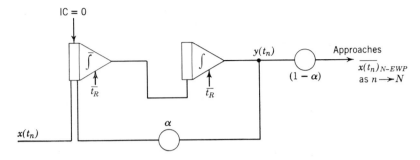

Figure 11.7 Circuit for estimating exponentially weighted past average for sampled data.

run times, equipment, etc. A convenient way of evaluating our estimate is in terms of probability of obtaining a result close to the expected value we desire to obtain by pure chance. The subject to design of experiments is a complex one and only one aspect will be treated here. Other aspects are given in the references.[5,7]

Assume that it is desired to find the value δ such that the true mean $E[x]$ of the process lies within the interval $[\overline{x_T} - \delta, \overline{x_T} + \delta]$ with a probability of .95 (i.e., the likelihood that we would obtain a result within $\pm \delta$ of the true mean by pure chance would only be .05). The limits $\pm \delta$ are referred to as the *confidence limits* at the .95 significance level. If we assume that our measurements have a Gaussian distribution and that we perform M measurements using Equation 11.19 with the circuit of Figure 11.6a and that equipment errors are negligible, then, the sample mean of our measurements is

$$\hat{x}_T = \sum_{i=1}^{M} \overline{x(t)_{Ti}} \qquad (11.27)$$

where $\overline{x(t)_{Ti}}$ represents the results of the i-th computer run. The sample variance is given by

$$\sigma_x^2 = \frac{1}{M-1} \sum_{i=1}^{M} [\overline{x(t)_{Ti}} - \hat{x}_T]^2 \qquad (11.28)$$

Thus, the confidence limits for the data are

$$\hat{x}_T - \frac{\beta \sigma_x}{\sqrt{M}} < E[x] < \hat{x}_T + \frac{\beta \sigma_x}{\sqrt{M}} \qquad (11.29)$$

where the coefficient β is obtained from tables of the Student t distribution with $(M - 1)$ degrees of freedom and the $t_{1-\alpha/2}$ percentile, if we desire a $(1 - \alpha)$ significance level.

EXAMPLE.† We estimate the mean of a process by making five runs T seconds long and measuring the time average using Equation 11.19. The sample mean, using 11.27, becomes

$$\hat{x}_T = 1.8 \text{ units}$$

The sample standard deviation becomes, using 11.28,

$$\sigma_x = 3.9 \text{ units}$$

To obtain the confidence interval, we find

$$t_{0.975} = 2.78$$

and, substituting in 11.29:

$$-2.1 < E[x] < 5.8$$

It is evident that this is a rather wide range. To reduce the spread $\pm \delta$ from the mean, we must either decrease the sample variance or take many more runs. Note, however, that for the same sample standard deviation σ_x, one hundred times more runs are needed to obtain a factor of 10 decrease in δ.

11.4 ESTIMATION OF CORRELATION FUNCTIONS

Correlation functions are among the most important quantities to be estimated in random process studies. They are important as tests of statistical dependence between two random functions, as tests for the presence of d-c and periodic components in a random process, as tools in system identification, etc.

A. Ensemble Average Measurements

The *auto-correlation function* of a random process $x(t)$ is defined as

$$\phi_{xx}(t_1, t_2) = E[x(t_1)x(t_2)] \qquad (11.30)$$

By analogy with the definitions of the ensemble average in the previous section, this expected value can be estimated by

$$\widetilde{\phi_{xx}(t_1, t_2)}_N = \frac{1}{N} \sum_{j=1}^{N} {}^j x(t_1) {}^j x(t_2) \qquad (11.31)$$

The *cross-correlation function* between two random processes $x(t)$ and $y(t)$ is defined as

$$\phi_{xy}(t_1, t_2) = E[x(t_1)y(t_2)] \qquad (11.32)$$

† This example problem is also solved by Levine.[10]

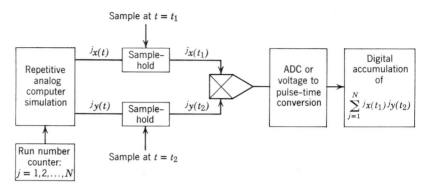

Figure 11.8 Block diagram of ensemble cross-correlator

which can be estimated from measurements on N sample functions of both processes as

$$\widehat{\phi_{xy}(t_1, t_2)}_N = \frac{1}{N} \sum_{j=1}^{N} {}^j x(t_1){}^j y(t_2) \qquad (11.33)$$

The circuit of Figure 11.5 can be readily adapted to the estimation of correlation functions by including a second sample-hold circuit and a multiplier as shown in Figure 11.8. From this figure it can be seen that since one analog computer run produces only one of the terms of the summation on the right of 11.33, N runs are needed to obtain *a single value* (one point) of the cross-correlation function. Consequently, if it is desired to estimate M points of the correlation function, MN analog computer runs (and MN multiplications) are required.

Note that

$$\phi_{xx}(t_1, t_1) = E[x^2(t_1)] \qquad (11.34)$$

is the ensemble mean squared value so that an auto-correlator can be used to estimate the mean square. When the process is stationary, the specific value of t_1 is immaterial. If we define $t_2 = t_1 + \tau$, it is clear that, for a stationary process, $\phi_{xx}(t_1, t_1 + \tau)$ is independent of t_1 and depends only on the interval τ.

B. Time Average Measurements from Continuous Data

When the process is stationary, correlation functions may be estimated from the finite time averages

$$\overline{\phi_{xx}(\tau)}_T = \frac{1}{T} \int_0^T {}^j x(t - \tau){}^j x(t)\, dt \qquad (11.35)$$

$$\overline{\phi_{xy}(\tau)}_T = \frac{1}{T} \int_0^T {}^j x(t - \tau){}^j y(t)\, dt \qquad (11.36)$$

11.4/ESTIMATION OF CORRELATION FUNCTIONS

where $^j x(t)$ and $^j y(t)$ represent a single record. These functions represent estimates of the correlation functions

$$\phi_{xx}(t - \tau, t) = E[x(t - \tau)x(t)]$$
$$\phi_{xy}(t - \tau, t) = E[x(t - \tau)y(t)]$$
(11.37)

A basic analog system for performing cross-correlation of a single record is shown in Figure 11.9, where the superscript j has been omitted for simplicity. There are two major problems with such devices:

1. Generation of time delays for analog signals is a difficult process. For low frequency signals and small delays, Padé approximations[11] may be satisfactory. On the other hand, tape recording and playback may be too costly, unless the functions being correlated are in fact stored on tape, in which case either head displacement or tape loops may be used to obtain delays. In at least one biomedical application,[12] time delay was obtained by analog-digital conversion, digital storage, and reconversion to analog form.
2. The second problem is that of open-loop integration. Here too exponentially weighted-past averaging (see Section 11.3B) may have advantages.

C. Time Average Measurements from Sampled Data

The cross-correlation function may also be estimated as the finite-time, sampled-data average

$$\phi_{xy}(k\,\Delta t)_N = \frac{1}{N} \sum_{n=0}^{N} {}^j x(t_n - k\,\Delta t)^j y(t_n)$$
(11.38)

where Δt is the sampling interval. If $x(t)$ and $y(t)$ are available as continuous signals, the circuit of Figure 11.8 can be adapted to the measurement of Equation 11.38 by sampling single records at the instants t_1,

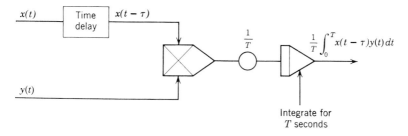

Figure 11.9 Analog cross-correlator.

$t_2, \ldots, t_N$. Note that a complete pass through the data is needed (with N multiplications) for an estimate of a *single value* of the correlation function at the time shift $k\,\Delta t$. A different value of the integer k will require a new run.

In many cases it may be more convenient to digitize both of the records and to perform the operations of multiplication and accumulation digitally.

D. Coarse Quantization Correlators

Early work of Widrow[13] and others, recently extended by Korn,[1] has led to the design of correlation devices where one or both input signals are quantized using only two quantization levels per signal. Thus, it can be shown that for Gaussian signals with zero mean, the cross-correlation of $x(t)$ can be evaluated from the relation

$$\phi_{xy}(t_1, t_2) = kE\{[\text{sgn } x(t_1)]y(t_2)\} \quad (11.39)$$

The advantage of this approach is that multiplication is replaced by switching, as indicated in the circuit of Figure 11.10. The constant of proportionality k depends on the root-mean-square of $x(t_1)$.

If estimates of the mean squared values are available, we can use one-bit quantization of both $x(t)$ and $y(t)$ by means of the relation

$$E[\text{sgn } x(t_1) \text{ sgn } y(t_2)] = \frac{2}{\pi}\arcsin\frac{\phi_{xy}(t_1, t_2)}{\sqrt{E[x^2(t_1)]E[y^2(t_2)]}} \quad (11.40)$$

This relation also applies only to Gaussian signals with zero mean. The advantage of this approach is that is requires only comparators and digital circuitry. A detailed discussion of correlation with quantized signals is given by Korn.[1]

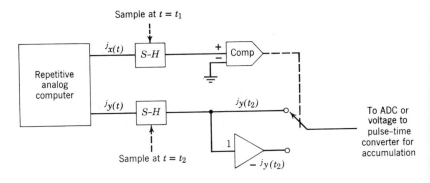

Figure 11.10 Cross-correlation with 1-bit quantization of $x(t)$.

11.5 ESTIMATION OF SPECTRAL DENSITY FUNCTIONS

Power spectral density functions (or simply power spectra) are frequency domain equivalents of correlation functions. They are primarily useful in the study of stationary random processes. For such a process the power spectrum may be viewed as an indication of the power contained by the process at various frequencies. This intuitive definition is the basis of the usual analog technique for evaluation of the spectral density functions. In recent years, digital techniques have been used almost exclusively in the estimation of power spectra, to the exclusion of both analog and hybrid methods. The three most common techniques for measurement of power spectra are: (1) analog filtering, (2) Fourier transforms of correlation functions, and (3) "fast Fourier transform" techniques.

A. Analog Spectral Analyzers

The power spectral density of a random process $x(t)$ is defined as

$$S_{xx}(\omega) = \int_{-\infty}^{+\infty} \phi_{xx}(\tau) e^{-j\omega\tau} \, d\tau \tag{11.41}$$

If $x(t)$ represents an electrical quantity measured in volts, then the units of $S_{xx}(\omega)$ are volts^2/rad/sec. Under appropriate restrictions, Equation 11.41 can be inverted to yield

$$\phi_{xx}(\tau) = \frac{1}{2\pi} \int_{-\infty}^{+\infty} S_{xx}(\omega) e^{j\tau\omega} \, d\omega \tag{11.42}$$

The pair of Equations 11.41 and 11.42 is termed the *Wiener-Khinchine* equations. Since the power spectral density represents (intuitively) the average power per unit frequency bandwidth, integration of the spectral density function yields the total power in the process:

$$\frac{1}{2\pi} \int_{-\infty}^{+\infty} S_{xx}(\omega) \, d\omega = \phi_{xx}(0) = \overline{[x(t)]^2} \tag{11.43}$$

which follows from the definition of the auto-correlation function.

For analog computer estimation of the power spectrum relation, Equation 11.43 is applied to compute the power dissipated in a 1 ohm resistor by components of $x(t)$ between a frequency f_0 and a frequency $f_0 + \Delta f$. With reference to Figure 11.11, if the filter output is denoted by x_0, then from 11.43 we obtain

$$\overline{[x_0(t)]^2} = 2 \int_{f_0}^{f_0+\Delta f} S_{xx}(\omega_0) \, df \cong 2 S_{xx}(\omega_0) \Delta f \tag{11.44}$$

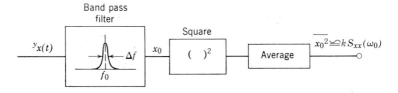

Figure 11.11 Analog spectral analyzer.

Evidently, the approximation inherent in this equation is also influenced by the nonideal filter characteristics as well as the errors of estimation inherent in finite time averages, as discussed in Section 11.3. In many practical cases, the implementation of a narrow-band filter by purely analog techniques is not feasible, especially at very low frequencies. In such cases it may be possible to obtain the analog signals on magnetic tape and play back the recorded data at higher speeds, thus multiplying the frequencies of interest by a suitable factor. In other cases,[14] heterodyne techniques make it possible to use low-pass rather than band-pass filters, as shown in block diagram form in Figure 11.12. Analog spectral analyzers based on the principle embodied in this figure have been used successfully for on-line power spectrum estimation in experiments involving human pilot responses.[15] It should be noted that circuits of the type of Figures 11.11 or 11.12 yield an estimate of the power spectral density at a single frequency ω_0. Thus, repetitive passes through the data or a large number of similar circuits are required to estimate the spectral density function over a band of frequencies.

B. Computation from Correlation Functions

The power spectral density can also be estimated by direct implementation of Equation 11.41 using digital computer techniques. The correlation function is first estimated from sampled data and then used as an input to a Fourier transform program.[16]

C. "Fast Fourier Transform" Techniques

In recent years an algorithm due to Cooley and Tukey has been used to obtain power spectral density estimates from sampled data with considerably shorter computation time than is needed when correlation functions are first computed.[17,18] Basically, the Cooley-Tukey method is based on the computation of the finite Fourier transform of the signal $x(t)$, defined as

$$A_T(f, x) = \int_{-T}^{T} x(t) e^{-j2\pi ft}\, dt \quad (11.45)$$

11.5/ESTIMATION OF SPECTRAL DENSITY FUNCTIONS

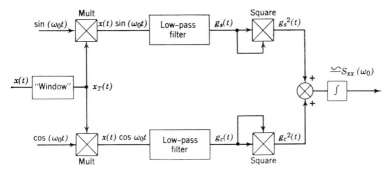

Figure 11.12 Block diagram of analog spectral analyzer suitable for low-frequency signals.

Then the rate of change of average power with frequency (or power spectral density) estimated for the finite length record is given by

$$G_T(f, x) = \frac{|A_T(f, x)|^2}{T} \quad (11.46)$$

Under appropriate conditions, as the record length $2T$ becomes longer, the quantity $G_T(f, x)$ approaches the spectral density function.[18]

The major advantage of the algorithms employing relationship 11.46 is that they do not require a prior evaluation of the correlation function. It can be shown that for sampled records involving n data points, the execution time using a correlation program is proportional to n^2, while the execution time using the Cooley-Tukey algorithm is proportional to the quantity $n \log n$.

D. Measurement Problems[16]

While power spectral density measurements are common in many industrial laboratories, such measurements are plagued with a large number of difficulties. Among these are the following:

The "Window Problem." Theoretical definitions of the spectral density functions are based on infinitely long integration times. In practice, we work with truncated time functions, say with a duration of T seconds. This is equivalent to the multiplication of a time function $x(t)$ by a window function $w(t)$

$$x_T(t) = x(t)w(t) \quad (11.47)$$

where
$$w(t) = 1, 0 \le t \le T$$
$$= 0, t > T \text{ and } t < 0 \quad (11.48)$$

The effect of this multiplication is to introduce the spectral characteristics of the window into the estimation process, thus causing a "spreading" in the frequency domain.

Biases. Both analog filtering and correlation techniques yield biased estimates of the power spectral density. In some cases, the effects of these biases can be removed if the spectrum of the input signal to the analyzer is as nearly flat (as close to "white noise") as possible. This can be achieved by a prefiltering process known as "prewhitening."

Aliasing. Power spectra of sampled signals contain harmonics at multiples of the sampling frequency. To avoid the introduction of erroneous components into the power spectral estimates, it is desirable to remove all frequencies above one-half the sampling frequency from the signal before attempting a spectral estimate.

These and other problems should be carefully examined before attempting power spectral analysis with analog, digital, or hybrid computers.

11.6 GENERATION OF NOISE SIGNALS IN HYBRID SYSTEMS

One advantage of hybrid computer techniques in the study of systems with random inputs, random initial conditions, or random parameters is in the wide choice of random noise sources available. It is possible to generate random noise using analog components and then obtain random sequences by digitizing the continuous noise signals. Conversely, we can generate pseudo-random sequences by appropriate digital computer algorithms and then obtain continuous random signals by digital to analog conversion and filtering. Finally, in a number of installations, special hybrid noise generators have been constructed.

A. Analog Noise Generators

Analog sources of noise are available as standard components from all manufacturers. They generally obtain their noise signal from a gas tube in a magnetic field whose output is then appropriately shaped. Ideally, an analog noise generator will have the following specifications:

1. Zero mean, i.e., $E[x(t)] = 0$.
2. Gaussian distribution.
3. Flat power spectrum to some maximum frequency f_{max}.

11.6/GENERATION OF NOISE SIGNALS IN HYBRID SYSTEMS

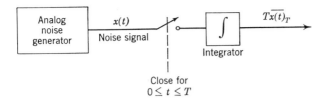

Figure 11.13 Measurement of noise generator output mean.

For hybrid computer purposes, the maximum frequency output should be relatively high; commercial analog noise generators are available with power spectra flat from nearly zero frequency to several thousand Hz.

The fact that analog noise generators contain energy at extremely low frequencies makes the verification of their specifications a difficult estimation problem. Consider for example the estimation of the mean value of the noise generator output which the manufacturer's specifications indicate to be zero. Assume that we simply integrated the noise generator output for T seconds, as shown in Figure 11.13. Then, from Equation 11.19, the sample mean is related to the output voltage by

$$V_T = \overline{Tx(t)}_T = \int_0^T x(t)\, dt \qquad (11.49)$$

The voltage V_T is a random variable which can be assumed to be normally distributed, with its mean and standard deviation both being functions of the integration time T. The true mean of the integrator output distribution is

$$\bar{V} = \int_0^T E[x(t)]\, dt = TE[x(t)] \qquad (11.50)$$

and the variance of the integrator output distribution is

$$\sigma^2 = \frac{N_0}{2}\int_0^T dt = \frac{N_0 T}{2} \qquad (11.51)$$

In this equation, N_0 represents the zero frequency spectral density of the noise in volts^2/Hz. As a concrete example, assume that we have a noise generator with a spectral density of 2.5 volts^2/Hz, that we use an integration period of four hours, and that the mean of our measurements yields, as an estimate of the noise generator mean, the value $\overline{x}_T = .05$

volts. Then the approximate 95% confidence limits are

$$.05 - \frac{2N_0}{T} < E[x(t)] < .05 + \frac{2N_0}{T}$$

or

$$.03 < E[x(t)] < .07$$

Evidently such long integration periods are not feasible using open-loop analog integrators; evaluation of averages from sampled data using digital techniques is required. The purpose of this example is to illustrate the fact that the specifications of analog noise generators are usually known only approximately, since serious measurement difficulties arise in attempting to obtain precise estimates. The situation is similar in attempting to estimate the spectral density or the output distribution.

B. Shift Register Noise Generators

An N-stage shift register produces a sequence of binary values (ones or zeroes) described by the relation

$$a_i = \sum_{k=1}^{N} c_k a_{i-k}, \quad i = 1, 2, \ldots, N \qquad (11.52)$$

where the coefficients c_k are either 0 or 1 and the summation is modulo 2, as illustrated in Figure 11.14a. If the coefficients c_i are carefully chosen,[19] it is possible to obtain a maximal length sequence of $2^N - 1$ bits before the pattern repeats. Furthermore, for many shift register lengths (i.e., for many values of N), maximal length sequences can be obtained with only two of the c_k equal to 1 and all others equal to 0. Furthermore, one of these c_k is always c_N. Therefore, a single exclusive OR gate is needed to provide the feedback. The theoretical correlation function of a maximal length linear shift register sequence is shown in Figure 11.14b. It can be seen that the auto-correlation function contains two triangular spikes separated by $2^N - 1$ clock periods and remains nearly zero in-between. Thus, for example, for a 23-bit shift register, we obtain over 8,000,000 uncorrelated bits before the pseudo-random sequence begins to repeat. For analog purposes, the shift register output is commonly shifted, amplified, and clipped in order to obtain a pseudo-random square wave as shown in Figure 11.14c. Low pass filtering of this square wave produces an analog voltage with an approximately Gaussian amplitude distribution. Shift register noise generators of the type described above have been built for the ASTRAC II computer at the University of Arizona[20] and elsewhere.

11.6/GENERATION OF NOISE SIGNALS IN HYBRID SYSTEMS

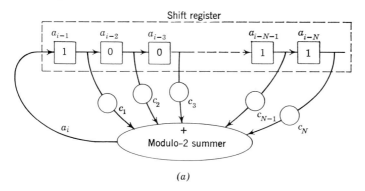

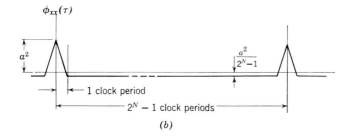

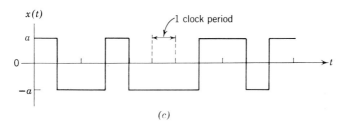

Figure 11.14 Shift register noise generation: (a) N-stage linear shift register, (b) approximate correlation function of maximal length shift register sequence and (c) pseudo-random square wave.

While shift register noise generators in theory have extremely attractive properties, in recent years they have come under increasing suspicion, because it has been demonstrated from measurements with practical devices that in many cases the output distributions are skewed.[21,22] As a result, in many hybrid computer laboratories shift register noise generators have been abandoned in favor of either analog noise generators or conventional digital computer algorithms.

C. Digital Algorithms[23,25]

The most common digital computer techniques for the generation of pseudo-random sequences are the mid-square method and a variety of so-called congruential methods. The mid-square method consists of generating each new random number by taking the middle n digits of the square of the previous n digit number. For example, if the starting number is $X_0 = 2189$ then $X_0^2 = 04791721$ from which one obtains as the next random number $X_1 = 7917$.

Mid-square generators have on the whole been superseded by congruential methods. These generators produce pseudo-random integers which can then be transformed into fixed point fractions or floating point numbers. The multiplicative congruential method is based on generating the k-th random number X_k from the $(k-1)$-st number by means of the relationship

$$X_k = cX_{k-1} \,(\text{mod } 2^r) \tag{11.53}$$

where r is the length of the computer accumulator in bits. The initial number X_0 can be taken as any odd number and the coefficient c is selected in accordance with rules given in the literature.[23] For example, with 36-bit machines the number $c = 5^{13}$ has been used. Equation 11.53 then indicates that the k-th number is obtained by multiplying $(k-1)$-st random number by a large coefficient c and then retaining only the least significant r bits.

The so-called "mixed congruential method" is described by the relation

$$X_k = [cX_{k-1} + d](\text{mod } 2^r) \tag{11.54}$$

where d is an odd integer.

An extensive literature on congruential random number generators exists and, in many cases, they have been shown to exhibit unusual properties which would not be expected of truly random sequences.[28,29] For this reason, in some installations random noise generator programs are not used and random digits are obtained from the highly respected RAND Corporation table of 1,000,000 random digits.[27]

D. General Considerations

From the standpoint of hybrid computer techniques, both analog and digital noise generation offers advantages. Digital techniques, whether arising from a random number generator program or from shift register sequences employed with appropriate caution, have the advantage that the sequence may be restarted from an arbitrary initial value. The succeeding numbers in the sequence will then be exact repetitions of previous runs. The advantage of such an approach for checkout of computer programs is obvious. On the other hand, all digital sequences are in fact pseudo-random sequences which eventually repeat and which may contain other hidden periodicities. For this reason, it may be advantageous, once checkout has been completed, to use analog noise generators and sampled outputs of analog noise generators to produce analog and digital noise, respectively.

11.7 LINEAR SYSTEM STUDIES

Among the most important computer techniques involving linear systems excited by random inputs are those concerned with the evaluation of certain correlation functions, shaping of spectral density functions, system identification from random inputs, and evaluation of mean square response. The purpose of this section is to provide a brief introduction to these topics.

A. Evaluation of Correlation Functions

Consider a time invariant linear system described by its weighting function $g(\tau)$. If such a system is excited with a deterministic input $x(t)$, its output is obtained from the convolution integral:

$$y(t) = \int_{-\infty}^{t} g(t - \tau)x(\tau)\, d\tau \qquad (11.55)$$

If the input to the system is a stationary random process with auto-correlation $\phi_{xx}(\tau)$, the input-output cross-correlation function is given by the Wiener-Lee relation:

$$\phi_{xy}(t) = \int_{-\infty}^{\infty} g(t - \tau)\phi_{xx}(\tau)\, d\tau \qquad (11.56)$$

By analogy with Equation 11.55, it can be seen that if a simulation of the linear system (say, on an analog computer) is excited by a signal whose waveform is proportional to the given auto-correlation function, then the system output will be proportional to the cross-correlation function $\phi_{xy}(t)$, as illustrated in Figure 11.15. Thus, if the auto-correlation function of the random process $x(t)$ has been previously evaluated and is stored in memory, the cross-correlation function between the input and output of a linear constant coefficient system can be evaluated with a single computer run.

The output auto-correlation function can be obtained from an additional computer run by using the expression

$$\phi_{yy}(t - t_0) = \int_{-\infty}^{\infty} g(t - \tau)\phi_{xy}(t_0 - \tau)\, d\tau \qquad (11.57)$$

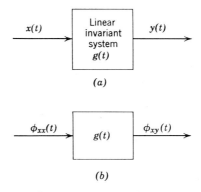

Figure 11.15 Linear system with weighting function $g(t)$: (a) evaluation of system response to input $x(t)$, and (b) evaluation of cross-correlation function $\phi_{xy}(t)$.

The input used in Equation 11.57 is the cross-correlation function computed using Equation 11.56 and run backwards in time.

B. Shaping of Spectral Density Functions

The input and output power spectral densities of the system of Figure 11.15a are related by

$$S_{yy}(\omega) = |G(j\omega)|^2 S_{xx}(\omega) \tag{11.58}$$

where $G(j\omega)$ is the frequency function corresponding to the weighting function $g(t)$. Equation 11.58 can be used to select the transfer function of a filter which will yield a desired power spectral density $S_{yy}(\omega)$.

C. System Identification

The identification of linear time invariant systems is greatly facilitated by the use of random inputs and the measurement of correlation or spectral density functions. Consider first the Wiener-Lee relation of Equation 11.56. If the input signal is a sample function of a process whose bandwidth sufficiently exceeds that of the system to be considered "white noise" (i.e., a noise whose power spectral density is constant for all frequencies), then

$$S_{xx}(\omega) = N_0 \tag{11.59}$$

which is a constant, and

$$\phi_{xx}(\tau) = N_0 \delta(\tau) \tag{11.60}$$

since the auto-correlation function of white noise is an impulse function.

Substituting 11.60 into 11.56 yields

$$\phi_{xy}(t) = \int_{-\infty}^{+\infty} g(t - \tau)\delta(\tau)\, d\tau$$
$$= N_0 g(t) \qquad (11.61)$$

In other words, input-output cross-correlation provides an estimate of the weighting function of an unknown system, as shown in Figure 11.16. The frequency domain equivalent of Equation 11.56 is the relation

$$S_{xy}(j\omega) = G(j\omega) S_{xx}(\omega) \qquad (11.62)$$

where $G(j\omega)$ is the frequency function (transfer function) of the system with weighting function $g(t)$, and $S_{xy}(j\omega)$ is the cross-spectral density between input and output of the system. By measuring both the spectral density of the input to a linear dynamic system and the cross-spectral density between input and output, it is then possible to estimate the system transfer function by solving Equation 11.62,

$$G(j\omega) = \frac{S_{xy}(j\omega)}{S_{xx}(\omega)} \qquad (11.63)$$

Since the cross-spectral density is a complex number at each frequency, this expression yields both the amplitude and phase of the system transfer function. Even in situations where the system is nonlinear, expression 11.63 provides a convenient technique for obtaining a linear approximation to the system behavior (the minimum mean-square error approximation). Such techniques have been used successfully in obtaining mathematical models for human operators in control systems for a number of years.[15]

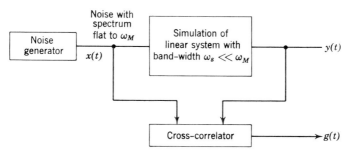

Figure 11.16 Evaluation of system weighting function $g(t)$ by cross-correlation.

The estimation of correlation functions and spectral density functions which is required in the implementation of the identification techniques described above can be performed by the techniques described in Sections 11.4 and 11.5, respectively.

D. Mean-Squared Response of Linear Systems

A number or techniques for the estimation of mean-squared values using either time or ensemble averages were described earlier (Section 11.3). However, all the techniques discussed require a sufficiently large number of runs to obtain the necessary accuracy in the estimates. It is, therefore, of interest to indicate a technique for obtaining mean-squared responses with a single computer run, thus eliminating the problems which arise in the use of the estimation techniques discussed previously. Such a technique is based on Parseval's theorem[5] which states, in effect, that the total energy at the system output is the same whether evaluated in the time or frequency domain. Applying this theorem, we can obtain a relationship between the weighting function and the transfer function, namely

$$\frac{1}{2\pi} \int_{-\infty}^{\infty} |G(j\omega)|^2 \, d\omega = \int_{-\infty}^{\infty} [g(t)]^2 \, dt \tag{11.64}$$

Assume now that the system input is white noise with a constant spectral density, as given by Equation 11.59. The system output spectral density is then given by Equation 11.58

$$S_{yy}(\omega) = N_0 |G(j\omega)|^2 \tag{11.65}$$

The mean-squared value of the output is obtained by the application of Equation 11.43:

$$\overline{[y(t)]^2} = \frac{1}{\pi} \int_0^\infty S_{yy}(\omega) \, d\omega \tag{11.66}$$

Substituting 11.65, we obtain for the mean-squared value

$$\overline{[y(t)]^2} = \frac{N_0}{\pi} \int_0^\infty |G(j\omega)|^2 \, d\omega \tag{11.67}$$

Finally, substituting Parseval's theorem, Equation 11.64, we obtain the expression

$$\overline{[y(t)]^2} = N_0 \int_0^\infty [g(t)]^2 \, dt \tag{11.68}$$

The operations involved in this expression are shown in Figure 11.17. It is important to note that the mean-squared value of the output is now

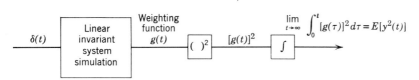

Figure 11.17 Measurement of mean square response to white noise with a single computer run.

obtained as a result of a single computer run, without actually using a random input and, consequently, without the need for averaging a large number of runs. In an actual simulation on an analog computer, the input indicated in Figure 11.17 is replaced by a unit initial condition on the first integrator. It is also apparent that an actual simulation of the linear system is not required. If the system weighting function is known, substitution into Equation 11.68 yields an estimate of the system mean-squared response to a white noise input. When responses to random inputs with spectral densities different from white noise are needed, the technique may still be applied by inserting between the white noise source and the simulated system an appropriate filter.[1]

When the linear system in question is time-varying, the system weighting function now depends on two variables, so that the convolution integral takes the form

$$y(t) = \int_{-\infty}^{t} g(t, \tau) x(\tau) \, d\tau \tag{11.69}$$

where $g(t, \tau)$ represents the response of the system at a time t to an impulse applied at time τ. In time-varying systems, it is evident that the mean-squared response is a function of the time of observation and can be computed from the expression

$$E[y^2(t_1)] = N_0 \int_0^{\infty} [g(t_1, \tau)]^2 \, d\tau \tag{11.70}$$

Thus, the mean-squared response at a time t_1 can still be computed from a single computer run if the weighting function is available as a function of the time of impulse application τ. This can be obtained either by use of the adjoint computing technique due to Laning and Battin[4,10,30] or by use of repetitive computing techniques with impulse inputs applied at successive values of τ.

If the mean-squared response of a nonlinear system is desired, the simplification technique described above is no longer applicable and we must apply the methods of Section 11.3. In other words, to estimate the mean-squared response of a nonlinear system, the system must be simulated, the random inputs must be applied, and the experiment must be repeated a sufficient number of times so that adequate confidence in the results is obtained. Such repeated experimentation with random inputs is known as the Monte Carlo method.

11.8 MONTE CARLO TECHNIQUES AND APPLICATIONS[31,32]

The Monte Carlo method refers to the performance of a large number of experiments in the study of systems with random initial conditions, random forcing functions, or random parameters in order to obtain estimates of response functions with appropriate accuracy. Monte Carlo studies have in fact become practical with the advent of high-speed hybrid computer techniques in which the system simulation can be performed repetitively on the analog portion of the system and the necessary statistics evaluated in the digital portion. Hybrid computer Monte Carlo techniques are at present the only practical methods for studies of complex, nonlinear, dynamic systems under the influence of random processes. Extensive investigation of hybrid Monte Carlo techniques has been performed by Korn and his associates.[2]

Monte Carlo techniques are also important in the hybrid solution of partial differential equations and in system optimization. These topics are discussed in Chapters 8 and 9, respectively.

In some cases the use of Monte Carlo techniques makes it possible to evaluate certain probability distributions by what may appear to be a "brute force" approach, namely by evaluation of a system response at a sufficiently large number of randomly selected points in an appropriate space. Nevertheless, such a technique is computationally feasible (using hybrid computers) and yields an efficient solution to a difficult parameter estimation problem, as illustrated by the work of McGhee and Walford.[33]

References

1. Korn, G. A., *Random-Process Simulation and Measurements*, McGraw-Hill, New York, 1966.
2. Korn, G. A., ed., "Hybrid Computers and Monte Carlo Techniques," Report EE S-9 Engineering Experiment Station, College of Engineering, Univ. of Ariz. 1965.

REFERENCES

3. Lee, Y. W., *Statistical Theory of Communication*, John Wiley, New York, 1960.
4. Laning, J. H., and R. H. Battin, *Random Processes in Automatic Control*, McGraw-Hill, New York, 1956.
5. Papoulis, A., *Probability, Random Variables and Stochastic Processes*, McGraw-Hill, New York, 1965.
6. Bendat, J. S., *Principles and Applications of Random Noise Theory*, John Wiley, New York, 1958.
7. Bendat, J. S., and L. Piersal, *Measurement and Analysis of Random Data*, John Wiley, New York, 1967.
8. Cameron, W. D., "Hybrid Computer Techniques for Determining Probability Distributions," *Proc. Int. Symposium on Analog and Digital Techniques in Aeronautics*, Liege, Belgium, 1963.
9. Brubaker, T., and G. A. Korn, "Accurate Amplitude Distribution Analyzer Combining Analog and Digital Logic," *Rev. Sci. Instruments*, **32**, 317–322, Mar. 1961.
10. Levine, L., *Methods for Solving Engineering Problems Using Analog Computers*, McGraw-Hill, New York, 1964.
11. Huskey, H., and G. A. Korn, *Computer Handbook*, McGraw-Hill, New York, 1962.
12. Macy, J. B., Jr., "Hybrid Computer Techniques in Physiology," *Annals N.Y. Acad. Sci.*, **115**, 568–590, 1964.
13. Widrow, B., "A Study of Rough Amplitude Quantization by Means of Nyquist Sampling Theory," *IRE Transactions on Circuit Theory*, Dec. 1956.
14. Seltzer, J. L., and D. T. McRuer, "Survey of Analog Cross-Spectral Analyzers," *WADC Tech. Rpt. 59-241*, Wright Air Development Center, Dec. 1959.
15. McRuer, D. T. et al., "Human Pilot Dynamics in Compensatory Systems," *Tech. Rpt. AFFDL-TR-65-15*, U.S. Air Force Flight Dynamics Research Laboratory, July 1965.
16. Blackman, R. B., and J. W. Tukey, *The Measurement of Power Spectra*, Dover, New York, 1958.
17. Cooley, J. W., and J. W. Tukey, "An Algorithm for the Machine Computation of Complex Fourier Series," *Math. of Computation*, **19**, 297–301, April 1965.
18. "Special Issue on Fast Fourier Transform and its Applications to Digital Filtering and Spectral Analysis," *IEEE Transactions on Audio and Electro-Acoustics*, **AV-15**, June 1967.
19. Tauseworth, R. C., "Random Numbers Generated by Linear Recurrence Modulo Two," *Math. of Computation*, **19**, 201, April 1965.
20. Hampton, R. L., "A Hybrid Analog-Digital Pseudo-Random Noise Generator," *Simulation*, **4**, 179–189, Mar. 1965.
21. Gilson, R. P., "Some Results of Amplitude Distribution Experiments on Shift-Register Generated Pseudo-Random Noise," *IEEE Transactions on Electronic Computers*, **EC-15**, 926–927, Dec. 1966.
22. White, R. C., Jr., "Experiments with Digital Computer Simulations of Pseudo-Random Noise Generators," *IEEE Transactions on Electronic Computers*, **EC-15**, 355–357, June 1967.
23. Hull, T. E., and A. R. Dobell, "Random Number Generators," *SIAM Rev.*, **4**, 230–254, 1962.
24. Hutchinson, D. W., "A New Uniform Pseudo-Random Number Generator," *Communications of the ACM*, **9**, 432–433, June 1966.
25. Hemmerle, W. J., *Statistical Computations on a Digital Computer*, Blaisdell, Waltham, Mass., 1967.

26. Hull, T. E., and A. R. Dobell, "Mixed Congruential Random Number Generators for Binary Machines," *J. ACM*, **11**, 31–40, 1961.
27. RAND Corp., *A Million Random Digits with 100,000 Normal Deviates*, Free Press, New York, 1955.
28. Peach, P., "Bias in Pseudo-Random Numbers," *J. Amer. Statist. Assoc.*, 56, 610–618, 1961.
29. Breenberger, M., "Method in Randomness," *Communications of the ACM*, **8**, 177–179, Mar. 1965.
30. Korn, G. A., "Adjoint Linear Systems in Analog Computation: A New Look," *Annals AICA*, Oct. 1962.
31. Meyer, H. A., ed. *Symposium on Monte Carlo Methods*, John Wiley, New York, 1956.
32. Hammersley, J. M., and D. C. Handscomb, *Monte Carlo Methods: Methuen's Monographs on Applied Probability and Statistics*, John Wiley, New York, 1964.
33. McGhee, R. B., and R. B. Walford, "A Monte Carlo Approach to the Evaluation of Conditional Expectation Parameter Estimates for Nonlinear Dynamic Systems," *IEEE Trans. on Automatic Control*, **AC-13,** *29–36, Feb. 1968.*

12
Hybrid Computation in Flight Simulation†

12.1 INTRODUCTION

It is generally known that the earliest widespread application of analog computers was in the area of flight simulation. This is because complete flight equations are so complex that solutions by analytic techniques in all but simplified cases are not possible, yet solutions are needed for many aspects of flight-vehicle design. In the late 1940's, analog computers and, in particular, electronic differential analyzers were the only type of computer able to solve flight equations in real time.

It is not surprising to find, therefore, that the first widespread application area of hybrid computers has been in flight simulation. In the mid-1950's, it was found that the computing precision required in calculating ballistic missile trajectories was not available in analog systems but the speed required in solving the rotational and control equations of motions was not available in digital computing systems. A hybrid computer offered the obvious solution, where the digital subsystem would be used to solve with high accuracy but limited speed the translational equations of motion as needed to simulate the trajectory and provide inputs for guidance simulation. The analog subsystem would be used to solve the rotational equations of motion and to simulate the control system dynamics where

† This chapter was contributed by Professor Robert M. Howe of the Department of Information and Control Engineering, University of Michigan.

accuracy requirements are limited but speed requirements are very high. With such a system it would be possible to solve the complete six-degree-of-freedom missile equations in real time, which was desirable in order to minimize overall computational time and essential if any actual flight hardware were used as part of the simulation. Unfortunately, in the early hybrid systems the digital-to-analog and, particularly, the analog-to-digital linkage hardware proved to be a sizeable problem area; the hybrid systems became operational much later than was originally anticipated. Nevertheless, the type of flight-simulation problem described above continues to represent a principal application even for current hybrid computer systems. Truitt has given a good summary of some of the early hybrid applications.[1] It is also not surprising to find that real-time flight simulation is currently one of the main application areas for all-digital simulation, particularly in flight-trainer systems, either for aircraft or space vehicles. Since currently available medium and large-sized general-purpose digital systems can solve flight equations in real time, we might reasonably ask why hybrid computers continue to be useful in flight simulation. One answer is that large numbers of faster than real-time solutions are often necessary or desirable for design purposes or to obtain statistical significance in the results when random variables are involved. Also, hybrid systems in many cases still preserve the excellent analog-type interface between the user and the computer system. In other words, a hybrid system generally permits better "hands-on" operation than a purely digital system. Finally, there are some real-time flight-equation problems where the problem frequencies are so high that a purely digital system is not adequate. These high frequencies may be present in control system loops, in elastic-structure modes, or in control-jet transients. Thus we find continued application requirements for hybrid computer systems in flight simulation.

In the next section we will present in block diagram form a suitable set of equations of motion for a complete six-degree-of-freedom flight simulation of vehicles in the atmosphere. We will point out why the equations as presented are particularly well suited to hybrid computation and how the equations should be divided between the analog and digital subsystems. In Section 12.3 we will present suitable equations for simulation of space-vehicle trajectories and point out some hybrid applications. Section 12.4 will discuss the general problem of digital and hybrid function generation and will present some useful methods for dynamic error analysis.

12.2 SIX-DEGREE-OF-FREEDOM AIRFRAME EQUATIONS

In this section we will be concerned specifically with the six-degree-of-freedom (three translational, three rotational) equations of motion of a flight vehicle such as an aircraft or missile flying in the atmosphere. Under these conditions the best choice of axis systems in which to write the equations depends on the predominant effect of the aerodynamic forces and moments. As we shall see in the next section, a completely different set of axes is appropriate for space-vehicle systems operating primarily out of the atmosphere.

A. Use of Flight-Path Axes vs. Body Axes

In connection with flight equations, it has been clear for many years that the use of flight-path axes (often called wind axes) as opposed to body axes for solving the translational equations makes much lower demands on computer accuracy and bandwidth.[2] Yet a number of current computer mechanizations continue to use body axes for solving the translational equations of motion. Figure 12.1 shows the interrelationships of the various axis systems. The body axes, x, y, and z, are a right-hand orthogonal set rigidly attached to the airframe. Stability axes, x_s, y_s, and z_s, differ from the body axes only by the angle of attack α. Aerodynamic force and moment data are often presented in terms of components along stability axes. The flight-path axes (frequently called wind axes), x_w, y_w, and z_w, differ from the stability axes by the angle of sideslip β. The x_w wind axis is aligned, by definition, in the direction of flight.

Let us assume that the air mass in which the vehicle flies is an inertial frame (this would be true for a flat earth with constant winds). We will let $\mathbf{V}_p$ be the vehicle velocity vector with respect to this frame. $\mathbf{V}_p$ will have

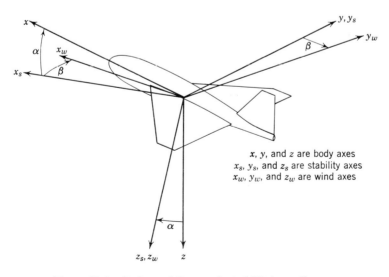

x, y, and z are body axes
x_s, y_s, and z_s are stability axes
x_w, y_w, and z_w are wind axes

Figure 12.1 Body, stability, and wind (flight-path) axes.

components U, V, W along the x, y, and z body axes, respectively. By definition, $\mathbf{V}_p$ will have components V_p, 0, and 0, along the wind axes x_w, y_w, and z_w, respectively. We will let $\mathbf{\Omega}$ be the vehicle angular velocity vector, with respect to the inertial frame. $\mathbf{\Omega}$ will have components P (roll rate), Q (pitch rate), and R (yaw rate) along the body axes x, y, and z respectively.

Let us denote external force components along the body axes by the symbols X_b, Y_b, and Z_b. These forces generally include contributions from gravity, propulsive and aerodynamic forces. If we sum forces along each body axis, we obtain the following three well-known equations:

$$m(\dot{U} - VR + WQ) = X_b \tag{12.1}$$

$$m(\dot{V} - WP + UR) = Y_b \tag{12.2}$$

$$m(\dot{W} - UQ + VP) = Z_b \tag{12.3}$$

where m is the mass of the vehicle. The inefficiency of these equations is immediately apparent when we consider the approximate size of the various terms. Let the vehicle be, say, a Mach 2 aircraft with $V_{\max} = 2000$ ft/sec. A reasonable upper limit on the pitch rate Q might be 2 radians/sec. Thus the term UQ in Equation 12.3 can get as large as 4000 ft/sec.² or 125 g's! On the other hand, Z_b/m, the normal acceleration due to external force (primarily aerodynamic lift) may have an upper limit of several g's. Hence, artificial accelerations, perhaps twenty to fifty times greater than the actual accelerations, are introduced because we are solving the translational equations of motion in terms of components along the rapidly moving body axes. In addition to scaling difficulties, we also have the high bandwidth angular rates P, Q, and R coupled inextricably with the translational equations; this is highly undesirable in a hybrid mechanization where the translational equations are to be solved digitally.

The use of flight-path axes eliminates these problems. Figure 12.2 shows a block diagram of the complete six-degree-of-freedom flight equations, where block 3 in the upper right-hand corner of the figure shows the translational equations of motion. Instead of U, V, and W, as in the case of the body axis equations, the velocity state variables become V_p, α, and β, all of which are directly needed for aerodynamic calculations. Equation 3.1 (in Figure 12.2) simply states that $\dot{V}_p$, the time rate of change of total vehicle velocity, is equal to X_w/m, where X_w is the force along the flight-path direction. For the case where the sideslip angle β is equal to zero, Equation 3.2 states that $\dot{\alpha}$, the time rate of change of angle of attack, is equal to the body axis pitch rate, Q, minus the flight-path axis pitch rate, $-Z_w/mV_p$, where Z_w is the external force along z_w in Figure 12.1.

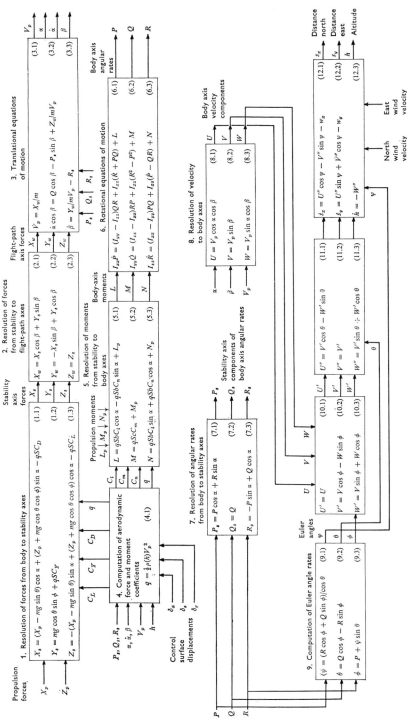

Figure 12.2 Block diagram of combined flight-path axis, body axis system for flat earth.

Finally, Equation 3.3 simply states that $\dot\beta$, the time rate of change of sideslip angle, is equal to the flight-path axis yaw rate, Y_w/mV_p, minus the body axis angular rate along z_w, namely R_s, where Y_w is the external force along y_w.

B. Six-Degree-of-Freedom Equations

Before describing the possible divisions between analog and digital computation in Figure 12.2, let us discuss briefly the various equations summarized in the figure. In block 1 the propulsion and gravity force components in body axes are added, resolved into stability axes, and summed with the aerodynamic forces. Block 2 shows the resolution of forces from stability to flight-path axes. As stated above, block 3 summarizes the translational equations. Block 4 represents the computations of the dimensionless aerodynamic coefficients and the dynamic pressure q. Note that in general the aerodynamic forces and moments are obtained by multiplication of the appropriate aerodynamic coefficient by qS, where S is the wing area. Block 5 shows the resolution of aerodynamic moments from stability to body axes and block 6 shows the rotational equations of motion. Here I_x, I_y, and I_z represent principal moments of inertia and I_{xz} is a product of inertia. We assume a flight vehicle that is symmetric about the xy plane. Block 7 shows the resolution of angular rates back to stability axes as needed for inputs to block 4. Block 8 resolves vehicle velocity into body axis components and block 9 represents the equations for computation of Euler angles ψ (heading), θ (pitch), and φ (bank angle). Finally, blocks 10, 11, and 12 resolve the velocity from body axis to earth axis components, so that distance north (s_x), distance east (s_y), altitude (h) can be computed. Here w_x and w_y are north and east wind components, respectively.

C. Hybrid System with Function Generation Performed Digitally

Block 4 in Figure 12.2 represents the computation of the aerodynamic coefficients. Very often these coefficients are complicated functions of a number of variables, including Mach number M, angle of attack α, sideslip angle β, altitude h, and control surface displacement δ. Numerous two, three, and even four variable functions are needed to represent the coefficients with high precision. All-analog generation of these functions is out of the question, so they are normally simplified to the point where analog function generation is feasible, or they are generated digitally in a hybrid computer system. Since the aerodynamic moments are involved in relatively high-frequency rotational-equation dynamic loops, the speed of digital function generation usually becomes an overall limiting factor in the speed of the hybrid simulation. Section 12.4 discusses quantitative

12.2/SIX-DEGREE-OF-FREEDOM AIRFRAME EQUATIONS

methods for estimating the sampling rates required to obtain acceptable accuracy.

The digital function generation is normally accomplished using table loop-up and linear interpolation. It is important to note that Mach number M and altitude h vary quite slowly compared with α, β, and δ in the multivariable functions. Thus it may be possible to interpolate with respect to M and h at a much lower rate. For example, a four variable function of M, h, α, and δ_a could be considered $L \times K$ functions of two variables α and δ_a, where L is the number of data points in α and K is the number of data points in δ_a. Perhaps the $L \times K$ two variable functions would only need to be updated, say, once per second as a result of interpolations with respect to M and h, whereas the final interpolation with respect to α and δ_a could be done at, say, one hundred times per second.

Note that in blocks 1 and 5 the analog multiplications of the form qSC and SbC are required, where q is the dynamic pressure, slowly varying and computed digitally and C is the aerodynamic coefficient. (S is a constant representing wing area and b is a constant representing wing span.) This multiplication in each equation is very nicely performed using MDAC's (multiplying digital to analog converters) where the dynamic pressure in analog form is applied to the analog input of each MDAC and the digital word representing the aerodynamic coefficient is applied to the register input of each MDAC. In this way, considerable equipment is saved and potentially greater accuracy is obtained over all-analog multiplication. Another alternative is to perform the multiplication digitally before conversion to analog form, in which case the digital computer outputs the three force components $-qSC_D$, qSC_Y, and $-qSC_L$, and the three moment components $qSbC_l$, $qScC_m$, $qSbC_n$, directly. This is not as accurate, however, as the MDAC scheme, since the total number of bits in the digital to analog converters will limit the dynamic range.

Added bandwidth can be obtained using hybrid function-generation techniques, one of which is discussed in Section 12.4C. Also, the mechanization of the coefficients themselves can be altered at times to produce better dynamic behavior. Consider, for example, the pitching moment coefficient C_m, which typically might be a function of Mach number M, angle of attack α, and elevator displacement δ_e. We can write C_m as follows:

$$C_m(M, \alpha, \delta_e) = [C_{M_\alpha}(M, \alpha, \delta_e)]\alpha + [C_{M_{\delta_e}}(M, \alpha, \delta_e)]\delta_e \quad (12.4)$$

Although we have created two three-variable functions where we had only one before, the new functions C_{M_α} and $C_{M_{\delta_e}}$ may be less strongly dependent on α and δ_e than C_M itself. If we perform the required multiplications,

$C_{M_\alpha}\alpha$ and $C_{M_{\delta_e}}\delta_e$, using MDAC's with α and δ_e the analog inputs, respectively, then analog bandwidth is maintained in the C_M calculation. It is analogous to computing a somewhat nonlinear spring force by multiplying the almost constant spring rate by the displacement, as opposed to generating the spring force directly. Clearly, if C_{M_α} and $C_{M_{\delta_e}}$ are strong functions of α and δ_e (which they frequently are), then little will be gained except for the case of small-motion transients in α and δ_e about equilibrium values.

D. Hybrid System with Translational Equations Performed Digitally

A next logical step in increasing the digital portion of the flight equations is to solve the translational equations of motion on the digital computer. In this case the equations shown in blocks 1 and 2 of Figure 12.2 are solved digitally, as well as Equation 3.1 and the Z_w/mV_p and Y_w/mV_p computation in Equations 3.2 and 3.3. Thus, $\dot{V}_p$ is integrated digitally to obtain V_p, while $\dot{\alpha}$ and $\dot{\beta}$ are integrated in the analog system to obtain α and β after the terms Z_w/mV_p and Y_w/mV_p are converted to analog signals. In this way the only terms with significant high-frequency content, namely $Q \cos \beta - P_s \sin \beta$ in Equation 3.2 and R_s in Equation 3.3, are always in analog form.

It may also be advisable to solve the equations in block 8 on the digital computer, as well as the coordinate conversion and integration indicated in blocks 10, 11, and 12. Here it should be noted that high-frequency components will be present in the Euler angles ψ, θ, and φ, so that the resolutions shown, when performed digitally, will contain dynamic errors. On the other hand, if we are dealing with an airframe in a near equilibrium cruise condition, these high-frequency errors will be minimal in effect and good trajectory accuracy will result on the average. Solution of the rotational equations in blocks 5, 6, and 9 would still be achieved by analog means; we might also consider mechanizing block 7 and the rotational stability derivatives C_l, C_m, and C_n in block 4 using the analog system. In this way, all of the high-frequency rotational degree-of-freedom loops would be devoid of significant digital elements.

E. Hybrid System with Only Control System Dynamics Simulated on the Analog Subsystem

Finally, we might wish to mechanize all of Figure 12.2 digitally which, as we pointed out earlier, may very well be feasible in real time. However, in most flight simulations there is a dynamic relationship between output state variables such as θ, φ, ψ, and h and the control surface

12.2/SIX-DEGREE-OF-FREEDOM AIRFRAME EQUATIONS

displacements δ_a, δ_e, and δ_r. This relationship is either established by a human operator or an autopilot. In the latter case the autopilot control surface servos may have natural frequencies well beyond the digital computer capability. In this case an analog simulation of the autopilot becomes the obvious choice. Control of the power plant thrust by an autopilot can still be performed digitally, since the time constants tend to be relatively long.

F. Miscellaneous Hybrid Configurations

Obviously there is a never ending list of alternative problem divisions between analog and digital subsystems, depending on the available hybrid computer and the simulation requirements. For example, we may be simulating an all-digital flight control system, in which case the digital computer should indeed be used for the portion of the control system which is digital in addition to any other portions of the system.[3]

G. Alternative Methods for Representing Angular Orientation

It is evident that conventional aircraft Euler angles become indeterminate when the pitch angle $\theta = \pm 90$ degrees. Under these conditions, it is necessary to go to a four-angle system,[4] direction cosines, or quaternions.[5] None of these systems is particularly desirable from a complexity point of view, and the direction-cosine scheme is the most complicated of all three if corrective loops must be utilized to prevent long-term drifts due to redundant integration.[5] Whether quaternions or some other four-parameter system may be best depends on the particular application, including control and display requirements.

If accurate computation of Euler angles is not required under highly unusual flight conditions, then wind axis (flight-path axis) Euler angles θ_w, φ_w, and ψ_w can be used to simplify the computations in blocks 8, 10, 11, and 12 in Figure 12.2. The following approximation formulas relate the body axis Euler angles θ, φ, and ψ to θ_w, φ_w, and ψ_w

$$\theta \cong \theta_w + \alpha \cos \varphi_w + \beta \sin \varphi_w \qquad (12.5)$$

$$\varphi \cong \varphi_w \qquad (12.6)$$

$$\psi \cong \psi_w + \alpha \sin \varphi_w - \beta \cos \varphi_w \qquad (12.7)$$

Here θ_w, φ_w, and ψ_w are obtained from formulas identical with Equations 9.1, 9.2, and 9.3 in Figure 12.2, except for subscripts w throughout. The wind axis (flight-path axis) angular rates P_w, Q_w, and R_w are obtained from the following formulas:[2]

$$P_w \cos \beta = P_s + Q_w \sin \beta \qquad (12.8)$$

$$Q_w = -\frac{F_{wz}}{mV_p} \qquad (12.9)$$

$$R_w = \frac{F_{wy}}{mV_p} \qquad (12.10)$$

Reference to block 3 in Figure 12.2 shows that Q_w and R_w have already been computed as part of the mechanization of $\dot\alpha$ and $\dot\beta$. Finally, the following formulas replace the equations in blocks 10, 11, and 12:

$$\dot s_x = V_p \cos \theta_w \cos \psi_w - W_x \qquad (12.11)$$

$$\dot s_y = V_p \cos \theta_w \sin \psi_w - W_y \qquad (12.12)$$

$$\dot h = V_p \sin \theta_w \qquad (12.13)$$

The approximations given in Equations 12.5, 12.6, and 12.7 are generally valid as long as less than three of the four angles α, β, θ_w, and φ_w are not large in magnitude simultaneously. For any but highly transient flight conditions, this represents a very good assumption. Note that mechanization of Equations 12.5 through 12.8 and 12.11 through 12.13 represents a considerable simplification over the equations given in blocks 8, 10, 11, and 12 in Figure 12.2.

H. Refinements in Trajectory Equations

In writing the equations given in Figure 12.2, we assumed a flat earth with constant surface winds. For a spherical earth that is rotating, the translational equations become much more complicated; thus it is no longer possible to write exact flight-path axis translational equations without introducing undue complexity. If we are willing to neglect the influence of centrifugal and coreolis acceleration on the trajectory, however, we can use the equations of Figure 12.2 along with the following formulas to obtain latitude L and longitude λ:

$$\dot L = \frac{\dot s_x}{r_0 + h}, \qquad \dot \lambda = \frac{\dot s_y}{(r_0 + h) \cos L} \qquad (12.14)$$

where r_0 is the radius of the earth.

12.3 SIMULATION OF SPACE VEHICLES

In this section we will consider the simulation by hybrid means of vehicles which spend most of their flight outside any planetary atmosphere. Under these conditions, a rather different axis system than the one suggested in the previous section is more efficient for the translational equations of motion.

12.3/SIMULATION OF SPACE VEHICLES

This is the H-frame axis system which utilizes horizontal velocity U_h, vertical velocity W_h (positive downward), and flight-path heading angle ψ_h as the three velocity state variables. This particular system has the advantage of especially good scaling for near circular orbits; it also allows use of an angular momentum integral to avoid an open-ended integration when the vehicle is in free-fall out of the atmosphere.[6]

A. Six-Degree-of-Freedom Equations Using the H-Frame

Figure 12.3 shows a block diagram of the six-degree-of-freedom flight equations using H-frame axes for the translational equations and body axes (as in Section 12.2) for the rotational equations. Note that body axes continue to represent the only practical axis system for the rotational equations, since the vehicle moments of inertia are fixed only in this system.

The equations in Figure 12.3 are exact for a rotating, oblate earth, except that gravitational perturbation forces from other planets have not been shown. These could easily be added, probably most conveniently to block 3. Notation is similar to that in Figure 12.2, where applicable, and is otherwise standard throughout. Block 1 shows the formulas for computing aerodynamic and jet reaction forces in body axes. Block 2 resolves these forces to E-frame components (i.e., components north, east, and downward). Here the noncentral force-field gravity terms g_x and g_z are added. Direction cosines $l_{1,2,3}$, $m_{1,2,3}$, and $n_{1,2,3}$ are utilized in this transformation. Block 3 resolves the forces into H-frame coordinates (horizontal and in the plane of the motion and downward). Finally block 4 summarizes the three translational equations, where the central force-field gravity term has been added to the $\dot{W}_h$ equation. Note that the equation for U_h, the horizontal velocity, has already been integrated and represents conservation of angular momentum. In the absence of an external horizontal force X_h in the plane of the motion, U_h becomes merely an algebraic function of radial distance r. This avoids an open-ended integration of $\dot{U}_h$ to obtain U_h. We will discuss further improvements in mechanization of these equations in Section 12.3D.

Block 5 resolves the H-frame velocities into velocity north, U_e, and velocity east V_e. Block 8 converts these to latitude L and longitude λ, where ω_e is the earth's spin rate. Block 6 resolves the E-frame velocity into body axes after adding wind components and earth spin components. The resulting vehicle velocity components U_{ab}, V_{ab}, and W_{ab} with respect to the air mass are used in block 7 to compute α, β, and total aerodynamic velocity V_a. Block 9 computes the aerodynamic and jet reaction moments in body axes. These are inputs to block 10, which gives the rotational equations of motion. Here P_b, Q_b, and R_b are pitch, roll, and yaw rates as viewed from an inertial frame. Block 11 corrects these to conventional

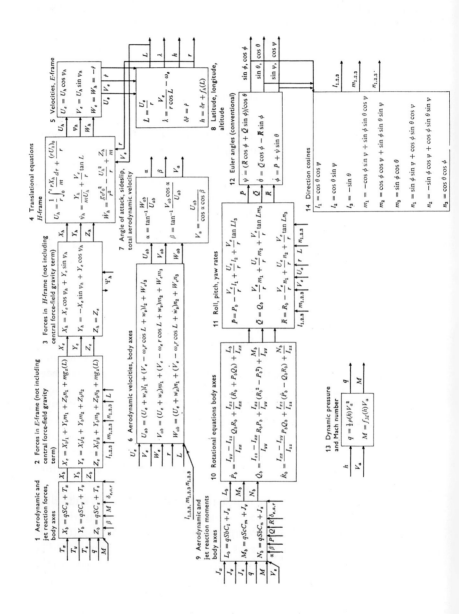

pitch, roll, and yaw rates as viewed from the E-frame (north, east, and downward). Block 12 shows the computation of conventional Euler angles from which the nine direction cosines in block 14 are computed. Finally, block 13 gives the equations for dynamic pressure and Mach number, M.

B. Possible Hybrid Configurations for Space-Vehicle Simulation

Because of the rather severe accuracy requirements which normally accompany space-vehicle trajectory calculations, it is natural to consider digital mechanization of the translational equations of motion in Figure 12.3. At a very minimum, this would include the computations in blocks 3, 4, 5, and 6. For flight out of the atmosphere, computation of the translational forces in blocks 1 and 2 by analog means would not yield significant errors, since the force levels are very low. In fact, when no power plant or aerodynamic forces are present, the analog voltage outputs representing those quantities could be switched to zero and the gravitational perturbation g_x and g_z could be computed digitally. We could probably also use digital mechanization of the computations in block 13, i.e., dynamic pressure and Mach number. Next, we should consider implementing on the digital computer all of the computations in blocks 1 and 2. This would cause high-speed dynamics in the direction cosines to be mixed into the translational force computations, but any resulting dynamic errors get smoothed effectively in blocks 4, 9, and 8 when the integrations are performed.

Blocks 6 and 7 represent a serious problem. Because of the enormous dynamic range in the velocity components during exit and reentry trajectories, we would like to implement the computation digitally. On the other hand, high-speed dynamics in the direction cosines will definitely affect the computation of α and β in block 7, and these represent primary contributions to aerodynamic moments in the rotational loops. If high computing speed is necessary, analog mechanization will have to prevail, with a subsequent loss in accuracy and dynamic range. If slow (e.g., real time or less) computing speed is possible, then digital mechanization should get the nod.

If it is decided to mechanize blocks 6 and 7 digitally, then digital mechanization of all the remaining blocks should probably be considered. This is because the digital computer has already been inserted into the high-frequency pitch, roll, and yaw loops due to the α and β computation. This still leaves room, however, for analog computation of autopilot dynamics relating θ, φ, and ψ with jet reaction control thrusts J_x, J_y, and J_z, as well as control surface displacements δ_a, δ_e, and δ_r.

If high-speed dynamics beyond digital capability must be preserved, then clearly blocks 6 through 12 and block 14 must remain analog, with the exception of any possible use of digital or hybrid function generation in computing C_l, C_m, and C_n.

Finally, we should note that the computation of jet reaction moments often involves a rather complicated set of logical operations to relate the turn-on and turn-off of individual control jets to pitch, roll, and yaw commands. This can, in many cases, be performed very nicely using the parallel patchable logic which is an integral part of current state-of-the-art analog subsystems.

C. Mechanization of Trajectory Computations

The translational equations presented in block 4 of Figure 12.3 can be further modified to increase computational accuracy whether using analog or digital systems. To make presentation of the equations simpler, let us define the following dimensionless variables:

$$\delta u_h = \frac{U_h}{\sqrt{g_0 r_0}} - 1, \quad w_h = \frac{W_h}{\sqrt{g_0 r_0}}, \quad \delta \rho = \frac{r}{r_0} - 1, \quad \tau = t \sqrt{\frac{g_0}{r_0}} \tag{12.15}$$

where g_0 is the central force-field gravity acceleration at radial distance r_0. Note that $\sqrt{g_0 r_0}$ is the circular orbit velocity for $r = r_0$ and that w_h is a dimensionless vertical velocity. δu_h is a dimensionless variation in horizontal velocity. $\delta \rho$ represent a dimensionless radial variation from r_0. τ is dimensionless time such that a circular orbit of radius r_0 has a period of 2π. We would normally select r_0 to be the mean radius of the expected trajectory.

In terms of these new coordinates, the H-frame translational equations of motion become the following:[7]

$$u_h = \frac{1}{1 + \delta \rho} [H - \delta \rho] \tag{12.16}$$

$$\frac{dw_h}{d\tau} = -\frac{\delta \rho/(1 + \delta \rho) + 2\delta u_h + (\delta u_h)^2}{1 + \delta \rho} + \frac{Z_h}{mg_0} \tag{12.17}$$

$$\frac{d\psi_h}{d\tau} = \frac{1}{1 + \delta u_h}\left[\frac{Y_m}{mg_0}\right] + \frac{(1 + \delta u_h) \sin \psi_h \tan L}{1 + \delta \rho} \tag{12.18}$$

$$\frac{d\delta\rho}{d\tau} = -w_h \tag{12.19}$$

$$\frac{dL}{d\tau} = \frac{(1 + \delta u_h) \cos \psi_h}{1 + \delta \rho} \tag{12.20}$$

$$\frac{d\lambda}{d\tau} = \frac{(1 + \delta u_h) \sin \psi_h}{(1 + \delta \rho) \cos L} - \omega_e \sqrt{\frac{g_0}{r_0}} \tag{12.21}$$

12.3/SIMULATION OF SPACE VEHICLES

where

$$H = \int_0^{\tau} (1 + \delta\rho) \frac{X_h}{mg_0} d\tau' + [\delta\rho + \delta u_h + \delta\rho \, \delta u_h]_{\tau=0} \quad (12.22)$$

Here H is a dimensionless angular momentum. It should be noted that the first term on the right side of Equation 12.17 represents directly the difference between centrifugal and central force-field gravity accelerations. This avoids the scaling problems inherent in computing centrifugal and gravity accelerations directly and then subtracting them. Because of this feature, along with the open-ended integration which has been avoided by use of the angular momentum integral, the translational equations shown can be used very effectively to mechanize accurate analog computation of satellite and space vehicle trajectories.[6,7] They have also proven to be extremely useful in all-digital mechanization, allowing a much larger integration step size for a given accuracy requirement or, conversely, much better accuracy for a given step size. The utilization of variation, $\delta\rho$, in dimensionless radius and variation, δu_h in dimensionless horizontal velocity permits up to a factor of two improvement in scaling. This is not only important in analog mechanization but also in special-purpose digital mechanizations where we might like to use fixed-point computation.

Note that in Equations 12.16 through 12.22 the noncentral force-field force components X_h, Y_h, and Z_h all appear as accelerations in units of g_0. In the solution of trajectory problems the vehicle angle of attack α and bank angle φ are normally considered to be control input variables when in the atmosphere. Thus the rotational degrees of freedom shown in Figure 12.3 do not need to be considered. Instead, the translational equations presented in this section can be used. The effect of α is introduced by calculating the resulting lift and drag forces acting on the vehicle. These are, in turn, revolved through the bank angle φ and flight-path pitch angle θ_w into horizontal in-plane force X_h, horizontal out-of-plane force Y_h, and downward force Z_h. In this way, accurate analog mechanization of ballistic and lifting reentry trajectories can be achieved.[6,7] This capability is very important in considering hybrid mechanization of trajectory optimization problems.

Additional refinement of the translational equations of motion presented in this section involves the inclusion of an energy constraint in the equations. Dimensionless total energy E_1 can be written in terms of the state variables. Thus

$$E_1 = \frac{(1 + \delta u_h)^2 + w_h^2}{2} - \frac{1}{1 + \delta\rho} \quad (12.23)$$

where the first term on the right side of Equation 12.23 is the kinetic energy and the second term is the potential energy. In the presence of additional external forces the following expression can also be used to calculate total energy:

$$E_2 = \int_0^\tau \left[(1 + \delta u_h) \frac{X_h}{mg_0} + w_h \frac{Z_h}{mg_0} \right] d\tau' + E_{\tau=0} \quad (12.24)$$

Ideally, E_1 should equal E_2, but due to computational errors this will not be true. We can define an error $\epsilon = E_2 - E_1$, compute this error continuously from Equations 12.22 and 12.23, and add correction terms on the right sides of Equations 12.17 and 12.19 proportional to the negative gradient of ϵ^2. Thus the correction term to be added to Equation 12.17 becomes

$$\Delta \frac{dw_h}{d\tau} = -K_w \epsilon \frac{\partial \epsilon}{\partial w_h} \quad (12.25)$$

and for Equation 12.19 becomes

$$\Delta \frac{d\delta}{d\tau} = K_\rho \epsilon \frac{\partial \epsilon}{\partial \delta \rho} \quad (12.26)$$

where K_w and K_ρ are constants.

This "steepest descent" method† of constraint enforcement has numerous applications.[8] In the particular case here, it has been shown[7] that the scheme works effectively even if only the correction given in Equation 12.25 is applied, i.e., if $K_\rho = 0$. Solving for $\partial \epsilon / \partial w_h$ and adding Equation 12.25 to the right side of Equation 12.17, we obtain[7]

$$\frac{dw_h}{d\tau} = \frac{1}{1 + \delta\rho} \left(\frac{\delta\rho}{1 + \delta\rho} + w_h^2 - 2E_2 \right) + K_w \epsilon w_h \quad (12.27)$$

where

$$\epsilon = E^2 - \frac{\delta\rho}{1 + \delta\rho} - \delta u_h - \tfrac{1}{2}(\delta u_h)^2 - \tfrac{1}{2}w_h^2 \quad (12.28)$$

By the use of this technique the exact closure of Keplerian orbits is assured and sizeable accuracy improvements are obtained when the space-vehicle trajectory is computed over very long times.

12.4 DIGITAL FUNCTION GENERATION‡

In Section 12.2C it was pointed out that a principal application of hybrid computers involves the use of the digital subsystem for multivariable

---

† See Chapter 9.
‡ See Chapter 7 for a further discussion of function generation software.

12.4/DIGITAL FUNCTION GENERATION 379

function generation and the analog system for integration and other operations. The use of digital function generation has proven particularly effective for functions of two or more variables. Typically, such multivariable functions are specified in terms of a table of function values at specified values of the independent variables. For example, a two variable function $f(x, y)$ would normally have f specified for L discrete x values, x_i, at each of K discrete y values y_j. The resulting $L \times K$ matrix of $f_{i,j}$ values ($f_{i,j} = f(x_i, y_j)$) is stored in digital memory starting with a specified location for $f_{0,0}$ and proceeding sequentially in a predetermined order, e.g., $f_{0,0}, f_{1,0}, f_{2,0} \cdots, f_{0,1}, f_{1,1}, f_{2,1}, \ldots$, etc.

If the data points x_i and y_j can be chosen with equal spacing, Δx and Δy, respectively, then it is not even necessary to store the x_i and y_j values in memory. Equal spacing also simplifies and, therefore, speeds up the computation of memory look-up locations as well as the required linear interpolation.[9] Note that the equivalent of arbitrary data point locations in original function variables $\bar{x}$ and $\bar{y}$ can always be obtained by generating two transformation functions $x = X(\bar{x})$ and $y = Y(\bar{y})$ such that the desired data points $\bar{x}_i$ and $\bar{y}_j$ produce equally spaced data points x_i and y_j, respectively.

Usually it is not worth the added complexity to implement higher than first-order (linear) interpolation. This is particularly true when the function to be represented is derived from empirical data, as is often the case with aerodynamic functions. The question frequently arises as to whether the function generation program should be written in a compiler language (e.g., FORTRAN) or in assembly language. Certainly, if the utmost in speed is the objective, then assembly language should be used. When this is done, there are a number of tricks which can be used to increase the speed further, such as choice of Δx to be an integral power of two with the result that divisions by Δx can be performed by simple shift operations.[9]

Also note that although there are frequently a large number of multivariable functions which must be generated in a complete flight simulation, they are typically functions of the same variables such as Mach number M, altitude h, angle of attack α, and control surface displacement δ. This makes digital function generation considerably faster than it would be if each multivariable function utilized different variables. It has already been pointed out in Section 12.2C that digital function generation can be speeded up by interpolating at multiple rates. Thus, interpolations with respect to the slow variables such as Mach number and altitude can be performed very much slower than interpolations with respect to α. In a digital system with an extended, medium access speed memory, such as a disk file, schemes like this can be used to reduce greatly the amount of high-speed memory required. The original data points

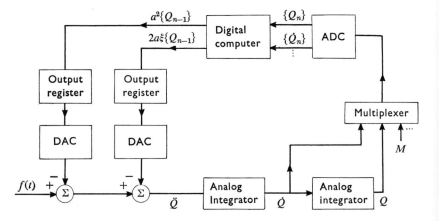

Figure 12.4 Block diagram of hybrid computing loop for second-order system.

representing all the variables are stored on the disk, but only those data points bracketing the current values of the slow variables M and h are read out into core to be used for high-speed interpolation.

The length of time required to generate each multivariable function depends obviously on the particular digital computer, the program that is used, whether the function variables are common with other functions, etc. A typical time for a two-variable function computed in current digital systems might range between 50 and 500 microseconds. For each added variable the time will roughly double.

A. Dynamic Error Analysis

If function generation in a hybrid flight simulation is to be performed digitally with the remainder of the computation to be performed with the analog system, then it becomes important to know quantitatively how much dynamic error is introduced because of the finite time delay and effective sample rate of the digital computation. Such error analysis is almost impossible to perform in general, but simple subloops in the hybrid system can be isolated, linearized, and analyzed as sampled-data systems using z-transforms as discussed in Chapter 4.[10]

For example, it can be shown that if small disturbances from an equilibrium symmetrical flight condition are assumed, the following linearized equation represents approximately the pitch-rate response Q to an elevator displacement δ_c:[11]

$$\ddot{Q} + 2a\xi\dot{Q} + a^2 Q = f(t) = K(\delta_c + \tau_s \dot{\delta}_c) \qquad (12.29)$$

where a is the undamped natural frequency of the short-period pitching motion and ξ is the damping ratio. In a hybrid mechanization of the flight equations the terms $2a\xi \dot{Q}$ and $a^2 Q$ would actually be nonlinear functions of other variables and hence would probably be computed digitally. Figure 12.4 shows a block diagram representative of the computation. The diagram indicates that both Q and $\dot{Q}$ get converted to data sequences $\{Q_n\}$ and $\{\dot{Q}_n\}$ by the multiplexer and analog-to-digital converter. These sequences are fed into the digital computer which feeds out two data sequences, each delayed by the function generation frame time T. The two data sequences are a^2 and $2a\xi$, respectively, times the two input data sequences. Each of the output data sequences is converted to analog form and subtracted from $f(t)$ to produce $\ddot{Q}$, as shown in the figure. The equivalent diagram in Figure 12.5 shows the blocks in terms of Laplace transforms when the outputs are continuous and z-transforms when the outputs are data sequences. The two switches represent the conversion of $Q(t)$ and $\dot{Q}(t)$ into data sequences with z-transforms $Q(z)$ and $Q_1(z)$, respectively. The block labeled $a^2 z^{-1}$ indicates multiplication by a^2 and delay by the sample period T. Similarly, the block labeled $2a\xi z^{-1}$ indicates multiplication by $2a\xi$ and delay by T seconds. The transfer function $H_e(s)$ represents a digital-to-analog converter with accompanying register to provide zero-order hold between updatings. To simplify the diagram, we have added the two data sequences and followed the sum with a single digital-to-analog converter rather than showing two separate converters. The results are entirely equivalent.

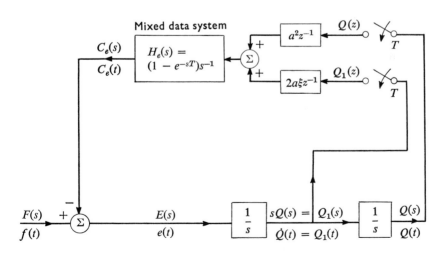

Figure 12.5 Representation of hybrid computing loop in terms of transforms.

Analysis of the system in Figure 12.5 using z-transforms shows that the real part σ and the imaginary part $\pm\omega$ of the equivalent characteristic roots are given by the following formulas when $\xi < 1$:[12]

$$\sigma \cong -a\xi + (\tfrac{3}{4} - 3\xi^2)aT, \qquad aT \ll 1 \qquad (12.30)$$

$$\omega \cong a\sqrt{1 - \xi^2}\left[1 + \frac{9\xi - 12\xi^3}{4(1 - \xi^2)}\,aT\right], \qquad aT \ll 1 \qquad (12.31)$$

The characteristic roots for the exact solution to Equation 12.29 are given by $\sigma = -a\xi$ and $\omega = a\sqrt{1 - \xi^2}$. By using Equations 12.30 and 12.31, we can compute the error in damping ratio and frequency which the digital feedback loop causes. Conversely, for a given maximum error requirement, we can specify the maximum frame period T which can be allowed for the digital function generation. For example, if $\xi = 0$ in the ideal system and we wish the hybrid loop to exhibit a ξ no more negative that $-.01$, then Equation 12.30 shows that $.75aT \leq .01$ or $(aT)^{-1} = 75$. This in turn implies 75 samples per radian or $2\pi(75) = 443$ samples per cycle of the short-period pitching motion represented by Equation 12.29.

It should be pointed out that the hybrid mechanization of Equation 12.29 as shown in Figure 12.4 does not bear quite a one-to-one correspondence with the mechanization that would result from Figure 12.2. However the results given by Equations 12.30 and 12.31 should be quite similar. We can also estimate the dynamic errors which occur for a forcing function $f(t)$ by comparing the ideal transfer function represented by Equation 12.29 for sinusoidal inputs with the hybrid loop z-transform.[10,12] As expected, the transfer-function error turns out to be of order aT.[12]

As indicated in Chapter 5, it can be shown that the zero-order extrapolation mechanized in an analog-to-digital converter is equivalent (to first-order) to a time delay of $T/2$ seconds.[10] This, along with the delay of T seconds in the digital computation, suggests a net equivalent feedback-loop delay of 1.5 seconds. Using the technique of Chapter 5, this can be compensated to first-order by supplying each analog to digital converter channel with the input variable plus $1.5T$ times the derivative of the input variable. Thus the multiplexer input Q in Figure 12.4 is replaced by $Q + 1.5\dot{Q}$, and the input $\dot{Q}$ is replaced by $\dot{Q} + 1.5T\ddot{Q}$. Since both $\dot{Q}$ and $\ddot{Q}$ are available explicitly in the simulation, the mechanization is easily accomplished. Care must be taken in selecting the multiplexer sampling instant for $\ddot{Q}$, since $\ddot{Q}$ is undergoing jumps whenever the digital-to-analog converter is updated. The sample must be taken just prior to digital-to-analog updating. Under these conditions, we can obtain the z-transform of the output Q from which the following equivalent characteristic roots

$\lambda = \sigma \pm j\omega$ are found:[12]

$$\sigma = -a\xi[1 - (3 + \tfrac{38}{3}\xi^2)a^2T^2], \qquad aT \ll 1 \quad (12.32)$$

$$\omega = a\sqrt{1-\xi^2}\left[1 + \frac{\tfrac{13}{24} - \tfrac{43}{6}\xi^2 + \tfrac{5}{2}\xi^4}{1-\xi^2}a^2T^2\right], \qquad aT \ll 1 \quad (12.33)$$

Note that the error in σ and ω varies as $(aT)^2$ rather than aT. Clearly the use of updating by $1.5T$ has decreased very significantly the error in equivalent characteristic roots for the hybrid computing loop.

For the case where $\xi = 0$, Gilbert has shown,[10] that $\sigma \simeq a^4T^3/2$ and $\omega \simeq a(1 + 13a^2T^2/24)$. The value of ω agrees with Equation 12.33 above, and the value of σ, of order T^3, is consistent with Equation 12.32. For the hybrid loop without updating and with $\xi = 0$, we found from Equation 12.30 that some 443 samples/cycle were needed to keep the effective damping ratio error below .01 in magnitude. Reference to Gilbert's results results shows that only 23 samples/cycle are needed for the same damping ratio error magnitude. More accurate but complex updating schemes can improve the accuracy even further.[11]

Although the results obtained in this section apply only to a linearized approximation to the longitudinal dynamic behavior of a flight vehicle, they are very indicative of what the actual behavior will be. Since the short-period motion represents invariably the highest frequency of any rigid body oscillatory mode for the flight vehicle, the results obtained so far in this section give an excellent insight into the digital-computer frame rate requirements for a full six-degree-of-freedom nonlinear flight simulation. For typical high-performance aircraft, the frequency a ranges between 1 and 10 radians/sec and ξ ranges from .1 to 1.

Before turning to a consideration of dynamic error analysis of the dominant lateral mode of flight vehicles, we should point out that in the case of the second-order hybrid loop without updating, as illustrated in Figures 12.4 and 12.5, there is a third characteristic root due to the one-period delay in the digital feedback which corresponds to a rapidly decaying exponential transient and is negligible in effect.[12] For the case where we use updating, there are two extra roots introduced; they correspond to a rapidly decaying oscillatory transient with a frequency equal to approximately one-fourth the sample frequency. Again the effect appears to be negligible.[12] A dominant lateral mode of the linearized flight equations can be represented by the equation[11]

$$\dot{P} + aP = f(t) = K\delta_a(t) \qquad (12.34)$$

where P is the roll rate and δ_a is the aileron displacement. This is known as the uncoupled rolling motion. The value of a ranges from 1 to 10 sec^{-1}

(equivalent to a time constant between .1 and 1 second) for typical high-performance aircraft. For a hybrid mechanization of Equation 12.34 similar to that shown in Figure 12.4 but with only one analog integrator in the forward loop, the equivalent characteristic root is given by:[12]†

$$\lambda = -a(1 + \tfrac{3}{2}aT), \qquad aT \ll 1 \qquad (12.35)$$

This should be compared with the ideal value of $\lambda = -a$. The extra root introduced by the feedback delay corresponds to a rapidly decaying oscillation at one-half the sample frequency. When updating is used ($P + 1.5T\dot{P}$ fed into the analog to digital converter), the equivalent characteristic root becomes[12]

$$\lambda = -a(1 - \tfrac{23}{12}a^2T^2), \qquad aT \ll 1 \qquad (12.36)$$

As expected, the error varies as T^2 compared with T in the case of no updating. There are two extra roots introduced when updating of the hybrid feedback loop is used. As in the case of the second-order system with updating, these two roots represent a rapidly decaying oscillation at one-fourth the sample frequency. The above formula can be used to estimate digital computer frame rate requirements when simulating a first-order system with digital function generation and analog integration. We have seen that by updating the input variables for digital function generation to take into account the time delay, rather dramatic improvement in overall computing accuracy can be obtained. In the cases considered here, the required time derivatives of the input variables were available explicitly. This may not always be true, in which case it may be necessary to differentiate the input variables, at least approximately, to implement the updating scheme. Before considering specific means for such differentiation, however, let us examine the problem of smoothing the zero-order hold output from the digital–analog converter channels. Frequently such smoothing is not necessary since the data may be fed directly to an integrator which accomplishes the smoothing itself. However, if smoothing is required, a low pass filter having a linear phase shift with frequency is highly desirable. The reason is that a linear phase lag with frequency corresponds to a pure time delay which can be compensated along with the other time delays in the function generator loop, using the input signal updating. For example, consider a second-order, low pass filter with natural frequency ω_n and a damping ratio ξ. The filter transfer operator $Y_F(p)$ is given by

$$Y_F(p) = \frac{1}{\dfrac{1}{\omega_n^2}p^2 + \dfrac{2\xi}{\omega_n}p + 1} \qquad (12.37)$$

† See also Chapter 4, Section 4.5.

12.4/DIGITAL FUNCTION GENERATION

For sinusoidal inputs the transfer function becomes

$$Y_F(j\omega) = \frac{1}{1 - \frac{\omega^2}{\omega_n^2} + j2\xi\frac{\omega}{\omega_n}} \tag{12.38}$$

The filter phase shift N is given by

$$N = \tan^{-1}\frac{2\xi\frac{\omega}{\omega_n}}{1 - \frac{\omega^2}{\omega_n^2}} \simeq -2\xi\frac{\omega}{\omega_n}, \qquad \omega \ll \omega_n \tag{12.39}$$

and the magnitude M is

$$M = \frac{1}{\sqrt{\left(1 - \frac{\omega^2}{\omega_n^2}\right)^2 + \left(2\xi\frac{\omega}{\omega_n}\right)^2}} \tag{12.40}$$

For $\xi = \sqrt{2}/2$ the ω^2 term in the radical vanishes and we have

$$M = \frac{1}{\sqrt{1 + \left(\frac{\omega}{\omega_n}\right)^4}} \tag{12.41}$$

Thus the magnitude of the filter transfer function remains very close to unity for input frequencies $\omega < \omega_n$. This is known as a second-order Butterworth filter and can be useful as an output-smoothing filter for digital-analog converters. The filter response to a sinusoidal input $Ae^{j\omega t}$ is given by

$$f_0 = Y(j\omega)Ae^{j\omega t} = Me^{jN}Ae^{j\omega t} \tag{12.42}$$

If the filter phase shift N is proportional to frequency, as is approximately true in Equation 12.39 for $\omega \ll \omega_n$, then we can write 12.42 as follows:

$$f_0 = MAe^{j\omega(t - 2\xi/\omega_n)} \tag{12.43}$$

Clearly if $M \simeq 1$, the filter output f_0 is equal to the input $Ae^{j\omega t}$ delayed in time by $2\xi/\omega_n$. But this time delay can be compensated by updating the function input variables by an additional $2\xi/\omega_n$ seconds. For $\xi = \sqrt{2}/2 = .707$, the filter transfer function approaches an ideal $M = 1$ characteristic for $\omega < \omega_n$ and yields a corresponding time delay of $.707/\omega_n$ seconds. A reasonable value of ω_n, the filter natural frequency, might range between .2 and .5 times the repetition frequency of the zero-order hold signal being filtered. A three-amplifier circuit for implementing the filter characteristic given by Equation 12.37 is shown in Figure 12.6, where the digital-analog

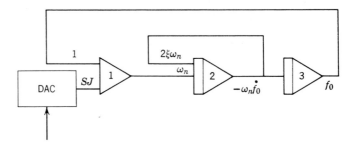

Figure 12.6 Example of second-order digital-analog smoothing circuit.

converter is terminated into the summing junction of an operational amplifier. One-amplifier circuits having the transfer operator of Equation 12.37 can also be implemented.

A simpler alternative for a smoothing filter is to use the first-order filter shown in Figure 12.7. Here the transfer operator is given by

$$Y_F(p) = \frac{1}{1 + \tau p}, \qquad \tau = RC \qquad (12.44)$$

where τ is the filter time constant. For sinusoidal inputs the filter magnitude and phase shift become, respectively,

$$M = \frac{1}{\sqrt{1 + (\tau\omega)^2}}, \qquad N = -\tan^{-1} \tau\omega \cong -\tau\omega, \qquad \omega \ll 1/\tau \qquad (12.45)$$

Clearly the filter time delay at low frequencies is τ which, as before, can be compensated by updating the function input variables by an additional τ seconds. Comparison of Equations 12.40 and 12.45 for filter magnitude M shows that the second-order filter provides more ideal filtering action. On the other hand, it is more complicated to implement. Obviously other low pass filter characteristics can be implemented, depending on the particular smoothing requirements.

Next consider the problem of differentiating the function input variables to allow updating. A circuit similar to that in Figure 12.4 can be used for this,

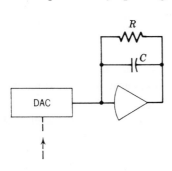

Figure 12.7 First-order digital-analog smoothing circuit.

where the input x is summed into amplifier 1 instead of the digital-analog converter current and where the output $\omega_n \dot{f_0}$ is obtained from amplifier 2. For $\omega \ll \omega_n$ this output will be equal to $\omega_n \dot{x}$ but delayed in time by $2\xi/\omega_n$. If $\dot{x}$ is not varying too rapidly, this time delay may be unimportant. Or, alternatively, the output f_0 from amplifier 3, which is equal to x for $\omega \ll \omega_n$ except for the same time delay $2\xi/\omega_n$, can be used as the input to the function generator in place of x. Thus the complete function generator input becomes $f_0 + \Sigma T \dot{f_0}$, where ΣT represents all time delays, including the delay $2\xi/\omega_n$ in obtaining f_0 and $\dot{f_0}$. By using the circuit of Figure 12.4 for differentiation, we assure that the frequency components in x above ω_n are attenuated proportional to ω.

B. Hybrid Function Generation

Rubin has suggested a scheme for function generation which combines the advantages of both analog and digital technology.[14] Basically, the technique implements the generation of a single-variable function $f(x)$ using the formula

$$f(x) = a_i x + b_i \qquad x_i \leq x \leq x_{i+1} \qquad (12.46)$$

where the computation is mechanized using an MDAC (multiplying digital to analog converter) and a conventional digital to analog converter terminated in a single amplifier, as in Figure 12.8. Here x_i and x_{i+1} represent data points in x at which the function is specified. Equation 12.46 implements linear interpolation where a_i is the slope and b_i is the intercept within the x interval shown. Whenever x enters a new region ($x > x_{i+1}$ or $x < x_i$), both of the digital to analog converters are updated to the proper slope and intercept for the new region. For an N-segment function representation there are N slopes and N intercepts which must be stored digitally or computed digitally from stored ordinates f_i.

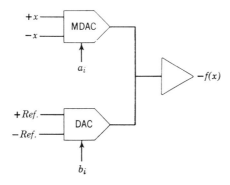

Figure 12.8 Hybrid function generator.

As long as x remains within a given segment, no digital output data is required because the MDAC and digital-analog converter settings remain fixed. Furthermore, the dynamic performance in generating $f(x)$ will be limited only by the performance of the MDAC, which should be excellent. Only when x enters a new segment region do the MDAC and digital-analog converter need to be updated. The net result is an enormously reduced load on the digital computer compared with an all digital mechanization having comparable dynamic errors.

For a two-variable function $f(x, y)$ the slope a_i and intercept b_i can be made linear functions of y using the same type of implementation. However, less equipment is needed if the following formula is used:

$$f(x,y) = a_{ij} + b_{ij}x + c_{ij}y + d_{ij}xy, \quad x_i \leq x < x_{i+1}, \quad y_j \leq y < y_{j+1}$$
(12.47)

Here there are two MDAC's and two digital-analog converters required for mechanization, along with an analog multipler to supply xy. Note that the same product xy can be used for other functions of x and y. The extension to functions of more than two variables is obvious.

This hybrid function generation technique has proven to be quite useful in flight simulation. Along with its obvious advantages, the main disadvantage is the jump discontinuity in the function generator output which occurs when crossing from one segment region to another. The size of the jump will depend on the slope change, the delay in updating, and the velocity of the input variable. For real-time flight simulation, this may not present a serious problem.

References

1. Truitt, T. D., "Hybrid Computation ... What is it? ... Who Needs it?" *IEEE Spectrum*, **1**(6), 132–146, June 1964.
2. Howe, R. M., "Coordinate Systems for Solving the Three Dimensional Flight Equations," *WADC TN 55-747*, June 1956.
3. Heartz, R. A., and T. H. Jones, "Hybrid Simulation of Space Vehicle Guidance Systems," *Simulation*, **6**, (1), 36–44, Jan. 1966.
4. Greenwood, D. T., "An Extended Euler Angle Coordinate System for Use with All Attitude Aircraft Simulators," *WADD Technical 60-372*, Aug. 1960.
5. Robinson, A. C., "On the Use of Quaternions in Simulation of Rigid Body Motion," *WADC TN 58-17*, Nov. 1957.

6. Fogarty, L. E., and R. M. Howe, "Flight Simulation of Orbital and Reentry Vehicles," *IRE Transactions on Electronic Computers*, **EC-11**, Aug. 1962.

7. Fogarty, L.E., and R. M.Howe, "Space Trajectory Computation at the University of Michigan," *Simulation*, **6**, (4), 220–226, April 1966.

8. Turner, R. M., "On the Reduction of Error in Certain Analog Computer Calculations by the Use of Constraint Equations," *Proc. Spring Joint Computer Conference*, San Francisco, 1960.

9. Paquette, G. A., "Progress of Hybrid Computation at United Aircraft Research Laboratories," *AFIPS EJCC Proc.* **26**, Part I, 695–706, 1964.

10. Gilbert, E. G., "Dynamic Error Analysis of Digital and Combined Analog-Digital Computer Systems," *Simulation*, **6**, (4), 241–257, April 1966.

11. Etkin, B. *Dynamics of Flight*, John Wiley, New York, 1962.

12. Howe, R. M., and L. E. Fogarty, "Error Analysis of Computer Mechanization of Airframe Dynamics," U.S. Air Force, Flight Dynamics Laboratory Tech. Rpt. (to be published).

13. Tutleman, D. M., "A Hybrid Computer Technique for Nonlinear Function Generation," *Simulation*, **6**(5), 308–322, May 1966.

14. Rubin, A. I., "Hybrid Techniques for Generations of Arbitrary Functions," *Simulation*, **7**(6), 293–308, Dec. 1966.

13

Man-Machine Systems

13.1 GENERAL REMARKS

A very important application of computer simulation is the study of systems in which a human being participates as an element of the system. Since mathematical models for human performance are not available except for extremely simple tasks, it is necessary to run such simulations in *real time*. It is the requirement for real-time simulation of sophisticated and complex man-machine systems which was largely responsible for the resurgence of interest in hybrid computer techniques. Training and research simulators in the aircraft industry have traditionally been constructed by the interconnection of a cockpit with an analog computer, with the computer being used to simulate aerodynamic phenomena and vehicle dynamics. The guidance and control functions were actually performed by the pilot using real or simulated control devices in the cockpit. As the complexity of piloted vehicles continued to increase, the difficulties of all-analog simulation became increasingly apparent. Thus, for example, the extensive simulation studies conducted during the design phase of the X-15 manned aerospace vehicle were considered by many to be the last all-analog simulation study of an advanced system.[1] Approximately four hundred analog amplifiers were involved in the simulation, interconnected with the cockpit and some actual hardware concerned with the motion of control surfaces. In connection with such vehicles as Gemini or Apollo, purely analog simulation is no longer possible primarily for the

13.2/HYBRID SIMULATION OF MANNED AEROSPACE VEHICLES

following reasons:

1. Extreme dynamic range requirements for vehicles operating from the earth's surface to a space environment.
2. Extreme duration of the mission which must be run in real time and thus is incompatible with drift of analog elements.
3. Accuracy requirements in the generation of trajectories.
4. The presence of digital computers on board the vehicles which must be simulated.

Primarily for the above reasons, many large simulators presently being constructed for the study of piloted space vehicles are based on hybrid computation techniques. The purposes of this chapter are to review the major features of hybrid simulation methods and to discuss briefly some recent results on hybrid models of human operator performance.

13.2 HYBRID SIMULATION OF MANNED AEROSPACE VEHICLES

Where a human pilot performs control or guidance functions in the operation of a system, some form of simulation is essential, both during the design phase and for purposes of training and evaluation. In the design of flight control and guidance systems, the simulation generally becomes some form of *physical simulation* in which there is an interrelationship between a human pilot, an actual or simulated portion of a vehicle control system (including manual controls, displays, dials, hand controllers, and so forth), and a general-purpose computer (analog, digital, or hybrid) which provides inputs to the cockpit and operator which represent the variation of environmental characteristics as well as vehicle dynamics during a particular flight mission[2,3,4] For purposes of illustration, consider the block diagram of the rendezvous and docking simulator illustrated in Figure 13.1. The three essential parts of a manned flight control system simulator are seen in this figure: (1) the pilot and cockpit (2) equipment required to provide input signals to the cockpit displays and process output signals from the operator, and (3) the computer. Where the pilot responds to simple dial movements, a computer may be adequate to provide the input signals. Sometimes, when a more realistic simulation of the external environment is required, films or television systems providing a simulated view of an approaching landscape or target area can be used. The diagram of Figure 13.1 shows a television system in which a camera movable in six degrees of freedom approaches a simulated target vehicle. The resulting television picture is displayed on a monitor located inside the cockpit.

A flow diagram listing the major elements of an actual rendezvous hybrid simulation performed by the McDonnell Aircraft Corporation and IBM[5]

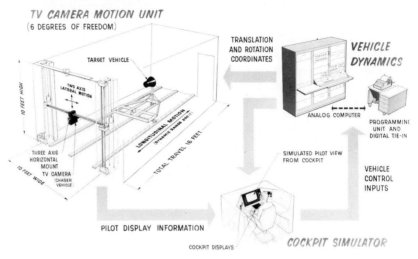

Figure 13.1   Rendezvous and docking simulator (courtesy TRW Systems Group).

is shown in Figure 13.2. The major portions of the simulation were the following:

1. The digital computer program consisted of three major parts. The first part was concerned with solving the translational equations of motion of the vehicle and the determination of vehicle attitude based on body rate information obtained from the analog computer. The second part was a simulation of the on-board digital computer. The third part was concerned with the computations necessary to drive the display devices in the crew station. The first part of this computation was termed the "environment program" as indicated in Figure 13.2. Among the detailed calculations required in the first part of this program were:
    a. Coordinate transformations including the location of the reference inertial frame relative to the equatorial plane and the North Pole, and the location of the body axes relative to the reference frame.
    b. Computation of gravity accelerations referenced to the inertial coordinate system.
    c. Aerodynamics programs.
    d. Thrust programs.
    e. Atmosphere characteristics programs.

13.2/HYBRID SIMULATION OF MANNED AEROSPACE VEHICLES

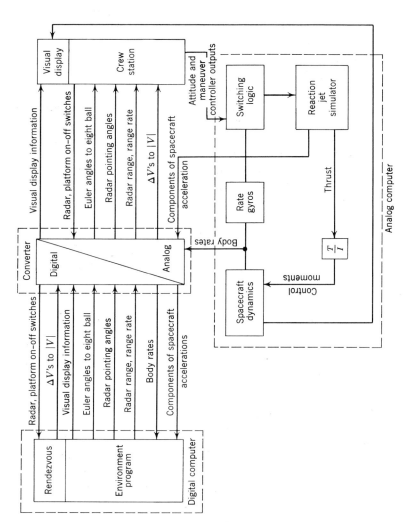

Figure 13.2 Rendezvous simulation flow diagram.[5]

2. The analog computer solves the differential equations of motion describing the vehicle rotation about the body axes. The computer receives information on the aerodynamic angular accelerations and pitch, roll, and yaw about the vehicle body axes from the digital computer; it sums them with the corresponding angular acceleration resulting from firing of the attitude control jets and then integrates them to produce spacecraft rotational rates. These rates are then used for simulation of the rate gyros and are transmitted to the digital program through analog-digital converters. Switching logic is used to obtain voltages corresponding to the firing of the jets based on the output of the simulated rate gyros.[6] The reaction jet simulator indicated as part of the program supplies such nonlinearities as hysteresis, on-off time delays, thrust rise, and decay times.
3. The crew station is provided with displays obtained from both analog and digital sources while its outputs (attitude and trajectory maneuvers) are fed as inputs to the analog computer program.

The simulation described briefly in the preceding paragraphs was also used to study the ascent and reentry phases of Gemini flight with considerable success.[5] The detailed equations solved by such a simulation are those required in the study of space vehicle trajectories and are described in detail in Chapter 12.

Hybrid computation has proven to be an extremely useful adjunct in the design and training phases of complex manned aerospace vehicles. However, it remains for the future to determine whether or not the increasing speed and increasing availability of digital displays will cause a gradual transition toward all-digital simulators.

13.3 SIMULATION OF THE HUMAN OPERATOR

As pointed out in the preceding paragraphs, one limitation of manned vehicle simulation is that it must be performed in real time due to the lack of adequate descriptions, in mathematical terms, of the human operator's performance. Considerable work is being devoted at present to the determination of mathematical models which describe the input-output characteristics of the human operator as an element of a closed loop system. If such models were available, preliminary design of manned systems could be performed considerably more rapidly and efficiently since the real-time limitation would be removed.

In recent years a number of such mathematical models have been proposed. The models, in general, are only adequate for presenting input-output characteristics of human operators in particular simple tasks, such

13.3/SIMULATION OF THE HUMAN OPERATOR

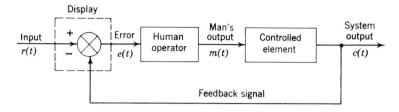

Figure 13.3   A human operator in a compensatory tracking loop.

as the tracking of low-frequency signals in one dimension, as illustrated in the block diagram of Figure 13.3. In this diagram the display can be an instrument or oscilloscope which represents the difference between a reference input and the controlled variable. The human operator is provided with a control stick or similar manual output device which provides input signals to a linear plant. It has been shown that when the input signal in this task contains no appreciable energy above .5 cps and when the input signal is random appearing, it is possible to represent the input-output characteristics of the operator by means of a describing function, i.e. a linear differential equation whose coefficients depend on the input bandwidth and on the controlled element dynamics. A number of such characteristics have been obtained experimentally and tabulated.[7]

Some recent evidence suggests that the operator's performance may in fact be closer to that of a hybrid rather than a continuous element. These studies indicate that the human operator in a control task samples the visual input at intervals of approximately one-third to one-half second and produces an output response from a reconstructed continuous signal. Mathematical models of human operator behavior for a restricted set of tasks have been constructed based on this assumption.[8] A typical model is shown in Figure 13.4 where the hybrid characteristics are immediately apparent. The operator is represented as a sampled data system where sampling is followed by a first-order hold and a linear continuous element

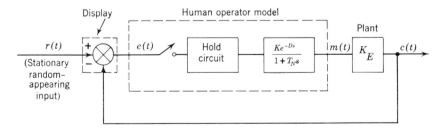

Figure 13.4   Sampled data model of the human operator.[8]

which approximately represents the time delay and the so-called neuromuscular lag. Experimental studies have been performed for the case where the controlled element is a simple gain; it has been shown that the model of Figure 13.4, which includes discrete as well as continuous operations, provides a closer fit to experimental data for that particular task than does a completely continuous model. Clearly, the simulation of the system of Figure 13.4 involves hybrid computing techniques. It is also possible that similar hybrid models will be required to represent human operator behavior in more complex tasks where the operator's attention must be time-shared between a variety of inputs and outputs.

The remainder of this chapter will be devoted to a discussion of two recent human operator models which involve hybrid concepts in order to illustrate that hybrid computation is not only a tool for the simulation and study of dynamic systems but may offer a point of view leading to new insight.

A. An Asynchronous Pulse-Amplitude, Pulse-Width Model of the Human Operator

The single-axis human operator model shown in Figure 13.4 contains a single periodic sampler. However, many studies have produced no evidence of periodic sampling behavior. The work of Merritt[9,10] was concerned with the development of a technique for synthesis and identification of mathematical models of human operator behavior which involve asynchronous sampling. The particular class of tasks considered was that involving control of a system with dynamics of second or higher order, which usually elicits pulse responses from human operators.[10] Such pulses were idealized by Merritt into triangular waveforms, each of which could be described by three numbers: a time of pulse initiation, a pulse amplitude, and a pulse width. A complete human operator model involving control of a simulated aircraft altitude in low level flight[9] is shown in Figure 13.5. The major portions of this diagram are the following:

1. A pulse-initiation decision element which indicates regions in the error-error rate phase plane in which the operator makes a decision to pulse. Note that a circular threshold region (in which no pulsing is initiated) is present. In addition, if the error trajectory is heading toward the origin, no pulse initiation takes place.
2. The initiation of an output pulse is a complex process which begins with the monitoring of error and error rate upon completion of an earlier pulse; it produces a decision to initiate a pulse when the error trajectory enters the preselected region of the phase plane.

13.3/SIMULATION OF THE HUMAN OPERATOR   397

3. Some time later (involving mental processes) the input vector is sampled and pulse amplitude and pulse width are computed.
4. Some time after the computation, a new output pulse is generated.

The coefficients in the model of Figure 13.5 were identified systematically by regression analysis techniques and showed excellent agreement with experimental data. Furthermore, the technique was extended to a simulation involving two instruments sufficiently separated so that the operator was required to move his eyes from one instrument to the other. Similar techniques were used to generate a discrete human operator model which could also predict to a reasonable accuracy the eye movements involved in the task. It is particularly interesting to note from an examination of Figure 13.5 that this model involves a combination of discrete and continuous operations. Thus the inputs to the pulse-initiation decision elements are continuous signals, but the ultimate output of a portion of the system is simply a discrete decision to start or terminate a pulse.

B. A Finite-State Model of Manual Control

The research study described in the preceding paragraph is based on asynchronous or aperiodic input-dependent sampling, but it assumes that the human operator receives continuous input information. In contrast to this approach, in the work of Angel and Bekey,[11] a class of human operator models has been proposed based on the assumption that the operator quantizes his input into a finite and very limited number of states by means of threshold gates. It is then assumed that the data processing is performed on asynchronous samples of this coarsely quantized input and that only a finite number of outputs is possible (i.e., that the human operator behaves as a finite-state machine). Finally, in order to obtain a continuous variation of output position, a hybrid actuator is used to provide a continuous output from discrete binary inputs. Such a model is shown in block diagram form in Figure 13.6 where it is applied to the control of a pure inertia plant. Both continuous and discrete signals are evident in the model. Furthermore, the model consists of three major elements: (1) threshold gates, which receive continuous inputs and generate binary outputs; (2) hybrid actuators, which receive binary inputs and generate continuous outputs; and (3) a combinational network which processes the discrete information as required to generate the required force program.

The threshold gate provides simply for a coarse quantization of the continuous error and error rate signals. They may involve only two or four levels, but these reference levels may in fact be adaptive in order to provide for a simulation of learning behavior on the part of the human operator.[12]

398 MAN-MACHINE SYSTEMS/13

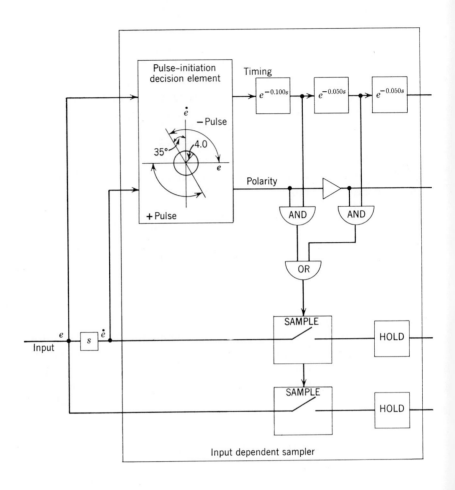

Figure 13.5 A pulse-modulation

13.3/SIMULATION OF THE HUMAN OPERATOR 399

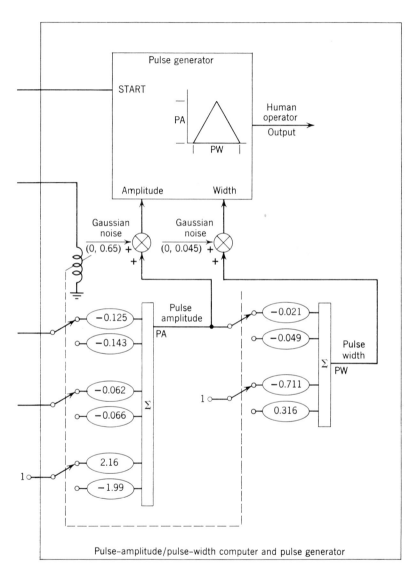

model of the human operator.[9]

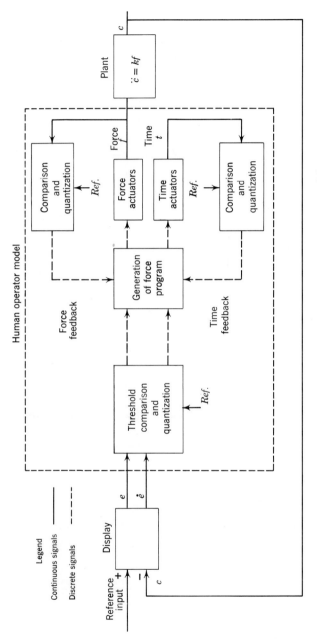

Figure 13.6 A finite-state model of the human operator.[11]

13.3/SIMULATION OF THE HUMAN OPERATOR 401

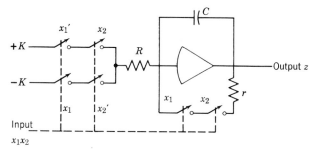

Figure 13.7 Hybrid (finite-state) actuator.

Transformation of discrete to continuous inputs was achieved in this model by means of a hybrid actuator, a combinatorial device which may be visualized with reference to Figure 13.7. Here digital control of an analog integrator provides the necessary features required in the hybrid actuator. The discrete inputs x_1 and x_2 control an analog integrator with two possible fixed input voltages $+K$ and $-K$. The integrator is assumed to introduce no sign change. Depending on the state of the discrete input, the output increases at a rate K, decreases at a rate K, or holds its value. A fourth input condition provides for reset to zero, as indicated in the input-output table, Table 13.1.

Prestored force programs involving bang-bang corrections provided the model with the capability of obtaining minimum-time position or velocity corrections. The response of the model to a ramp input is shown in Figure 13.8, which strongly resembles typical human operator response.

While both of the models described in the preceding paragraphs are basically feasibility studies, they illustrate the power of hybrid techniques in leading to a new synthesis of experimental data in the form of entirely new models of system behavior. An application of finite-state techniques to the development of a hybrid model of animal locomotion is discussed in Chapter 14.

Table 13.1 Finite-State Actuator Characteristics

Input (x_1x_2)	Output z	$\dfrac{dz}{dt}$
0 0	Constant	0
0 1	Increases	$+K$
1 0	Decreases	$-K$
1 1	Resets	$-z/rC$

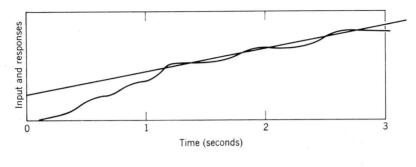

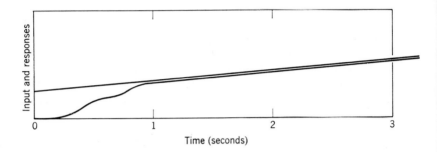

Figure 13.8   Response of human operator model to increasing ramps.

References

1. Cooper, N., "X-15 Analog Flight Simulation—Systems Development and Pilot Training," *Proc. Western Joint Computer Conference*, pp. 623–638, 1961.
2. Thomas, O. F., "Analog-Digital Hybrid Computers in Simulation with Humans and Hardware," *Proc. Western Joint Computer Conference*, pp. 639–644, 1961.
3. Connelly, M. E., "Computers for Aircraft Simulation," M.I.T. Electronic Systems Laboratory, Report 7591-R-2, Dec. 1959.
4. Sadoff, M. E., and C. W. Harper, "A Critical Review of Piloted Flight Simulator Research," presented at IAS Meeting on "The Man-Machine Competition," Seattle, Aug. 1962.
5. Jacobsen, C. A., "The Application of the Hybrid Computer in Flight Simulation," *Proc. IBM Scientific Computing Symposium on Computer-Aided Experimentation*, pp. 206–244, 1966.
6. Landauer, J. P., "Simulation of Space Vehicle Reaction Jet Control Systems," Electronic Associates, Inc., Hybrid Computation Application Study 3.4.1h, 1963.

REFERENCES   403

7. McRuer, D., et al., "Human Pilot Dynamics in Compensatory Systems," U.S. Air Force, Flight Dynamics Laboratory Tech. Rpt. AFFDL-TR-65-15, July 1965.

8. Bekey, G. A., "The Human Operator as a Sampled-Data System," *IRE Transactions on Human Factors in Electronics*, **HFE-3**, 43–51, Sept. 1962.

9. Merritt, M. J., "Synthesis and Identification of Mathematical Models for the Discrete Control Behavior of Human Operators", Ph.D. Dissertation, University of Southern Calif., Dept. of Elec. Eng., June 1967.

10. Merritt, M. J., and G. A. Bekey, "An Asynchronous Pulse-Amplitude, Pulse-Width Model of the Human Operator," *Proc. 3rd Annual NASA-University Conference on Manual Control*, NASA SP-144, pp. 225–240, 1967.

11. Bekey, G. A., and E. S. Angel, "Asynchronous-Finite State Models of Manual Control Systems," *Proc. 2nd Annual NASA-University Conference on Manual Control*, NASA SP-128, pp. 25–38, 1966.

12. Angel, E. S., and G. A. Bekey, "Adaptive Finite-State Models of Manual Control Systems," *IEEE Transactions on Man-Machine Systems*, **MMS-9**, 15–20, Mar. 1968.

14

Biological Systems

14.1 COMPUTATION AND DATA PROCESSING IN BIOMEDICAL SCIENCES

The simulation of biological systems and the processing of biological signals offer potentially fruitful areas for hybrid computer techniques. However, the overwhelming majority of such simulations is being performed by digital techniques; it is expected that the trend toward digital computer utilization will continue. Nevertheless, there are a number of areas in which hybrid techniques have found important applications and others where possible applications may result.

Perhaps the most important application of hybrid techniques involves analog preprocessing of biological signals. The analog output is then converted to digital form and sent to a digital computer for further analysis. In terms of the spectrum of applications discussed in Chapter 1, this is evidently not a balanced application. However, it is one of great importance since the signals recorded from living organisms are generally in analog form, including such variables as blood pressure, blood flow, temperature, respiration rate, and a variety of bioelectric signals such as those obtained from electrocardiograms, electromyograms, or electroencephalograms. In a number of biomedical computing applications such variables are measured, converted to electrical form, digitized, and fed directly into digital computers. However, in other applications considerable savings in digital computer time and complexity of programs can be obtained if the measured variables are first processed by means of simple analog techniques. In Section 14.3, a detailed illustration of this

14.1/COMPUTATION AND DATA PROCESSING IN BIOMEDICAL SCIENCES

technique, as applied to analog processing of electrocardiograms, will be discussed.

While analog processing may offer certain advantages, it is also important to note that the nature of biological signals makes such processing quite difficult in certain other applications. Biological systems are characterized by their variability. Responses to the same stimulus may vary significantly not only from animal to animal or subject to subject but also for successive stimulations of the same preparation. It has become customary for the processing of such responses to obtain signal enhancement by the averaging of a number of such transient responses obtained on successive stimulations. Evidently, the averaging of a series of transient records requires some form of storage. Small special-purpose digital computers such as the "Computer of Average Transients" (CAT), manufactured by Technical Measurement Corporation,[1] digitize the analog records at fixed intervals of time, accumulate the sampled variables in memory, and generate an oscilloscope display of the average response, as illustrated in the block diagram of Figure 14.1. In this type of experiment the original responses are analog in form, digital techniques are used for averaging, and an analog display is obtained.

A limited number of biomedical installations possess balanced hybrid systems. Among these are the Latter Day Saints Hospital in Salt Lake City, the Sloan-Kettering Institute in New York City, and the Albert Einstein College of Medicine in New York City. In such installations a broader spectrum of hybrid computer techniques can be applied to the processing of biological data. Several examples from such applications will be given in Section 14.4.

Mathematical models of biological systems offer a fertile ground for hybrid computer techniques. The techniques for parameter optimization

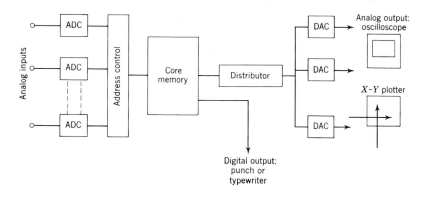

Figure 14.1 Block diagram of average response computer.

discussed in Chapter 9 can be applied to biological as well as inorganic systems. Furthermore, certain biological systems have characteristics which, at least in principle, make them ideally suited for hybrid computation. For example, transmission of information in the nervous system is carried on by means of pulse-frequency modulation, at least in part. Here the pulses are discrete events, while their frequency is an analog variable. Mathematical models of eye movements developed by Young and others[2] contain both continuous and discrete channels. A "pursuit" channel provides continuous eye movements while a "saccadic" channel provides for the discrete jumps observed in visual tracking of moving targets. These two examples alone illustrate that such systems logically lend themselves to hybrid simulation. On the other hand, there may be overriding considerations leading to the use of digital simulation.

Mathematical models of biological systems in recent years have been simulated on both analog and digital computers.[3,10] Extensive steady-state chemical equilibrium processes have been studied by means of digital computers.[4] On the other hand, cardiovascular system dynamics have been studied most extensively by means of analog computers.[5] A special-purpose hybrid computer (basically a digitally-controlled analog) has been constructed for the simulation of certain biochemical processes.[6] Only a limited number of general-purpose hybrid simulations has been performed[7,8,9,17] and, consequently, this remains an area for future research and investigation. Finally, it is interesting to note that hybrid techniques may offer significant advantages in the design of artificial limbs and other prosthetic devices. In the synthesis of such devices, it is necessary to accomplish the reverse process from that required in the processing of biological data, namely to generate a control strategy by means of digital computation, and then transform it to a suitable analog output. In view of the potential importance of this application, a detailed example is discussed below in Section 14.5.

14.2 THE NATURE OF BIOLOGICAL SIGNALS

It was noted above that digital computation and digital preprocessing are commonplace in biomedical installations. Among the reasons which account for this phenomenon are the following:

1. Biological signals, in general, are of relatively low frequency. Many biological variables contain little information above 10 to 20 Hz and only a few contain significant information above a few hundred Hz. Compared to the range of frequency in such areas as electromagnetic radiation, it is evident that the frequencies are indeed low. The significance of this fact is that relatively inexpensive analog to digital

conversion may provide adequate sampling rates for biological signals, thus rendering them suitable for digital computation.
2. Biological signals are known for their variability and, thus, require averaging for signal enhancement, as pointed out above. The operation of the signal averaging computers is based on the assumption that the random disturbances in successive responses are independent. In that case, if the response to the i-th stimulus is denoted by $f_i(t)$ and its variance by var $\{f(t)\}$, then the average response is given by

$$\overline{f(t)} = \frac{1}{N} \sum_{i=1}^{N} f_i(t) \tag{14.1}$$

and the variance of the average is

$$\text{var}\,\{\overline{f(t)}\} = \frac{1}{N} \text{var}\,\{f(t)\} \tag{14.2}$$

Thus the r.m.s. fluctuation of the random component of the responses decreases as $1/\sqrt{N}$ if N responses are averaged.
3. Biological experimentation and diagnostic display of computer variables often require the storage of a large amount of information but relatively simple computational procedures. Here again digital computers excel and analog techniques are deficient.
4. As may be expected, analog techniques find their major advantage in certain nonlinear modeling applications and in the processing of certain biological signals. In addition, a number of small-scale analog computers are used in biomedical laboratories to provide rapid and simple generation of biological stimuli and certain on-line data reduction capabilities. On the other hand, biological signals may exhibit both discrete and continuous characteristics. For example, information transmission in certain areas of the nervous system is carried on by frequency modulated pulse trains.[11] The shape of each pulse (the nerve impulse) may be invariant, but the firing frequency is modulated. Over a given range of the stimulus, the pulse frequency may be a continuous function of the input. Thus, a faithful simulation of this process would require both discrete and continuous elements.

14.3 HYBRID PROCESSING OF ELECTROCARDIOGRAM (ECG) DATA

The electrocardiogram is a recording of the electrical activity of the heart.[23] The activity of the heart, like that of other muscles, is accompanied by electrical activity associated with the depolarization of the membrane

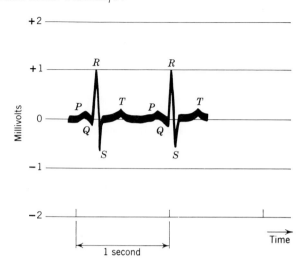

Figure 14.2   Typical normal electrocardiogram.

surrounding the muscle fibers. Powerful contractions of the heart muscle make it possible to record this electrical activity on the surface of the skin. To take into account the spatial orientation of the heart within the body, it is necessary to record this manifestation of electrical activity at a sufficient number of points in order to obtain the components of the "heart activity vector." A number of electrode systems are available; they are sometimes referred to by means of the number of leads used in the recording process. Among the common techniques are the "three-lead electrocardiogram" and the "twelve-lead electrocardiogram." For this discussion let us assume that three components of the heart vector, measured at approximately *orthogonal* orientations, will be obtained. The typical form of any of these recordings is shown in Figure 14.2. Examination of this record shows that the wave of electrical activity associated with each heartbeat is characterized by five pulses, or "waves" as they are commonly called, which are associated with particular phases of the heart cycle. The waves are normally denoted by the letters P, Q, R, S, and T as indicated in Figure 14.2. For example, the T wave is associated with the depolarization of the atria before contraction. The QRS complex is due to currents generated by depolarization of the ventricles just prior to their contraction. The clinical importance of the electrocardiogram arises from the fact that the regularity of the waves and their timing are destroyed in the presence of certain diseases. For example, the P wave may not be visible at all, the interval between successive R waves may

14.3/HYBRID PROCESSING OF ELECTROCARDIOGRAM (ECG) DATA

change drastically, distortion in various forms can be present, etc. Thus, interpretation of the electrocardiogram is based on a study of the amplitudes, durations, and frequencies of the waves. Typical parameters which are required are the amplitudes of P, Q, R, S, and T, durations of the intervals PR, QT, and RR, the average duration of the ST interval in a series of heartbeats, and so forth. Basically, the desired information consists of some 15 data values per beat.

The interpretation of the data requirements discussed above indicates immediately the possible usefulness of analog preprocessing. If 1 minute of a record from a single electrocardiogram lead is digitized, assuming that useful information is in the frequency range from 0 to 100 Hz and that .1% resolution (10 bits) is desired, some 600,000 bits of information are obtained. On the other hand, if we obtain only 15 data points by analog preprocessing from each heartbeat and 120 heartbeats in a minute are assumed so that only 1800 points per minute must be digitized, we obtain less than 20,000 bits of information. In other words, analog preprocessing may result in the discarding of some 95% of the input data, thus resulting in significant savings in the required amount of digital memory, at the expense of only minimal analog complexity.

The simplest approach, as suggested by Wortzman et al.,[12] is based on the use of a series of level detectors operating on the output of a differentiating circuit whose input is the low-pass-filtered electrocardiogram. By using a series of positive and negative levels with respect to ground, the occurrence of each wave can be detected. The output of the level detector is a logic signal indicating a need to sample the electrocardiogram. To compensate for the delay in the filtering and level detection process, Figure 14.3 indicates that the original analog signal must be delayed prior to sampling. The basic purpose of this circuit is to reduce the amount of information supplied to the digital computer.

More sophisticated techniques of ECG analysis are based on the use of peak detection and interval measurement on averaged records, rather than on single records, in order to decrease the effect of drifts and variations occurring during individual heartbeats. The baseline of the electrocardiogram record shifts during the recording process due to a combination of amplifier drift in the instrumentation system and an artifact due to respiration. To decrease the effect of this variation, the onset of a new ECG cycle must be detected; an average response computer can then be used to sum individual cycles beginning from an appropriate triggering point. It is evident that the choice of triggering point must be made carefully in order to eliminate "time jitter" in the summed records. Several triggering techniques have been suggested in the literature. Rautaharju[13,14] has suggested a triggering system based on the detection of a rising signal with a slope

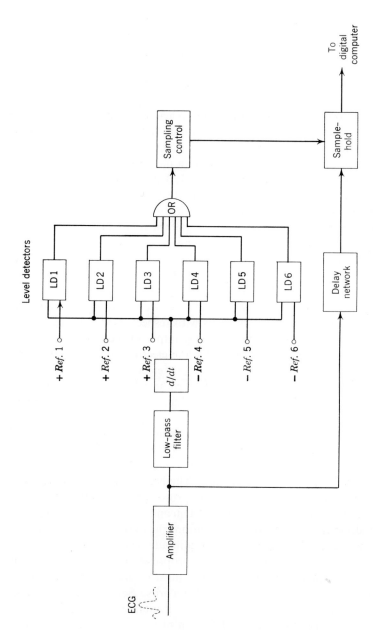

Figure 14.3 Analog preprocessing of ECG data.[12]

exceeding a preset threshold followed by an adjustable time interval and later by a negative slope exceeding another preset threshold. Furthermore, in order to minimize the effect of heart position, the triggering signal is generated from operations performed on the sum of the squares of three orthogonal ECG leads, as shown in Figure 14.4. Evidently in this system analog preprocessing is used only to obtain the necessary triggering signals as inputs into an average response computer.

Mitchell and Cady[7,24] have also used a combination of level detection, triggering, and signal averaging to extract data from the electrocardiogram. Their system is shown in block diagram form in Figure 14.5. In this particular system the feature extraction is performed by means of analog circuits, obtained by reconverting the outputs of the average response computers into analog form. The advantage in this technique, suggested by the authors, is that relatively inexpensive serial memory may be used for function accumulation and averaging, without the necessity of tying up the general-purpose digital computer. The triggering signal indicated at the bottom of Figure 14.5 is obtained by the detection of the point of maximum negative slope on the electrocardiogram, i.e., the steepest falling point following the peak of the R wave.

The extraction of the relevant data points from the cardiogram (such as intervals between successive peaks or detection of given peak levels) is accomplished by standard analog techniques when these are used.[15,16] However, in all probability, future developments for ECG instrumentation will use digital feature extraction due to the decreased cost and increasing availability of digital computers. Nevertheless, the use of analog prefiltering and possibly analog triggering will continue to play an important part in the efficient extraction of information from electrocardiographic recordings.

14.4 ON-LINE HYBRID COMPUTER APPLICATIONS

As an illustration of the usefulness of hybrid computer techniques in conjunction with biological experimentation, two examples from the range of applications developed by Dr. Josiah Macy, Jr., at the Albert Einstein College of Medicine[17] will be reviewed here.

Consider first an application of the system to studies in cardiac physiology. During the course of an experiment in the laboratory, the electrocardiogram and blood pressure signals are sent to the computer directly in analog form. On receipt of a signal (obtained from a push button) in the laboratory, a number of events are initiated. The $R - R$ interval, indicating the duration of the cardiac cycle during which the push button

412 BIOLOGICAL SYSTEMS/14

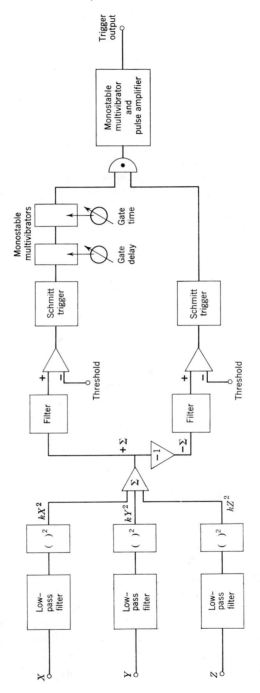

Figure 14.4 Hybrid preprocessing of ECG data.[13]

14.4/ON-LINE HYBRID COMPUTER APPLICATIONS 413

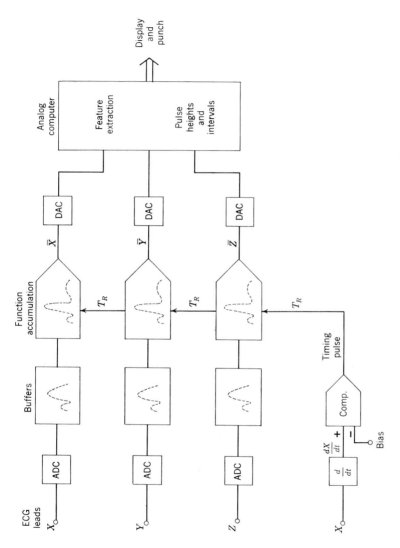

Figure 14.5 Hybrid processing of ECG data.[7]

signal was delivered, is measured by analog techniques, stored in a sample-hold amplifier, and multiplied by a percentage set by the laboratory. If the duration of the i-th cardiac cycle is denoted by T_i and the percentage set by the laboratory by p, then at a time given by

$$T_s = pT_i$$

a stimulus pulse is delivered along an appropriate analog line back to the laboratory. By this means a stimulus can be delivered at an appropriately selected phase of succeeding cardiac cycles. The effect of the stimulus can be monitored by observation of the aortic blood pressure waveforms or ventricular pressure waveforms in the animal. For example, it may be desired to observe the average pressure during a short portion of the cardiac cycle known as the isovolumetric phase (during which pressure increases but the volume of the ventricle remains approximately constant). This phase can be detected by differentiating the blood pressure curve, detecting its peak, digitizing a short portion (say 10 msec) of the pressure waveform,

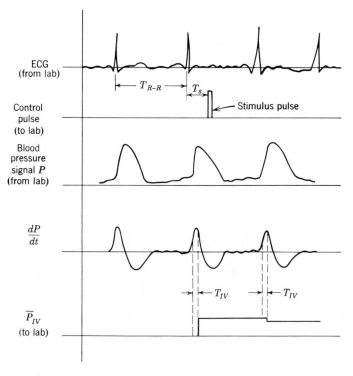

Figure 14.6   On-line hybrid processing of blood pressure signals.[17]

averaging in the digital computer, storing the results, and on request sending an analog display back to the experimenter. The relevant waveforms are indicated in Figure 14.6.

As another example, consider the determination of cardiac output as measured by the so-called indicator dilution technique. In this experiment a dye of known optical concentration is injected into the bloodstream; at an appropriate point in the arterial system, blood samples are run through a densitometer. By using appropriate calibrations, the densitometer readings will then correspond to known dye concentrations. A typical dye concentration curve from the laboratory is shown in Figure 14.7, illustrating that it is a typical noisy biological record. Analog techniques are used to remove most of the noise by low pass filtering, thus producing the curve in Figure 14.7b. An analog function generator is used to produce a curve proportional to the logarithm of the concentration as indicated in Figure 14.7c. The logarithmic curve is now sent to the digital computer in order to compute the area of the shaded portion. The shaded portion can be converted to an estimate of the cardiac output during a given cycle by previously stored scalefactors and conversion constants. The digital computer performs the extrapolation to the baseline (which is necessary for the computation of the area) by means of a least-squares fit of the best straight line which fits the first portion of the descending part of the curve. The resulting computation of the area is returned to the laboratory for display.

These two examples illustrate the applicability of on-line hybrid computation in conjunction with biological experiments.

14.5 A HYBRID APPROACH TO THE STUDY OF ANIMAL LOCOMOTION

Perhaps the most outstanding example of the application of hybrid concepts to the study of biological systems is their usefulness in the study of animal locomotion. Much of the original work in this field is due to Tomovic who first suggested the use of a shift register representation to the study of creeping movements[18] and later developed a general approach to the study of self-moving machines.[20] Tomovic and Karplus[19] published a simulation study of animal locomotion. The applications of finite-state automata theory to the analytical study of quadruped locomotion and the design of prosthetic devices have been investigated by Tomovic and McGhee.[21,22] The following paragraphs are extracted primarily from the latter references.

Two approaches to the study of leg movements involved in animal locomotion are possible. One approach, which may be termed the "analog approach," consists of describing the trajectory of each leg in space by means of the appropriate differential equation. A second approach, as

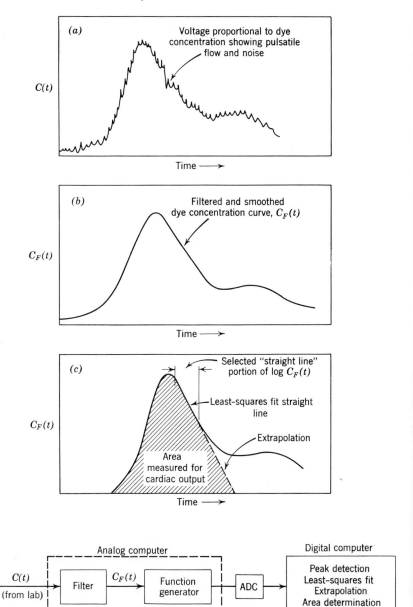

Figure 14.7 On-line hybrid determination of cardiac output.[17]

14.5/A HYBRID APPROACH TO THE STUDY OF ANIMAL LOCOMOTION

suggested by Tomovic and McGhee,[21] is based on the assumptions that the animal's central nervous system provides only a minimum of control information to the limbs and that each leg then behaves as a preprogrammed device under control of the spinal cord. This principle, termed the "Principle of Maximal Autonomy," suggests that the legs should be represented as hybrid devices with discrete inputs and continuous outputs. In order to make possible a description of a continuum of output positions for each leg, the limbs are considered as levers linked together at joints and powered by opposing muscle groups. By proper excitation of the agonist-antagonist muscle groups, a joint may be rotated in either direction. It is also possible to lock a joint or to release it so that it swings freely. Thus, at least four states are required to describe the behavior of a given joint. A device capable of accepting discrete inputs and generating the continuous outputs described above has been termed a "cybernetic actuator"[21] and is described by Table 14.1. From a computer point of view, it is evident that the cybernetic actuator is in fact an appropriately coded digital-to-analog converter, with a binary input and a continuous output.

To use the concept of the joint described in the preceding paragraph in the description of a complete leg movement cycle, it is also necessary to characterize the feedback provided by the sensory capabilities of the biological element. For example, there are evidently limits to the angular excursion permitted around knee joints, hip joints, and ankle joints. Upon reaching a particular limit, feedback signals must be provided for control. To describe such feedback by means of simple concepts, threshold gates can be used. Such gates are again hybrid elements with continuous inputs and binary outputs. If we define the input to such a gate as an error signal given by the difference between a signal and a reference (for example, a reference angle and the actual excursion of the leg), the behavior of the threshold gate can be described by Table 14.2.

The description of the leg movements during locomotion then involves the use of a small digital computer to provide a sequence of discrete inputs which control the states of the various joints. Each joint, idealized as a

Table 14.1 Definition of a Cybernetic Actuator

Input	Actuator State	Output
0 0	0	Free
0 1	1	Decreasing
1 0	2	Increasing
1 1	3	Locked

Table 14.2 Behavior of a Threshold Gate

Error Signal†	Output
Positive	1
Zero	0
Negative	0

† s = signal input; r = reference input; and e = error signal

$$e = s - r$$

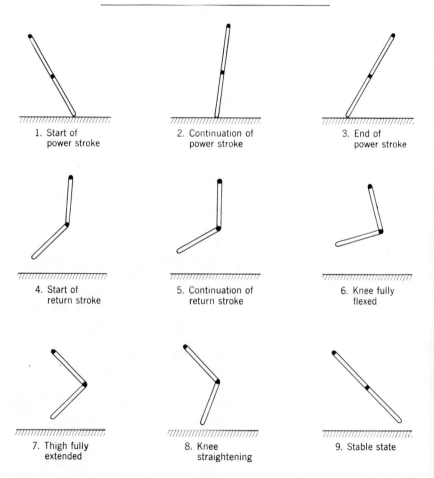

Figure 14.8 Sequence of events during movement of a mono-stable leg.[26]

14.5/A HYBRID APPROACH TO THE STUDY OF ANIMAL LOCOMOTION 419

Column number	Corresponding foot
1	Left front
2	Right front
3	Left rear
4	Right rear

(a)

State of legs				Duration (seconds)
1	1	1	0	$\frac{3}{16}$
1	0	1	0	$\frac{6}{16}$
1	0	1	1	$\frac{3}{16}$
1	0	0	1	$\frac{6}{16}$
1	1	0	1	$\frac{3}{16}$
0	1	0	1	$\frac{6}{16}$
0	1	1	1	$\frac{3}{16}$
0	1	1	0	$\frac{6}{16}$

(b)

Figure 14.9 Sequence of states associated with a quadruped walk: (a) correspondence of feet to columns of state sequence and (b) sequence of states.[26]

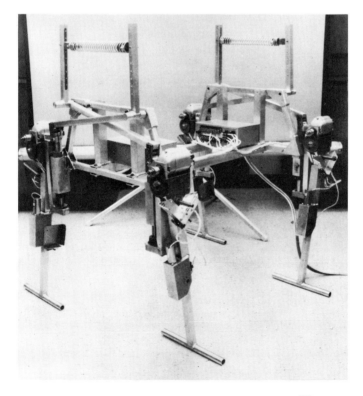

Figure 14.10 Artificial quadruped prototype[26]

cybernetic actuator, then transforms the discrete input into a continuous motion. The threshold gates represent idealizations of feedback elements which transform the continuous output to discrete control signals. The control of the legs of a quadruped during a stable symmetrical walk can then be idealized as indicated in Figure 14.8.

The use of the principles described in the preceding paragraph has made possible the design and construction of an artificial quadruped.[26] The sequence of events for one leg of such a quadruped is indicated in Figure 14.9, which shows that only hip and knee joints were considered in its realization for simplicity. A photograph of the prototype machine is shown in Figure 14.10.

The description of locomotion as a finite-state machine has been included to illustrate the fact that when discrete and continuous processes are considered (rather than either one to the exclusion of the other), then a range of conceptual tools for the description of biological systems, as well as the engineering synthesis of devices which simulate biological systems, becomes possible.

References

1. Clynes, M. E., "The CAT-Computer of Average Transients," *Med. Electronic News*, pp. 5–7, June 1962.
2. Young, L., and L. Stark, "Variable Feedback Experiments Testing a Sampled Data Model for Eye Tracking Movements," *IEEE Transactions on Human Factors in Electronics*, **HFE-4**, 38–51, Sept. 1963.
3. Clymer, A. B., and G. E. Graber, "Trends in the Development and Applications of Analog Simulations in Biomedical Systems," *Simulation*, **2,** 41–59, April 1964.
4. DeLand, E. C., and G. B. Bradham,"Fluid Balance and Electrolyte Distribution in the Human Body," RAND Corp. Memorandum RM-4347-PR, Santa Monica, Calif., Feb. 1965.
5. Noordegraaf, A., G. N. Jager, and N. Westerhof, eds., *Circulatory Analog Computers*, North Holland Pub., Amsterdam, 1963.
6. Higgins, J. J., "A Hybrid Computer for the Investigation of Chemical Reactions: The Johnson Foundation Electronic Computer, Mark II," *Annals N.Y. Acad. Sci.*, **115,** 1025–1037, 1964.
7. Mitchell, B. A., and L. D. Cady, "Hybrid Computing Techniques Applied to EKG Analysis," *Annals N.Y. Acad. Sci.*, **128,** 850–860, 1966.
8. Ellis, M. E., et al., "Hybrid Computer Simulation of the Mammalian Cardiovascular System," *Proc. Annual Conference on Eng. in Medicine and Biology*, **8,** 312, 1966.

9. Siler, R. W., "Hybrid Computation in Bioscience," in *Computers and Biomedical Research*, (R. W. Stacy and B. Waxman, eds.), Vol. 1, Academic Press, New York, pp. 87–108, 1965.
10. Stacy, R. W., and B. Waxman, eds., *Computers and Biomedical Research*, Academic Press, New York, 1965.
11. Jones, R. W., et al., "Pulse Modulation in Physiological Systems," *IRE Transactions on Biomed. Eng.*, **BME-8**, 59–67, Jan. 1961.
12. Wortzman, D., B. Gilmore, H. D. Schwetman, and J. I. Hirsch, "A Hybrid Computer System for the Measurement and Interpretation of Electrocardiograms," *Annals N.Y. Acad. Sci.*, **128**, 876–899, 1966.
13. Rautaharju, P. M., "Hybrid and Small Special Purpose Computers in Electrocardiographic, Ballistocardiographic and Pulse Wave Research," *Annals N.Y. Acad. Sci.*, **126**, 906–918, 1965.
14. Rautaharju, P. M., "Deterministic Type Waveform Analysis in Electrocardiography," *Annals N.Y. Acad. Sci.*, **128**, 939–954, 1966.
15. Electronic Associates, Inc., "Hybrid Computer Analysis of Electrocardiographic Data," EAI Applications Study 4.4. 1h, 1964.
16. Licher, R. M., and R. Nesbit, "Electrocardiogram Analysis by the Beckman/SDS Integrated Computing System," Beckman Instruments, Inc., Computer Operations Report, 1964.
17. Macy, J. B., Jr., "Hybrid Computer Techniques in Physiology," *Annals N.Y. Acad. Sci.*, **115**, 568–590, 1964.
18. Tomovic, R., "A General Model of Creeping Displacement," *Cybernetica*, **4**, (2), 98–107, 1961.
19. Tomovic, R., and W. J. Karplus, "Land Locomotion-Simulation and Control," *Proc. 3rd AICA Conference on Analog Computation*, Opatija, Yugoslavia, pp. 385–390, Sept. 1961.
20. Tomovic, R., "On the Synthesis on Self-Moving Automata," *Automation and Remote Control*, **26**, 297–304, Feb. 1965.
21. Tomovic, R., and R. B. McGhee, "A Finite State Approach to the Synthesis of Bioengineering Control Systems," *IEEE Transactions on Human Factors in Electronics*, **HFE-7**, 65–69, June 1966.
22. McGhee, R. B., "Abstract Quadruped Locomotion Automata," Univ. of Southern Calif. Tech. Rpt., USCEE 130, April 1965.
23. Guyton, A. C., *Textbook of Medical Physiology*, 3rd ed., Saunders, 1966 (Chapter 14: "The Normal Electrocardiogram").
24. Zeitlin, R. A., L. D. Cady, L. J. Tick, and M. A. Woodbury, "Combined Analog-Digital Processing of Cardiograms," *Annals N.Y. Acad. Sci.*, **115**, 1106–1114, July 1964.
25. Warner, H. R., "Simulation as a Tool for Biological Research," *Simulation*, **3**, 57–63, 1964.
26. McGhee, R. B., "Finite State Control of Quadruped Locomotion," *Simulation*, **9**, 135–140, Sept. 1967.
27. Wortzman, D., N. J. King, and B. G. Gilmore, "Analysis and Pattern Recognition of Electrocardiogram Waveforms Using a Simulated On-line Hybrid Computer System," *Proc. 16th Conference on Eng. in Medic. and Biol.*, **5**, 181–191, 1963.
28. Stern, J. A., F. Bogdanoff, and G. M. Hethcote, "Hybrid Computer Systems for the Analysis of Electroencephalographic Alpha Desynchronization," *Proc. 16th Conf. on Eng. in Medic. and Biol.*, **5**, 70–71, 1963.
29. Cochrane, I., "Hybrid Computer Analysis of Chromosomal Patterns," *Bio-Medical Engineering*, **3**, 58–65, Feb. 1968.

15

Solution of Integral Equations†

15.1 INTRODUCTION

The mathematical description of many problems of engineering interest contains integral equations. Typical of a large class of such problems is the Fredholm integral equation of the second kind,

$$y(x) = f(x) + \lambda \int_a^b K(x, t) y(t)\, dt \tag{15.1}$$

where $f(x)$ and the kernel $K(x, t)$ are given functions, a and b are constants, λ is a parameter, and $y(x)$ is to be found. From a computational point of view, equations of this type may be considered as problems in two dimensions, where one dimension (t) is the dummy variable of integration. For digital computer solution, both variables must be discretized. For analog computer solution, it is possible to perform continuous integration with respect to the variable t for a fixed value of x and perform a scanning process to obtain step changes in the second variable. In either case, the solution is iterative and results in a sequence of functions $\{y_n(x)\}$, $n = 1, 2, \ldots$, which, under certain conditions, converge to the true solution $y(x)$ as n increases.

It is evident that such a sequential solution, with a two-dimensional array $K(x, t)$, may be extremely time consuming for pure digital solution.

† Some of the material in this chapter was adapted from a paper previously published in the *Proceedings of the AFIPS Computer Conference*, **31**, 143–148, 1967, under the title "Solution of Integral Equations by Hybrid Computation," by G. A. Bekey, R. Tomovic, and J. C. Maloney.

On the other hand, the scanning and iteration procedures, which require storage of the successive approximation to the solution, do not lend themselves to pure analog computation. Rather, the problems require a hybrid combination of high speed, repetitive integration, memory, and flexible control logic.

The advantage of using such hybrid computational methods for the solution of integral equations was realized quite early. A special-purpose computer for solution of integral equations was proposed by Wallman in 1950.[1] Basically, the computer technique consisted of replacing the integration with respect to two independent variables by scanning at two different rates. These original proposals for iterative solution of integral equations were based on the classical or Neumann method.[2] In 1957, M. E. Fisher proposed an iteration technique for high-speed analog computers equipped with a supplementary memory capacity which resulted in considerably faster convergence than the classical technique.[3] However, little practical experience with his method is available due to the complexity of the function storage and playback apparatus.[4] One test of Fisher's method was made using a repetitive analog computer in which the computer operator manually adjusted a set of potentiometers in a special function generator at the end of each iteration cycle.[5]

The purpose of this chapter is to examine hybrid computer solution of integral equations by both the Neumann and Fisher method.

15.2 THE NEUMANN ITERATION METHOD

The classical or Neumann iteration procedure for the solution of Equation 15.1 specified by

$$y_{n+1}(x) = f(x) + \lambda \int_a^b K(x, t) y_n(t)\, dt \tag{15.2}$$

with $y_0(t) = 0$. Under conditions discussed by McKay and Fisher,[4] the process converges to a limit $y_\infty(x)$ which is the solution of Equation 15.1.

From a computer point of view, an integration over the whole range of t must be made for each particular selected value of x. If the range of x is divided arbitrarily into I segments of length Δx, the function is represented by the values of $y(x)$ at the midpoint of each segment, i.e., $y(x_i)$, $i = 1, 2, 3, \ldots, I$. It is clear that a total of I integrations in the t domain must be made before a single change in $y_n(t)$ is made in Equation 15.2. Such an integration in t for a single value of $x = x_i$ will be called a *minor cycle*. In order to increase the index n, i.e., to derive the next approximating function $y_{n+1}(t)$, one has to complete I minor cycles. This group of minor cycles will be called a *major cycle*. The complete solution theoretically requires an infinite number of major cycles. However, practical experience[4] has demonstrated that accuracies of the order of 1% are attainable in about 20 major cycles using Neumann's method.

It should be noted that digital computer implementation of the strategy defined by 15.2 requires that the variable t also be discretized and that an appropriate numerical integration formula be used. For example, if Euler or rectangular integration is used, Equation 15.2 becomes

$$y_{n+1}(x_i) = f(x_i) + \lambda \sum_{j=1}^{J} K(x_i, t_j) y_n(t_j) \Delta t \qquad i = 1, 2, \ldots, I \quad (15.3)$$

where $\Delta t = $ constant is the integration step size and J is the total number of steps in the interval $(b - a)$. Since t is only a dummy variable, the total range in t must equal the range of x; it is possible to choose the number of steps in t and x to be equal, i.e., let $I = J$. More sophisticated numerical procedures do not change the need for minor and major integration cycles. Equation 15.3 requires only the algebraic operations of addition and multiplication and is well suited to digital computation.

For hybrid computer solution, each minor cycle may be performed continuously (using analog integration) and the resulting values stored. Assignment of other specific computational functions to the analog or digital portions of a system may significantly affect overall accuracy, as will be discussed later in this chapter. Thus, the generation of the functions $f(x)$ and $K(x, t)$, as well as the multiplication under the integral sign in 15.2 may be performed either in the analog or digital computer. A flow chart illustrating the programming of Neumann's method is shown in Figure 15.1. A stopping criterion given in the flow chart is based on reducing the difference between successive approximations to a sufficiently small value.

15.3 THE FISHER ITERATION METHOD

An examination of Equation 15.2 and Figure 15.1 reveals that Neumann's method requires the storage of $y_{n+1}(x_i)$ at the end of each minor cycle (i.e., for $i = 1, 2, \ldots, I$). At the end of I minor cycles the entire vector $[y_n(t)]$ under the integral sign is replaced and a major cycle has been completed.

Fisher's method[3,4] for the solution of the same problem requires that one element of the vector $[y_n(t)]$ be updated at the end of each minor cycle. Consequently, the Fisher version of the digital process of Equation

15.3 / THE FISHER ITERATION METHOD

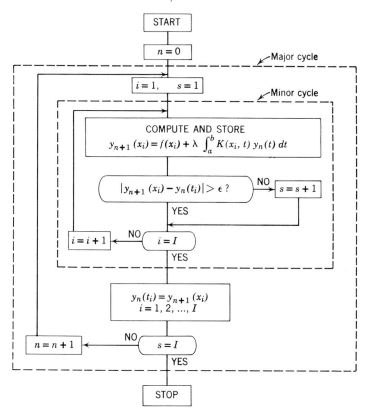

Figure 15.1 Flow diagram of Neumann's method.

15.3 becomes

$$y_{n+1}(x_i) = f(x_i) + \lambda \sum_{j=1}^{J} K(x_i, t_j) y_{n,i-1}(t_j) \Delta t \qquad (15.4)$$

$$i = 1, 2, \ldots, I, \qquad j = 1, 2, \ldots, J, \qquad I = J$$

Note that $y(t)$ under the summation sign now carries a double subscript. The idea is to replace at the end of each minor cycle the existing value of $y_n(x_i)$ in the memory with the newly obtained value of $y_{n+1}(x_i)$. The notation (n, i) $i = 1, 2, \ldots, I$ implies that during each major cycle the unknown function $y(x_i)$ is gradually adjusted as the index i is increased, not waiting, as in the Neumann method, until all minor cycles are completed. In other words, Fisher's method is based on using each piece of new information as soon as it is available so that the adjustment from $y_n(x)$ to $y_{n+1}(x)$ proceeds gradually, rather than being performed all at once after I minor cycles.

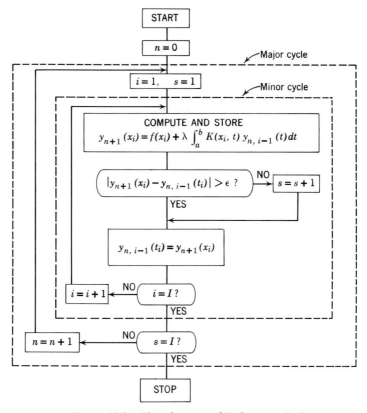

Figure 15.2 Flow diagram of Fisher's method.

The hybrid computer version of Fisher's method takes the form

$$y_{n+1}(x_i) = f(x_i) + \lambda \int_a^{x_i - \Delta x/2} K(x_i, t) y_{n+1}(t)\, dt$$

$$+ \lambda \int_{x_i - \Delta x/2}^{b} K(x_i, t) y_n(t)\, dt \qquad i = 1, 2, \ldots, I \quad (15.5)$$

The first integral on the right side of Equation 15.5 contains the results which have been obtained during the previous minor cycles of the present major cycle. A flow chart showing the hybrid computer implementation of this strategy is shown in Figure 15.2.

15.4 ILLUSTRATIVE EXAMPLES

Fisher[4] has shown that for symmetric kernels his algorithm converges whenever the Neumann algorithm does and, in certain cases, also when the classical method fails. (A symmetric kernel is characterized by $K(x, t) = K(t, x)$). Furthermore, using a simple example, Fisher has demonstrated that this method may speed up convergence significantly. In order to obtain practical results concerning this comparison, several problems were solved using a small hybrid computer (IBM 1620 digital computer and Beckman 2132 analog computer).

To facilitate the evaluation of the two methods, solutions $y_n(t)$ were compared with exact analytical solutions, $z(t)$, by means of a root sum square criterion, defined by

$$F_n = \sqrt{\sum_{i=1}^{I} [z(x_i) - y_n(x_i)]^2} \quad (15.6)$$

EXAMPLE 1. The following equation was solved:

$$y(x) = \tfrac{2}{3} + 2\int_0^1 (x - t)y(t)\, dt \quad (15.7)$$

with the initial approximation $y_0(t) = \tfrac{2}{3}$. The step size was chosen as $\Delta x = .1$ and the multiplication was performed on the analog computer. The solutions obtained by the Neumann and Fisher methods are compared in Figure 15.3 using the criterion F_n defined in 15.6. The considerably faster convergence of Fisher's method is clearly illustrated.

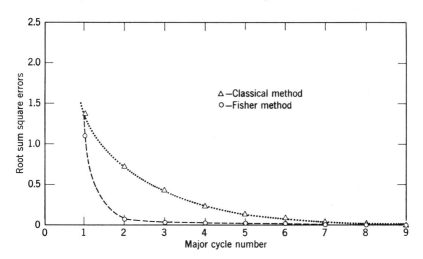

Figure 15.3 Comparison of hybrid solutions using classical and Fisher methods:

$$y(x) = \tfrac{2}{3} + 2\int_0^1 (x - t)y(t)\, dt$$
$$y_0(x) = \tfrac{2}{3}, \quad \Delta x = 1$$

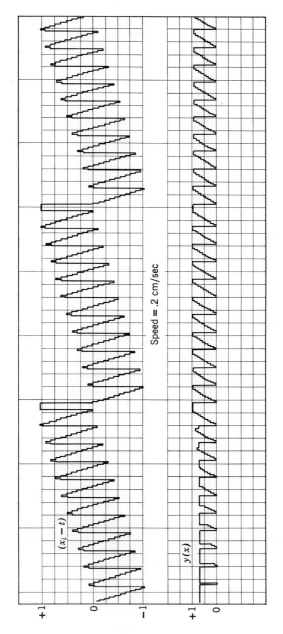

Figure 15.4 Time history of Fisher's solution to

$$y(x) = \tfrac{2}{3} + 2\int_0^1 (x-t)y(t)\,dt$$

$$y_0(x) = \tfrac{2}{3}$$

15.4/ILLUSTRATIVE EXAMPLES

Time histories of the kernel $K(x, t) = x - t$ and the functions $y_n(x_i)$ for several major cycles are shown in Figure 15.4.

EXAMPLE 2. The following equation was solved:

$$y(x) = 1.5x - \tfrac{7}{6} + \int_0^1 (x - t)y(t)\, dt \tag{15.8}$$

To illustrate the effect of choice of initial conditions, the equation was solved once with $y_0(t) = 0$ and once with $y_0(t) = \tfrac{2}{3}$. The results are shown in Figure 15.5 for $\Delta x = .01$. It is evident that a poor choice of initial approximation may lengthen the convergence process.

EXAMPLE 3. Examples 1 and 2 used kernels which could be considered functions of a single variable. Consider now the equation

$$y(x) = 1 + \int_0^1 e^{-xt} y(t)\, dt \qquad y_0(t) = 0 \tag{15.9}$$

The results are shown in Figure 15.6. Once again the superiority of Fisher's method is evident.

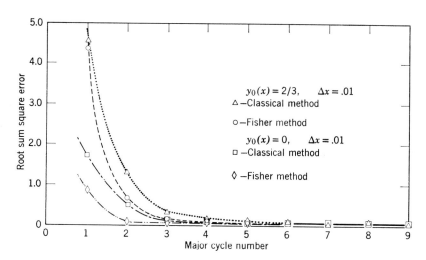

Figure 15.5 Comparison of hybrid solutions using classical and Fisher methods:

$$y(x) = \frac{3x}{2} - \frac{7}{6} + \int_0^1 (x - t)y(t)\, dt$$

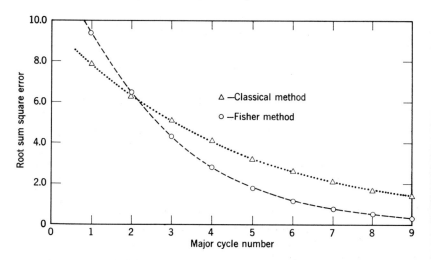

Figure 15.6 Comparison of hybrid solutions using classical and Fisher methods:

$$y(x) = 1 + \int_0^1 e^{-xt} y(t)\, dt$$
$$y_0(x) = 1, \qquad \Delta x = .01$$

15.5 DISCUSSION OF ERRORS

The major errors which enter into the hybrid solutions of integral equations are the following: (a) truncation errors (due to the fact that the functions have been quantized), (b) analog-digital and digital-analog conversion errors, (c) other analog computer errors, and (d) phase shifts due to digital execution time.

Truncation errors arise from the quantization of the variables x and t in Equation 15.1. In order to test the importance of the quantization interval, Example 1 was solved using 10 intervals ($\Delta x = .1$) and 100 intervals ($\Delta x = .01$). A comparison of the root sum squared errors for both interval sizes using the Fisher method is shown in Figure 15.7. It can be seen that convergence is speeded up by the choice of a smaller interval. After a sufficiently large number of major cycles, there is no apparent advantage to the small interval size. In fact, an oscillation in the final error criterion values ensues with the small Δx, which may be due to a random compounding of round-off errors from the large number of minor cycles.

The effect of quantization also enters into the integration process since the approximating function $y_n(t)$ is reconstructed from the samples

$y_n(x_i)$, $i = 1, 2, \ldots, I$. In the present study this reconstruction was accomplished using simple zero-order holds, i.e.,

$$y_n(t) = y_n(t_i) \quad \text{for} \quad t_i - \frac{\Delta t}{2} \leq t < t_i + \frac{\Delta t}{2} \tag{15.10}$$

It can be shown[4] that with both zero-order and first-order reconstruction the errors are proportional to $(\Delta t)^2$. Second-order interpolation formulas will reduce the error to $0(\Delta t)^3$, but the additional computation required to achieve it may not be justifiable.

Analog-to-digital and digital-to-analog conversion errors were extremely important in the solution of the example problems. However, it should be noted that both the Fisher and Neumann techniques are stable processes as long as the solution to the problem is analytically convergent.[4] Thus, random errors (which may enter the problem from converter or multiplier inaccuracies) during any one iteration (major cycle) will be corrected in subsequent iterations. Consequently, random errors delay the convergence process and may also cause a final indeterminacy region in the solution.

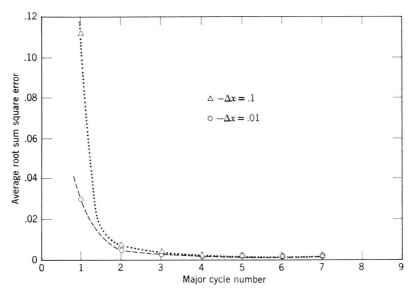

Figure 15.7 Comparison of hybrid solutions using Fisher method:

$$y(x) = \tfrac{2}{3} + 2\int_0^1 (x - t)y(t)\, dt$$
$$y_0(x) = \tfrac{2}{3}$$

Alternative mechanization of kernel generation and multiplication were also investigated. The resulting effects on solution accuracy are clearly a function of the quality and precision of available analog multipliers and function generators, as well as conversion errors. It should be noted, however, that analog generation of the kernel $K(x, t)$ has the advantage of producing a program which has a solution time independent of kernel complexity and whose digital portion is universally applicable.

While the actual times taken for a given solution clearly depend on the particular digital computer being utilized, comparisons between hybrid and all digital solutions are of interest. A speedup factor of 300 to 1 was obtained by using hybrid computation over digital computation, with comparable final accuracy.

15.6 EXTENSION TO OTHER TYPES OF INTEGRAL EQUATIONS

The above discussion has been devoted entirely to equations of the Fredholm type. However, extension of the technique to many other types of equations is possible. Consider, for example, the Volterra equation

$$y(x) = f(x) + \lambda \int_a^x K(x, t)y(t)\, dt \qquad (15.11)$$

This equation differs from the Fredholm equation in that the upper limit of integration is variable. The algorithms of Figures 15.1 and 15.2 are still applicable if the kernel is redefined such that

$$K(x, t) = 0 \quad \text{for } t > x \qquad (15.12)$$

A simple digitally controlled switch for implementing 15.12 can be used in conjunction with the Fredholm equation program to solve Volterra-type equations.

Wallman[1] and Fisher[3] have also indicated possible extensions of hybrid techniques to the solution of multidimensional integral equations, integrodifferential equations, and certain more general functional equations. Such extensions have yet to be proved in practice.

15.7 CONCLUSION

Hybrid computation techniques, involving fast repetitive analog integration and function generation and digital storage and control, are well suited to the solution of integral equations. Hybrid techniques lead to a substantial reduction in solution time when compared to all digital methods. Further, the examples solved in this study substantiate the faster convergence of Fisher's iteration scheme when contrasted with the classical Neumann technique. Extensions to other areas of application appear promising but remain to be tested.

References

1. Wallman, H., "An Electronic Integral Transform Computer and the Practical Solution of Integral Equations," *J. Franklin Inst.*, **250**, 45–61, July 1960.

2 Hildebrand, F. B., *Methods of Applied Mathematics*, Prentice-Hall, Englewood Cliffs, N.J., 1952.

3. Fisher, M. E., "On the Continuous Solution of Integral Equations by an Electronic Analogue," *Proc. Cambridge Phil. Soc.*, **53**, 162–173, 1957.

4. McKay, D. B., and M. E. Fisher, *Analogue Computing at Ultra-High Speed*, John Wiley, New York, 1962.

5. Tomovic, R., and N. Parezanovic, "Solving Integral Equations on a Repetitive Differential Analyzer, *IRE Transactions on Electronic Computers*, **EC-9**, 503–506, Dec. 1960.

Problems*

CHAPTER 2

2.1 Determine the change in the analog output of a DAC with reference voltage $E = 100$ volts corresponding to a change in the digital input from $x_D = .0001111111$ to $x_D = .1111100000$. Use any of the digital-to-analog conversion techniques discussed in the chapter.

2.2 Find an equation that gives the change in the analog output of the DAC of Problem 2.1 if the reference input is not fixed but replaced by $y_A = 100t$, $0 \leq t \leq 1$.

2.3 Assume that a 5-volt step is to be converted to digital form using both 10-bit successive approximation and incremental converters. Calculate and plot the error (ADC output minus 5 volts) at each time step for the first 20 steps with each converter, assuming that in both cases one l.s.b. $= 0.1$ volt.

2.4 Assume that an 8-bit DAC has been calibrated such that a full-scale analog output is obtained with the reference voltage N equal to 100 volts. If the *actual* reference signal during the calibration was $(N + \delta)$ volts, what is the error in volts at the analog output for: (a) a setting of .10000000 and (b) a setting of .01010100?

* Certain problems are marked with an asterisk, indicating that computers are required for their solution. However, many of the remaining problems, which require preparation of analog schematics or digital flow charts, also lend themselves well to laboratory assignments.

2.5 What analog-to-digital conversion technique is potentially the fastest? The most accurate? The least sensitive to noise? Explain.

2.6 Discuss the problems and feasibility of designing a 32-bit analog-to-digital converter of the: (a) incremental type and (b) successive approximation type.

2.7 A sinusoid ($x = 100 \sin \omega t$) is converted to digital form and then reconverted to analog form by using a DAC with a hold at its output. Plot one cycle of the output of the hold element (corresponding to one cycle of the input) if the ADC, storage, and DAC times are negligible and the following specifications apply to the converters:
(a) Sampling rate = 10 samples/cycle. Word length = 6 bits.
(b) Sampling rate = 10 samples/cycle. Word length = 12 bits.
(c) Sampling rate = 4 samples/cycle. Word length = 10 bits.

2.8 A positive analog voltage x_A is converted to a digital quantity x_D, using a 12-bit ADC. The digital computer forms the quantity

$$y_D = \sqrt{x_D}$$

and y_D is sent to a very accurate DAC whose output is the analog voltage y_A. Assume that $0 < x_A < 100$ volts. If y_A is scaled to full-scale, what is the magnitude of the jumps in y_A corresponding to the quantization in the ADC?

2.9 Which of the following digital codes is most desirable for use in the analog-digital channels of a hybrid system: (a) ordinary binary code, (b) BCD, (c) Gray code, (d) error-correcting code, or (e) error-detecting code? Explain by identifying advantages and limitations of each code for this application.

2.10 An ADC is being used to convert a sine wave, $x(t) = 100 \sin \omega t$, into pure binary code including 11 bits plus sign. The converter requires 10 μsec per bit. If the sampling is periodic and amplitude errors are not to exceed .1%, what is the highest frequency that can be handled if the converter is of the: (a) successive approximation type, and (b) incremental type.

CHAPTER 3

3.1 Most of the integration formulas in Chapter 3 are based on using a fixed step size h. Assume that sets of values $\{x_i\}$ and $\{y_i'(x_i)\}$ are obtained from the analog computer asynchronously, so that the step size $\Delta x_i = x_i - x_{i-1}$ is not constant but varies in some way.

Which integration formulas can be used to solve the differential equation

$$y' = f(x, y),$$
$$y = y_0 \quad \text{at} \quad x = x_0$$

with the irregularly spaced values indicated above? Give flow charts of any applicable formulas and discuss their features in detail.

3.2 Set up the following differential equations for numerical solutions by each of the methods discussed in the chapter:

(a) $\dfrac{dy}{dx} + 17.3y = f(x), \quad y(0) = 3.0$

(b) $\dfrac{d^2y}{dx^2} + 0.3(1 - y^2)\dfrac{dy}{dx} + y = 0, \quad y(0) = 1.0, \quad y'(0) = 0$

(c) $\dfrac{dy}{dx} + e^y = u(x), \quad y(0) = 0$

where $u(x)$ is the unit step function

(d) $\ddot{\theta}_1 + b\dot{\theta}_1 + k(\theta_1 - \theta_2) = 0$
$\ddot{\theta}_2 + c\dot{\theta}_2 + k(\theta_2 - \theta_1) = 0$
with $\theta_1(0) = \theta_2(0) = 0, \ \dot{\theta}_1(0) = +1, \ \dot{\theta}_2(0) = -1, \ b = 3, \ c = 2$, and $k = 4$.

(e) $\dot{\mathbf{x}} = A\mathbf{x} + B\mathbf{u}$ where

$$A = \begin{bmatrix} a_{11} & a_{12} \\ a_{21} & a_{22} \end{bmatrix}, \quad B = \begin{bmatrix} b_1 \\ b_2 \end{bmatrix}, \quad x(0) = \begin{bmatrix} x_{10} \\ x_{20} \end{bmatrix}$$

*3.3 Solve the following differential equation on the digital computer by using Euler, fourth-order Runge-Kutta, and predictor-corrector-modifier formulas:

$$\ddot{x} + 3\dot{x} + 40x = 2 \sin 7t \quad \text{(P3.3)}$$
$$x(0) = \dot{x}(0) = 0$$

(a) Use at least two different step sizes, and (b) compare your solution with the analytical solution of Equation P3.3.

3.4 The following two equations are given:

$$y' + y = 1 \quad \text{(P3.4a)}$$

and

$$y' - y = 1 \quad \text{(P3.4b)}$$

Assume that the initial condition of both equations is zero. (a) Solve the two equations numerically using Euler (rectangular) integration for 10 time steps, using $h = 0.2$ second, and compare the solutions with the analytical solutions. Which solution has the greater error? Why? (b) Assume that a fixed error ϵ is added to the solution of each time step. How do the errors accumulate in the solution of Equation P3.4a? P3.4b?

*3.5 Repeat Problem 3.4a by actually solving the equations on a digital computer, using Euler, predictor-corrector, and Runge-Kutta formulas of your choice.

3.6 The equation $y' = f(x, y) = -5y + g(x)$ is to be solved by the following procedure:

$$y_{n+1} = y_{n-1} + 2hy_n'$$

$$y_{n+1}' = f(x_{n+1}, y_{n+1})$$

$$y_{n+1} = y_n + \frac{h}{2}(y_{n+1}' + y_n')$$

Is the procedure stable? How does it compare with other formulas?

3.7 Draw a flow chart for solution of the following equation by the trial-and-error method, using: (a) third-order Runge-Kutta integration and (b) Adams-Bashforth 3-point predictor integration:

$$y''' + 2y'' + 3y' + t^2 y = f(t)$$

$$y''(0) = b_1, \quad y'(0) = b_2, \quad y(t_1) = b_3, \quad 0 \leq t \leq t_1$$

3.8 Discuss the following statements in detail: (a) Unstable integration formulas should never be used. (b) Round-off errors are much more important than truncation errors in present-day hybrid systems. (c) Digital integration is always more accurate than analog integration.

CHAPTER 4

4.1 Using Table 4.1 in the chapter, solve the difference equation

$$f(n + 1) + 3f(n) = 1 \tag{P4.1}$$

using z-transforms and the initial condition $f(0) = 1$.
Answer: $f(n) = \frac{1}{4} + \frac{3}{4}(-3)^n$.

4.2 The Heun integration method may be described as follows: If the equation to be integrated is

$$y' = f(y, x), \quad y(x_0) = y_0 \tag{P4.2a}$$

then the $(i + 1)$st value of y will be given by

$$y_{i+1} = y_i + \left(\frac{f_i + f_{i+1}}{2}\right)T \qquad \text{(P4.2b)}$$

The slope f_{i+1} is computed by using rectangular integration on equation P4.2a. Apply this method to the specific equation

$$y' = -y, \qquad y(0) = 1 \qquad \text{(P4.2c)}$$

(a) Find a recursion equation of the form P4.2b for this problem such that the right-hand side of the equation depends only on y_i and not y_{i+1}. (b) Show that the z-transform of the solution is given by

$$X(z) = \frac{x_0}{1 - [1 - T + (T^2/2)]z^{-1}}$$

(c) Study the stability and accuracy of the solution as a function of T.

4.3 The equation

$$\dot{y} + 5y = x(t), \qquad y(0) = 0$$

is to be solved by using rectangular integration, as in Section 4.5A. (a) Derive the transform expression $Y(z)/X(z)$. (b) Plot the T-root locus and find the range of stability. (c) Plot the gain and phase of $Y(z)/X(z)$ with the substitution $z = e^{j\omega T}$ and $T = 1$.

4.4 Derive expressions for the z-transforms of: (a) a trapezoidal integrator and (b) Simpson's $\frac{1}{3}$ rule integrator

$$y_2 = y_0 + \frac{T}{3}(x_2 + 4x_1 + x_0)$$

and (c) Simpson's $\frac{3}{8}$ rule integrator

$$y_3 = y_0 + \tfrac{3}{8}T(x_3 + 3x_2 + 3x_1 + x_0)$$

4.5 The equation $\dot{y} = f(y, t) = -5y + r(t)$ is to be solved by the following procedure:

$$y_{n+1} = y_{n-1} + 2T\dot{y}_n$$
$$\dot{y}_{n+1} = f(t_{n+1}, y_{n+1})$$
$$y_{n+1} = y_n + \frac{T}{3}(\dot{y}_{n+1} + \dot{y}_n)$$

Is the procedure stable?

4.6 Show that if $\mathscr{Z}\{f(nT)\} = F(z)$, then
$$\mathscr{Z}\{nf(nT)\} = -z\frac{d}{dz}F(z)$$

4.7 In order to save the rare gogo bird from extinction, the University Birdlovers Club has purchased a small island and released N birds on the island. Every year they come back to count and release more birds. The discover that every year *half* of the birds from the previous year have died and they release 4 new birds every year. Ignoring such minor mathematical problems as fractional birds, (a) write a difference equation describing the n-th year bird population, and (b) find the steady-state population, using the final value theorem of z-transforms:
$$\lim_{n\to\infty} f(nT) = \lim_{z\to 1} (1 - z^{-1})F(z)$$
Answer: 8 birds.
(c) What is the effect of the initial population N on the steady-state population?

CHAPTER 5

5.1 Design a filter that removes both the first-order and second-order terms of the power-series expansion of the time-delay operator $e^{-\tau s}$ in Equation 5.3. Under what conditions would such a compensation scheme be practical? (**Hint:** Let $F(s) = 1 + a_1 s + a_2 s^2$ and substitute in Equation 5.3a.)

5.2 Consider the hybrid system shown in Figure 5.2 with $\tau_1 = 10$ msec, $\tau_2 = 20$ msec, $T = 40$ msec, and $D(s) = 1$. (a) Using only the first two terms of the power-series 5.3, obtain an expression for the error $e(t)$ where
$$e(t) = y(t) - x(t)$$
(b) If a value of τ which is in error by 10% is used in a time-delay compensation scheme, what is the error?

5.3 Prove the following statement which appears in Section 5.4: "Consider, for example, the sampling, at a sampling rate of 100 samples per second, of two sinusoidal signals, one having a frequency of 60 cycles per second and the other a frequency of 40 cycles per second. It can be shown that at every sampling instant the magnitude of these two sine waves is identical."

5.4 Prove that the mean-square value of the quantization error is given by
$$\overline{e_q^2} = \sqrt{q^2/12}$$

5.5 An 8-channel multiplexer is used to connect the outputs of 8 identical analog sine-cosine generators to an analog-digital converter. If all the outputs are given by $x_i(t) = 100 \sin 628t$, $i = 1, 2, \ldots, 8$, and if they are addressed sequentially, what is the minimum speed of the converter (in conversions per second) if the slewing error is not to exceed 1° between any two sine waves?

5.6 Equation 5.8b indicates that if a sampled sine wave is to be reconstructed with a zero-order hold in such a way that the *maximum* instantaneous error does not exceed .1% of full-scale, about 6280 samples per cycle are needed. How many samples per cycle would be required if the sampling is nonperiodic and the sampling times are such that the error in *each* sampling inverval is .1%?

5.7 The output of a fractional-order hold circuit is given by

$$y(\tau) = y(nT) + \frac{k\tau}{T}\{y(nT) - y[(n-1)T]\}, \quad 0 \leq \tau < T \quad \text{(P5.7)}$$

where $0 < k < 1$. Note that this expression describes a zero-order hold if $k = 0$ and a first-order hold if $k = 1$. Develop an analog computer mechanization of expression P5.7 using track-and-hold amplifiers, assuming that the sequence of values $\{y(nT), n = 1, 2, \ldots\}$ is available at the output of a digital-analog converter.

5.8 Develop: (a) an analog computer program and (b) a FORTRAN program to compute the sensitivity coefficients $\partial x/\partial a$ and $\partial x/\partial b$ for the equation

$$\ddot{x} + ax^2\dot{x} + x^3 = f(t) \quad \text{(P5.8)}$$
$$\dot{x}(0) = 0, \quad x(0) = b$$

Use the method of sensitivity equations described in Section 5.8 and show all initial conditions and inputs.

5.9 Sensitivity coefficients can also be defined by using finite differences, as

$$u(t, q_0) \cong \frac{x(t, q_0 + \Delta q) - x(t, q_0)}{\Delta q} \quad \text{(P5.9)}$$

Develop: (a) an analog computer program and (b) a FORTRAN program for computation of the sensitivity coefficients in Problem 5.8, using the definition in Equation P5.9.

5.10 A sine wave, $x(t) = 64 \sin \omega t$ is sampled at 16 samples per cycle, converted to digital form, reconverted to analog form, and reconstructed with a zero-order hold. (a) Compute the maximum instantaneous amplitude error due to the sampling and data reconstruction, neglecting ADC and DAC conversion times.

(b) Select a word-length for the output of the ADC such that the maximum instantaneous quantization error is no greater than one-tenth of the error of part a, assuming the two errors to be independent.

5.11 It is desired to solve the equation

$$\ddot{x} + \dot{x} + 4x - 17\dot{x}^3 = 0$$

on a hybrid computer. Suggest various possible allocations of computing functions between the analog and digital portions of the system and compare them on the basis of probable errors.

5.12 The chapter describes data reconstruction using zero-order and first-order hold circuits. A second-order hold is described by the transfer function

$$H_2(s) = T(Ts + 1)\left(Ts + \frac{1}{2}\right)\left(\frac{1 - e^{-Ts}}{Ts}\right)^3$$

During a sampling interval, its output is given by:

$$x(nT + \tau) = x(nT) + \left(\frac{\tau}{T}\right)[x(nT) - x((n-1)T)]$$

$$+ \frac{1}{2}\left(\frac{\tau}{T}\right)\left(1 + \frac{\tau}{T}\right)[x(nT) - 2x((n-1)T) + x((n-2)T)]$$

where $\{x(nT)\}$ is the sequence of numbers being converted to analog form.

(a) Draw the reconstructions of a sampled sine wave $x(nT) = \sin \omega_0 nT$ obtained with zero-order, first-order, and second-order holds if the sampling occurs 8 times per cycle ($\omega_s = 8\omega_0$). Use different colors for the various outputs.

(b) Draw a scaled analog computer diagram to generate second-order hold outputs from the outputs of a standard DAC which includes a zero-order hold.

(c) Obtain the amplitude and phase characteristics of a second-order hold and compare them with those of Figure 5.10. Describe the major advantages and limitations of each hold from this frequency-domain data.

(d) Calculate the number of samples per cycle required to limit the maximum instantaneous amplitude error to .1% when reconstructing a sampled sine wave. Compare your results to those for zero-order and first-order holds discussed in Section 5.7.

5.13 Consider that a sine wave is applied to a quantizer which uses a total of four (4) bits to cover the entire -100 volt to $+100$ volt dynamic range. Plot the quantization error versus phase angle $\theta = \omega t$. In what part of the dynamic range is the error most significant?

5.14 The following equation is solved on a hybrid computer:

$$\ddot{y} + a(1 - y^2)\dot{y} + y = 0$$

$$y(0) = b, \quad \dot{y}(0) = 0$$

The digital computer samples y and $\dot{y}$ from the analog and computes the quantity $a(1 - y^2)\dot{y}$. Everything else is done on the analog. ADC time is 2 msec per conversion. Digital computer execution time is 2 msec after obtaining y and $\dot{y}$. DAC and zero-order hold updating interval (desampling interval) is 4 msec. The multiplexer first samples $\dot{y}$ and then, 1 msec later, samples y. It is desired to compensate for the most significant portion of the time-delay error by introducing lead compensation into the analog portion of the system. Draw the complete analog diagram, including numerical values of all potentiometer settings. The interface and digital computer should be drawn simply as blocks.

5.15 Professor Hyppolates J. Bridd (known as "Hy" Bridd among his friends) has proposed the following inequality as characterizing hybrid computers:

$$\frac{\pi f_m(T + 2T_e)}{\phi} \leq 1 \qquad \text{(P5.15)}$$

where $T = 1/f_s$ is the sampling interval, T_e is digital computer execution time, f_m is the maximum analog signal frequency, and the maximum allowable phase shift of a sine wave $\sin(2\pi f_m t)$ due to digital processing and data reconstruction is ϕ radians. (a) Prove or disprove Equation P5.15. (b) Discuss the following statement, derived from Equation P5.15 ("Hy Bridd's First Law"):[†] "The faster the digital computer, the less often one has to sample." Evaluate the validity (if any) of this "law," its assumptions, limitations, etc.

[†] Bridd's law was first suggested to the authors by Elbert Hartsfield of TRW Systems, Redondo Beach, California.

CHAPTER 6

6.1 Investigate (using manufacturers' literature, published papers, trade journals, etc.) various interrupt structures for digital computers and comment on their relative advantages and disadvantages for hybrid computation.

6.2 Discuss the relative merits of word-length, memory cycle time, and priority interrupt in the choice of a digital computer for hybrid operation.

6.3 You have been asked to design a hybrid computer which will drive a car in an optimum manner. The computer inputs will be obtained from optical, acoustic, electromagnetic, or any other sensors you wish. The computer outputs will control actuators for steering, braking, acceleration, etc. Prepare a block diagram of your proposed design and discuss such problems as data acquisition, interfaces, selection of continuous or discrete signals, number of multiplexers, sense lines, and memory requirement. Estimate the computer cycle time. Indicate the allocation of computing functions between analog and digital portions of the system and justify it.

6.4 From manufacturers' literature, locate and study unusual features of currently available digital computers and discuss their possible applicability in hybrid computers.

6.5 It has been suggested that an ideal hybrid computer would contain both analog and digital computation elements which could be addressed by the system software. Evidently, such a system would require automatic interconnection of analog elements with DAC's and ADC's ("automatic patching"). Estimate the number of possible connections which might be needed in an analog subsystem consisting of 10 amplifiers, 20 potentiometers, and 4 multipliers. Assume that every amplifier can be operated as a summer, integrator, or high-gain amplifier.

CHAPTER 7

7.1 It is desired to use the digital computer in a hybrid system to perform diagnostic tests of the analog computer, in order to test the normal functioning of all analog components. (a) Prepare a flow chart of a digital computer program to perform diagnostic maintenance of the analog computer. (b) Prepare a schematic diagram of the analog patching required to test normal operation of all elements. The special connections you prepare will be used during the diagnostic test.

7.2 Define the following terms: (a) assembler, (b) compiler, (c) monitor, (d) supervisor, (e) utility routine, (f) subroutine, (g) executive, (h) interpreter, (i) priority interrupt, (j) real-time package, and (k) sense line. Are all the above needed with hybrid software systems?

7.3 Function-generation routines for hydrid systems usually consist of stored tables of $f_i(x_i)$, $i = 1, 2, \ldots, N$. The increment in the independent variable, $\Delta x_i = x_i - x_{i-1}$, is usually fixed. However, considerable economies of high-speed memory are possible if the data are not equally spaced in the x domain. Prepare a flow chart and discuss a function storage and retrieval system for a hybrid computer which will function as follows:

(a) A function $f(x)$ is generated by the analog computer and available at the input to an ADC.

(b) The function and the independent variable are sampled and stored whenever $|f(x) - f_i(x_i)| = C$, where C is an appropriate constant.

(c) The entire function must be retrievable during the following iteration on the analog computer.

7.4 In connection with the asynchronous sampling and function storage scheme of Problem 7.3, compute the saving in words of core memory which may be possible, as compared with synchronous sampling, if the function $f(t) = 100 \sin t$ is sampled, stored, and retrieved such that: (a) data reconstruction is done by connecting retrieved points with straight lines and (b) the maximum error between $f(t)$ and the reconstructed function does not exceed .1 volt with either synchronous or asynchronous sampling.

Don't forget that with synchronous sampling, you must also store the increment or step-size Δt; with asynchronous sampling it *may* be necessary to store the double array $\{f_i, t_i\}$.

7.5 Prepare a flow chart of a system for verifying the correctness of an analog set-up of a given set of m ordinary n-th-order differential equations.

7.6 You have been asked to develop software for a special-purpose hybrid computer where the analog portion solves *only* second-order, linear, ordinary differential equations. Hence, the analog portion is hard-wired (has no patchboard) and only the coefficients and initial conditions are entered as part of the data input in the digital program. Prepare a flow chart of a program which will automatically scale and set the potentiometers for the analog portion of this system if N differential equations must be solved. Scaling should be such that the maximum values of all amplifier outputs lie between 50 and 100 volts, if possible.

CHAPTER 8

8.1 Prepare analog computer schematics for solution of the following equations by the DSCT method: (a) the wave equation (8.5) and (b) the biharmonic equation (8.6).

8.2 The partial differential equation, which may represent the one-dimensional diffusion of Los Angeles smog on a windy day, is

$$\frac{\partial^2 \phi}{\partial x^2} = v_x \frac{\partial \phi}{\partial x} + \frac{\partial \phi}{\partial t} \qquad (P8.2)$$

where v_x is the wind velocity (a constant) and ϕ is the concentration of the diffusing pollutant. Assume that the field in the x direction is to be discretized into 10 sections with nodes $x_0, x_1, \ldots, x_{10}$.

(a) Set up a finite-difference equation that approximates the continuous variation of ϕ at node 5.

(b) Assuming ϕ_4 and ϕ_6 are available, show an analog computer circuit for solving the equation of part a.

(c) Discretize time in the equation of part a and write an implicit DSDT approximation to Equation P8.2.

8.3 Prepare a flow chart for the digital portion of the hybrid DSCT method discussed in Section 8.8 for the solution of the one-dimensional diffusion equation.

8.4 Prepare a flow chart for the digital portion of the hybrid CSDT method discussed in Section 8.9 for the solution of the diffusion equation.

8.5 Prepare a flow chart for extending the techniques of either Problem 8.3 or 8.4 (whichever you consider most suitable) to solution of the two-dimensional diffusion equation:

$$\frac{\partial^2 \phi}{\partial x^2} + \frac{\partial^2 \phi}{\partial y^2} = f(x, y) \frac{\partial \phi}{\partial t}$$

8.6 The following finite-difference approximation to the diffusion equation is proposed:

$$\frac{1}{\Delta x^2}(\phi_{x+\Delta x, t} - \phi_{x, t+\Delta t} - \phi_{x, t-\Delta t} + \phi_{x-\Delta x, t}) = \frac{K}{2\Delta t}(\phi_{x, t+\Delta t} - \phi_{x, t-\Delta t})$$

Note that the term $\phi_{x,t}$ does not appear. (a) By comparing this approximation to others in the chapter, would you consider this an *explicit* or *implicit* approximation? Explain. (b) Suggest a hybrid method for solution, using this approximation. (c) Compare it to others with respect to stability, ease of implementation, computation time, equipment requirements, etc.

8.7 The following equations are said to govern the diffusion of boron in certain semiconductors:

$$\frac{\partial V}{\partial t} = K_1 \frac{\partial^2 V}{\partial y^2} + K_2 \frac{\partial C}{\partial t}$$

$$\frac{\partial C}{\partial t} = K_3 \left(\frac{\partial V}{\partial y}\right)\left(\frac{\partial C}{\partial y}\right) + K_4 V \frac{\partial^2 C}{\partial y^2}$$

where V is the concentration of vacancies (holes) in the material, C is the concentration of boron, y is linear distance, t is time, and K_1 through K_4 are constants.

(a) Show how to solve this set of equations on an analog computer, using a DSCT approach. Clearly indicate the initial and boundary conditions needed to obtain a solution.

(b) Propose and describe in detail a hybrid method of solution.

CHAPTER 9

9.1 Consider the boundary value problem

$$\ddot{x} + 0.5\dot{x} + x = 0 \tag{P9.1a}$$
$$x(0) = 0, \quad \dot{x}(0) = q, \quad x(1) = 1$$

The initial condition q is unknown. To insure that the solution passes through the second boundary, we select a criterion function

$$F(q) = [1 - x_q(1)]^2 \tag{P9.1b}$$

where $x_q(t)$ is the solution of P9.1a with a particular value q for the unknown initial vote. It is desired to minimize $F(q)$ by steepest descent.

(a) Derive the sensitivity equation for $u_q \triangleq \partial x/\partial q$.

(b) Show that the initial conditions for the sensitivity equation are $u(0) = 0$, $\dot{u}(0) = 1$.

(c) If continuous steepest descent can be used in the solution of this problem, let

$$\dot{q} = -K \frac{\partial F}{\partial q}$$

and show an unscaled analog schematic for solution of the problem. If it cannot be used, explain.

(d) Draw an iterative analog or hybrid diagram for solution of the problem where the j-th initial condition will be given by

$$q^{(j)} = q^{(j-1)} - K \frac{\partial F(q^{(j-1)})}{\partial q}$$

9.2 Use the technique in Section 9.4 to prepare a detailed analog schematic for solution of the following equations by a steepest descent method, with the criterion functions indicated:

$$2x_1 + 3x_2 = 5$$
$$x_1 + 5x_2 = 7$$

(a) Let $f = e_1^2 + e_2^2$, and (b) let $f = |e_1| + |e_2|$, where e_1 and e_2 are defined by Equation 9.30.

9.3 Show that the matrix expression in Equation 9.29 is equivalent to the steepest descent equations in Equation 9.38.

*9.4 Use continuous steepest descent to find the four real roots of the polynomial

$$P(x) = x^4 + 10x^3 + 35x^2 + 50x + 24 = 0$$

Let $F(x) = [P(x)]^2$. (a) Use the analog computer to solve the problem and compare your results with analytical solution for the roots. (b) Prepare a digital program for solution of the problem by a discrete gradient method.

9.5 Section 9.5 describes an approximate continuous steepest descent technique suitable for analog mechanization. An alternate mechanization, sometimes known as the "equation error method" (references 8, 16, and 17 of Chapter 9) may be used for parameter identification in the following way.

Let the system to be modeled have the form shown in Figure 9.10; that is,

$$k_1 \dot{y}_D + k_2 y_D = x \qquad (P9.5a)$$

We assume that both y_D and $\dot{y}_D$ are measurable. Using the initial guesses at the parameter values, namely α_1 and α_2, we form the quantity

$$\alpha_1 \dot{y}_D + \alpha_2 y_D - x = \epsilon \qquad (P9.5b)$$

Note that ϵ, the "equation error," is an algebraic function of the parameters; hence the partial derivatives $\partial \epsilon / \partial \alpha_1$ and $\partial \epsilon / \partial \alpha_2$ are well defined.

(a) Develop the necessary equations to minimize the function $f(\epsilon) = \epsilon^2$ by steepest descent by continuous adjustment of α_1 and α_2.

(b) Draw the corresponding analog computer diagram.

*(c) Select values of k_1 and k_2 and obtain an actual solution on the analog computer.

***9.6** Use the mechanization of Figure 9.10 to optimize the parameters α_1 and α_2 of Problem 9.5 by approximate steepest descent. Compare your results with those of Problem 9.5c.

9.7 Draw a flow chart of the digital portion of the hybrid optimum gradient method of Figure 9.19, using sensitivity equations rather than finite differences and neglecting the quadratic interpolation of Figure 9.18. Let all differential equations be solved by the analog computer.

9.8 Draw the digital flow chart and an unscaled analog diagram for hybrid mechanization of the optimization technique described in Section 9.7. Apply it to a third-order system of your choice, with 3 adjustable parameters α_1, α_2, and α_3, with the criterion function

$$F(\alpha_1, \alpha_2) = \int_0^T (y - y_D)^2 \, dt$$

where y_D is a desired response and y is the output of your system.

9.9 Repeat Problem 9.8 for the relaxation method described in Section 9.8.

***9.10** Program and run a hybrid optimization by random search of the problem described by Ung in reference 43 of Chapter 9.

9.11 Modify the algorithm illustrated in Figure 9.25 in order to accomplish the following: (a) Instead of "absolute biasing," use the strategy of Equation 9.100 to modify the mean of the step distribution, and (b) include a subroutine that decreases the variance of the distribution as the local optimum is approached.

9.12 The random search algorithms of Figures 9.25 and 9.26 use sampled noise from an analog noise generator. Study the techniques of digital noise generation discussed in Chapter 11 and modify the algorithms as necessary to use at least two different digital noise generation methods. Be particularly careful to note the distribution of the pseudo-random sequence available with your computer software. Which noise generation method do you think is best suited for hybrid random search optimization?

9.13 You have an analog computer with several multipliers and no dividers. It has been suggested that (continuous) steepest descent can be used to produce a divider as follows: Let the desired quotient be $z = x/y$ and the error be $e = zy - x$. Choose a criterion function $f(e)$, develop the detailed mechanization, and discuss its advantages and limitations. Will it divide two signals regardless of their polarity?

CHAPTER 10

*10.1 Program an analog computer for solution of the linear time-optimal problem described in Section 10.4. Let

$$\dot{\mathbf{x}} = A\mathbf{x} + Bu \qquad (\text{P10.1})$$

where

$$A = \begin{bmatrix} 0 & 1 \\ 2 & 3 \end{bmatrix}, \quad B = \begin{bmatrix} 1 \\ 2 \end{bmatrix}, \quad \mathbf{x}(0) = \begin{bmatrix} 1 \\ 1 \end{bmatrix}$$

and

$$|u(t)| \leq 4$$

It is desired to find the control function $u(t)$ which will transfer the system to the origin of the state space in minimum time. Set up the problem and find the initial conditions of the adjoint equations by trial and error.

10.2 Write a FORTRAN program for digital solution of Problem 10.1.

*10.3 Use iterative analog methods to solve the example problem described in Section 10.4c (Equations 10.34 and following). Let $\omega = 1$, $c_1 = c_2 = .25$.

10.4 Prepare a flow chart for digital solution of the chemical reactor optimization problem described in Section 10.6B.

*10.5 Read the paper of reference 35 in Chapter 10 and prepare a critique of the proposed method. Use computer techniques to try out the sample problem described in the paper. If your solution disagrees with that of the paper, explain.

10.6 Prepare a detailed flow chart and analog diagram for solution of a trajectory optimization problem by the method of Wingrove and Raby (Section 10.9). Select any coefficients and numerical values you wish, but let the system be of second-order or higher.

CHAPTER 11

11.1 Develop an all-analog mechanization of the probability distribution analyzer shown in Figure 11.2, using diodes to develop slicers, gates, etc.

*11.2 Use the circuit of Problem 11.1 to test the probability density of the random noise generator on your computer.

11.3 Prepare a flow chart and FORTRAN program for a digital technique of estimating the probability density function of sampled analog data stored in memory.

11.4 Develop a detailed circuit analogous to Figure 11.2 for estimating the joint probability distribution defined in Equation 11.12.

*11.5 Use the circuits of Figures 11.6a and b to estimate the time average of the output of an analog noise generator, using 100-second samples of data. (a) Compare the results of the two methods. (b) Using whatever reference books you wish, try to calculate confidence limits for your estimates.

11.6 Consider two random analog signals $x(t)$ and $y(t)$ with no significant energy above 50 Hz. Choose a sampling frequency and estimate the *digital* computer time needed for calculation of the cross-correlation function by means of the technique in Section 11.4C.

11.7 An analog noise generator produces a random signal with zero mean, Gaussian amplitude distribution, and constant spectral density from 0 to 40 Hz. (a) How can the output distribution be modified to a uniform distribution? (b) How can the spectrum be modified to

$$S_{yy}(\omega) = \frac{K}{a^2 + \omega^2}$$

*11.8 It is desired to estimate the mean-squared noise response of a system described by the differential equation

$$\ddot{y} + 3\dot{y} + 2y = x$$

where $x(t)$ is assumed to be white noise. (a) Prepare digital flow charts and/or analog schematics for three different ways of estimating $E[y^2(t)]$. (b) Program and run the method of Figure 11.17 and compare the results with at least one other method.

11.9 It is desired to estimate the mean-squared response of the time-varying system

$$\ddot{y} + t\dot{y} + e^{-t}y = x$$

to a white noise input. (a) Prepare an analog schematic for solution of the problem by the "adjoint method" (references 4, 10, and 30 of Chapter 11). (b) Prepare the necessary analog schematics and digital flow charts for solution by a direct method where impulse inputs are applied at successive times to obtain the system weighting function (see Equation 11.70).

CHAPTER 12

12.1 Three alternative hybrid mechanizations of the six-degree-of-freedom airframe equations are described in Sections 12.2C, D, and E. (a) Draw a detailed flow chart for the digital portion of each proposed mechanization. (b) Draw an unscaled analog schematic for the analog portion of each proposed mechanization.

12.2 A number of alternative hybrid mechanizations of the equations needed for space-vehicle simulation are discussed in Section 12.3B. (a) Draw a detailed flow chart for the digital portion of each proposed mechanization. (b) Draw an unscaled analog schematic for the analog portion of each proposed mechanization.

12.3 Compare the alternative mechanizations of Problems 12.1 and 12.2 in terms of: (1) analog equipment requirements, (2) estimated digital memory requirements, and (3) interface requirements. Based on your analysis, which method would you recommend for solution of: (a) atmospheric flight equations and (b) space flight equations. Justify and explain your answers.

CHAPTER 13

*13.1 Construct a human operator control task of the type shown in Figure 13.3. Let the operator watch an oscilloscope screen where only a vertical line appears. The reference input $r(t)$ can be a sum of three nonharmonic sine waves or low-frequency noise passed through at least two stages of filtering to insure that there is negligible energy above .5 Hz. As a controller, use a potentiometer with a handle located such that control movements right and left (in a vertical plane) correspond to line displacements on the screen. For simplicity, let the controlled element be simply a gain.

(a) Practice tracking (nulling the display error) by moving the controller such that the line stays as close as possible to the center of the screen.

(b) Construct a model of yourself on the analog computer by using the expression

$$H(s) = \frac{Ke^{-Ds}}{1 + T_N s} = \frac{\bar{M}(s)}{E(s)}$$

where $\bar{m}(t)$ will represent the model output. Drive the model with the tracking error and adjust K, D, and T_N until $m(t)$ and $\bar{m}(t)$ are as close as possible.

(c) Replace the human operator model by the one shown in Figure 13.4 and again adjust the parameters (including the sampling interval T) for best fit. A first-order hold is described by

$$x_{\text{FOH}}(t) = x(nT) + \{x(nT) - x[(n-1)T]\}\frac{(t - nT)}{T}$$

for the interval $nT \leq t < (n + 1)T$. Which model fits your tracking performance better? Why?

(d) Optimize the parameters of your model by using steepest descent or similar optimization techniques to minimize

$$f = [m(t) - \bar{m}(t)]^2$$

or

$$F = \int_0^{t_f} [m(t) - \bar{m}(t)]^2 \, dt$$

13.2 You have been asked to design a hybrid computer-based simulator for the study of new methods of automobile control. Study data requirements, driver computer interfaces, computation requirements, advantages of analog versus digital generation of displays, and processing of steering wheel and pedal outputs, etc.

CHAPTER 14

14.1 Design: (a) an analog and (b) a hybrid technique for measurement of heart rate. The input will consist of audio signals obtained from a microphone strapped to your body; heart rate will be defined as

$$HR(t_i) = \frac{1}{T_i}$$

where T_i is the i-th interval between heartbeats.

*14.2 Build and test your device by computer simulation in the laboratory. Plot the heart rate transients which follow: (a) taking a deep breath, (b) exhaling, (c) holding your breath, and (d) standing up suddenly from a sitting position.

14.3 Draw detailed analog schematics and digital flow charts for the cardiac-output determination method illustrated in Figure 14.7.

14.4 You have been asked to design a computer-based technique to analyze and evaluate the electrocardiograms of all the inhabitants of a city with a population of 100,000. It is desired to do this on-line and in real time, if possible. Devise a scheme and discuss its advantages and limitations. (For example, you could use analog preprocessing as in Figure 14.3 and compare the quantities obtained with normal and pathological levels for the population. Or, you could use a pattern-recognition scheme of some kind, or . . . .)

14.5 A large hybrid computer has been installed in your university hospital. Design (in block-diagram form) the following applications for the computer:

454 PROBLEMS

(a) Generating patient-status displays for a remote nurses station, with such quantities as temperature, blood pressure, and heart rate being monitored, displayed, compared to reference levels with necessary alarms, etc.

(b) Generating on-line displays during surgery to assist the surgeons and anesthetists.

For each application, discuss data acquisition, interface requirements, analog equipment accuracy requirements, digital memory, word-length, cycle time, etc.

CHAPTER 15

15.1 Prepare analog schematics and digital flow charts for solution of the equation

$$y(x) = e^{-x} + 0.3 \int_1^x \frac{1}{x^2 + t^2} y(t)\, dt$$

with at least two different allocations of computing tasks between analog and digital computers using (a) Neumann's method and (b) Fisher's method.

15.2 Discuss the applicability of the techniques of this chapter to the numerical evaluation of the Laplace transform of a function $f(t)$ defined as

$$F(s) = \int_0^\infty f(t) e^{-st}\, dt$$

Do not forget that s is a complex quantity, defined as $s = \sigma + j\omega$. Show flow charts, block diagrams, etc.

15.3 Apply the techniques of this chapter to the solution of differential equations. Given the equation

$$\dot{y} = f(y, x, t), \; y(0) = y_0$$

note that its solution can be written as

$$y(t) = y_0 + \int_0^t f(y, x, \tau)\, d\tau$$

(a) select a specific linear equation and (b) select a specific nonlinear equation. Prepare all necessary flow charts and schematics.

Author Index

Anderson, M. D., 311, 322, 330
Andrews, J. M., 20
Angel, E. S., 397, 403
Athans, M., 303, 328–329

Balakrishnan, A. V., 328–329
Battin, R. H., 359, 361
Beckenbach, E. F., 299
Bekey, G. A., 150, 299–300, 397, 403, 422
Bendat, J. S., 361
Bihovski, M. L., 135, 138, 149
Blackman, R. B., 361
Bogdanoff, F., 421
Boltyanskii, V. C., 329
Bradham, G. B., 420
Bradley, R. E., 20
Breenberger, M., 362
Brennan, R. D., 207
Brooks, S. H., 284, 300
Brubaker, T., 361
Brunner, W., 280, 298, 316, 328
Bryson, A. E., Jr., 323, 329

Cady, L. D., 411, 420–421
Cameron, W. D., 335, 361
Carr, J. W., III, 89
Chandler, W. J., 207
Chapelle, W. J., 20
Chernoff, H., 298

Chestnut, H., 298
Chuang, K., 288, 243
Clancy, J. J., 20
Clymer, A. B., 299, 420
Clynes, M. E., 420
Cochrane, I., 421
Collatz, L., 89
Connolly, M. E., 20, 149–150, 402
Connolly, T., 299
Cooley, J. W., 348, 361
Cooper, N., 402
Courant, R., 323, 328
Crane, F. D., 329
Crockett, J. B., 298
Curry, H. B., 298

Darcy, V. G., 328
de Backer, W., 318, 328
Debroux, A., 20, 207
D'Hoop, H., 20, 207
DeLand, E. C., 420
Denham, W. F., 323, 329
Dobell, A. R., 361–362
Duersch, R. R., 298

Elkind, J. I., 298
Ellis, M. E., 420
Etkin, B., 389
Eykhoff, P., 299

AUTHOR INDEX

Falb, P., 303, 329
Fancher, P. S., 322, 328
Farrenkopf, R., 300
Favreau, R. R., 284, 300
Federoff, O., 150
Feldbaum, A. A., 299, 303, 329
Fifer, S., 20
Fineberg, M. S., 20
Fisher, M. E., 423, 426–427, 432–433
Fitsner, L. N., 299
Fogarty, L. E., 389
Fowler, M. E., 112
Franks, R. G., 284, 300
Frew, A., 300
Frost, P. A., 329

Gaines, W. M., 298
Gamkrelidze, R. V., 329
Gelman, R., 20, 149–150
Genna, J. F., 20
Gilbert, E. G., 91, 110–111, 313, 328, 382–383, 389
Gill, S., 89
Gilliland, M. C., 20, 112
Gilmore, B. G., 421
Gilson, R. P., 361
Graber, G. E., 420
Gran, M., 300
Graupe, K. K., 299
Green, C., 20, 207
Green, D. M., 298
Greenwood, D. T., 388
Gupta, S. C., 311, 322, 330
Gurin, L. S., 300
Guyton, A. C., 421

Hahn, W. R., Jr., 20
Halbert, P. W., 316, 329
Hammersley, J. M., 362
Hamming, R. W., 74, 89
Hampton, R. L., 361
Handler, H., 241–243
Handscomb, D. C., 362
Haneman, V. S., 215, 242
Hannen, R. A., 328
Harper, C. W., 402
Hartsfield, E., 443
Heartz, R. A., 388
Hemmerle, W. J., 361
Henrici, P., 112

Hethcote, G. M., 421
Higgins, J. J., 420
Hildebrand, B., 68, 84, 89, 433
Howard, D. R., 329
Howe, R. M., 101, 111, 215, 242, 363, 368, 389
Hull, T. E., 361–362
Humphrey, R. E., 299
Hung, J. W., 112
Hurney, P. A., Jr., 20
Huskey, H., 361
Hutchinson, D. W., 361

Iwata, J., 149, 243, 322, 330

Jackson, A. S., 20
Jacobson, C. A., 402
Jager, G. N., 420
Jones, R. W., 421
Jones, T. N., 388
Jury, E. S., 112
Jury, S. H., 216–217, 242

Karplus, W. J., 20, 149–150, 234, 242–243, 415
Kazda, L. F., 243
Keitel, G. H., 79, 89
Kelley, H. J., 323, 329
Keludjian, G., 149
King, C. M., 20
King, N. J., 421
Knapp, C. H., 329
Korn, G. A., 47, 289, 300, 318, 329, 331, 335, 339, 346, 360–362
Korn, T. M., 47
Kovacs, J., 207
Kumar, K. S. P., 300
Kunz, K. S., 89
Kuo, B. C., 112
Kuo, S. S., 89

Landauer, J. P., 316, 329, 402
Laning, J. H., 359, 361
Lee, E. S., 329
Lee, Y. W., 361
Leitmann, G., 303, 329
Levine, L., 253, 299–300, 361
Lew, A. Y., 179, 207
Licher, R. M., 421
Liebmann, G., 219, 242

AUTHOR INDEX 457

Linebarger, R., 207
Little, W. D., 241–243
Lubin, J. F., 20, 207

McGhee, R. B., 179, 207, 277–278, 299–300, 360, 362, 415, 417, 421
McKay, D. B., 423, 433
McRuer, D. T., 361, 403
Macy, J. B., Jr., 361, 411, 421
Maloney, J. C., 422
Margolis, M., 299
Maybach, R. L., 311, 319–320, 329
Meissinger, H., 135, 137–138, 149, 266, 299–300, 316, 329
Merritt, M. J., 91, 396, 403
Meyer, H. A., 362
Miller, K. S., 135, 149
Milne, W. E., 71, 87, 89
Mischenko, E. F., 329
Mitchell, B. A., 284, 300, 329, 411, 420
Miura, T., 119, 149, 229, 243, 322, 330
Monroe, A. J., 112
Moser, J. H., 300
Munson, J. K., 284, 299
Murray, F. J., 135, 149

Nesbit, R., 421
Neustadt, L. W., 300, 312–314, 328–329
Noordegraaf, A., 420
Norkin, K. B., 299

Ornstein, G. N., 299

Paiewonsky, M., 313, 316, 329
Palevsky, M., 20
Papoulis, A., 361
Paquette, G. A., 389
Parezanovic, N., 433
Peach, P., 362
Piersal, L., 361
Pontryagin, L. S., 303, 329
Potts, T. F., 299
Pyne, I. B., 299

Raby, J. S., 326, 329–330
Ralston, A., 75, 82–83, 89
Rastrigin, L. A., 290, 300
Rautaharju, P. M., 409, 421
Reed, C. O., 300
Rekasius, Z. V., 329

Robinson, A. C., 388
Rogers, A. E., 299
Romanelli, M. J., 64
Rose, R. E., 300
Rubin, A., 284, 299, 389

Sabroff, A. E., 300
Sadoff, M. E., 402
Saucedo, R., 112
Schmid, H., 20, 39, 47
Schwetman, H. D., 421
Sellars, H. L., 300
Seltzer, J. L., 361
Shannon, C. E., 124
Shaw, J. C., 207
Sherrill, H., 204
Shridhar, R., 300
Shubin, A. B., 299
Siler, R. W., 421
Skramstad, H. K., 20
Stacy, R. W., 421
Stakhovskii, R. I., 299
Stark, L., 420
Steinmetz, H. L., 321, 329
Stern, J. A., 421
Strauss, J., 207
Susskind, A. K., 43, 47
Sze, T. W., 112

Tauseworth, R. C., 361
Thomas, O. F., 402
Tick, L. J., 421
Tiechroew, D., 20, 207
Tomovic, R., 20, 135, 149–150, 228, 243, 415, 417, 421–422, 433
Tompkins, C. B., 299
Tou, J. T., 112, 303, 329
Truitt, T. D., 364, 388
Tsuda, J., 322, 330
Tukey, J. W., 348, 361
Turner, R. M., 389
Tuttleman, D. M., 389

Ung, M. T., 177, 297, 300
Urban, W. D., 20

Vichnevetsky, R., 234, 243, 299
Vidal, J., 149–150

Wait, J. V., 112
Walford, R. B., 360, 362

Wallman, H., 423, 432–433
Warner, H. R., 421
Waxman, B., 421
Wertz, H. J., 299
White, R. C., Jr., 361
Widrow, B., 346, 361
Widrow, D., 127, 149
Wilde, D. J., 300
Wilf, H. S., 89
Williams, E. J., 299
Windeknecht, T., 243

Wingrove, R. C., 326, 329–330
Witsenhausen, H. S., 234, 243, 299, 316, 329
Wong, A., 300
Woodbury, M. A., 421
Wortzman, D., 409, 421

Young, L., 406, 420

Zeitlin, R. A., 421

Subject Index

Adage, Inc., 33
Adams-Bashforth method, 68
  z-transform analysis of, 96
Adams-Moulton method, 70
  z-transform analysis of, 96
Addalink, 155
Addaverter, 155
Adjoint computing technique, 359
Adjoint variables, 305, 308, 317, 323–324
Administration of hybrid facilities, 160–162
Albert Einstein College of Medicine, 405, 411
Algebraic equations, 258–260
Amplitude errors in sample-hold units, 130–133
Analog-computer-oriented systems, 114
Analog computers, general characteristics, 4
Analog-digital converters, 20, 30, 42
  closed-loop type, 39
  incremental operation, 37–39
  in flight simulation, 364
  sample-hold unit for, 31–33
  simultaneous type, 35
  successive-approximation, 37–39
  time-interval, 41–42
Analog hold, 23
APACHE, 8
Aperture error, 31–32, 115, 125

Applied Dynamics, Inc., 6, 155
Arizona, University of, 9, 312
ASTRAC II, 241, 242, 312, 352
Autocorrelation function, 343
Average response computer, 405

Beckman Instruments, Inc., 155, 290
Biharmonic equation, 212
Biological signals, 404, 406–407
Biological systems, 406, 411–415
Blood pressure, hybrid processing of, 414
Boundary-value problems, 49, 83, 85, 211, 242, 308
Brown Engineering Co., 33

Calculus of variations, 301
CASPRE, 194–199
CAT, 405
CLASH program, 165–170
Closed-loop analog-digital converter, 36–39
Closed-shop operation, 156–157
Comcor/Astrodata, Inc., 6, 134
  CI-5000, 104, 165–169
Common-mode signals, 128
Computational stability, 58, 72, 74
  in digital finite-difference methods, 220–223
  in hybrid CSDT method, 233–234
Computer Control Corp., 9

SUBJECT INDEX

Confidence limits, 341–342
　of analog solutions, 293
Constraints, in linear programming, 261
　on control functions, 302
Continuous-space-continuous-time, 213–214
Continuous-space-discrete-time, see CSDT method
Control Data Corp., 10, 33, 153, 164, 165–169, 170
　CDC 6400, 165–169
Control Equipment Corp., 33
Control lines, 190
Control vector, 304
Convair Astronautics, 154, 155
Corrected inputs, method of, 149
Correlation functions, 343-346
　from continuous data, 344–345
　from ensemble measurements, 343
　in linear system studies, 355–356
　using coarse quantization, 346
Costate vector, 305
Cost functional, 304, 323
Criterion function, 251–252
Criterion functional, 263, 302
Cross-bar multiplexers, 47, 127
Crosstalk errors, 128
CSCT method, 213–214
CSDT method, 213
　analog, 216–218
　hybrid, 229–234

Debugging software, 194–199
Demultiplexer, 127–129, see Multiplexers
Diffusion equation, 212
Digital-analog converter, 23–29
　errors, 129–130
　general organization, 25
　in function generation, 387–388
　ladder network type, 29–30
　multiplying, 26, 369, 387
　weighted resistor type, 26–29
Digital computer, general characteristics, 5
Digital-computer-oriented systems, 114
Digital differential analyzers, 8
Digital Equipment Corp., 9, 33, 47, 153
Digital hold, 22
Discrete-space-continuous-time, see DSCT
Discrete-space-discrete-time, see DSDT
Distance function, 252

Distributor, errors in, 127–129
DSCT method, 213
　analog, 214–216
　hybrid, 228–229
DSDT computer, 9
DSDT method, 213
　analog, 218–220
　hybrid, 234–239
Dynamic Instrumentation Co., 33
Dynamic System Electronics, Inc., 33

EAI, see Electronic Associates, Inc.
Electrocardiogram, hybrid processing of, 407–411
Electronic Associates, Inc., 6, 8, 20, 155–156, 165–168, 170–177, 207, 243
　690 computer system, 173–176
　8900 computer system, 162–173, 180–207
　CSDT method, 231–233
Electronic Development Corp., 33
Electronic Engineering Corp., 34
Ensemble average, 333–334, 337–338
Errors, amplitude, 130–133
　analog, 293
　analyzed by z-transform techniques, 99–109, 381–382
　classification of, 115–117
　first-order hold, 129–135
　in flight simulation, 378, 380–382
　in optimal control problem studies, 308
　in solution of integral equations, 430–433
　in Taylor series, 58
　multiplexer, 127–129
　per-step, 143–147
　phase, 130–133
　round-off, 50
　sampler, 146
　skewing, 129
　slewing, 128
　time delay, see Time-delay errors
　truncation, 50–58, 78–83
　zero-order hold, 129–135
Euclidean norm, 275
Euler angles, 368, 371
Euler integration, 58, 200
　z-transform analysis, 94, 95
EURATOM, 8
Expected values, 337–341

SUBJECT INDEX 461

Exponentially weighted-past average, 340

Field-effect transistor, 43, 127
Filter, smoothing, 385, 386
Finite-state models, of manual control, 397–401
  of quadruped locomotion, 417
First-order hold circuit, errors in, 125–139
Flight simulation, 363–364
  analog, 390
  atmospheric, 364–372
  body axes for, 365
  flight path axes for, 365
  hybrid computers in, 363–364
  of manned vehicles, 390–394
FORTRAN for hybrid computers, 181–183
Forward integration method, 52, 67–70
Function generation, digital, in flight simulation, 368–369, 378–387
  errors, 380
  hybrid, 200–204, 387–388
  three-variable, 369
Function space, 323
  gradient in, 322–323
Fundamental matrix, 312

Gaussian random noise, 350
Gaussian random sequence, 286–287, 352
Gill integration method, 64–67, 200
Gradient method, continuous, 256–258, see Steepest descent methods
Gradient method, discrete, 270–280
  convergence and stability, 277
  digital computer implementation, 277
  geometrical interpretation, 272
  hybrid implementation, 279
  optimum, 275
  step size determination, 273
Gradient method in functional optimization, 322–326
Gradient vector, 249, 257, 322
  approximate, 272
Gray code, 127

Hamiltonian, 305–306, 309, 318, 321–324
H-frame, 373–374
HODAD, 167
HOI, 181, 183, 186

HSL, 176, 181, 186–189
Human operators, simulation of, 267–270, 394–402
Hybrid actuator, 401, 417
HYDAC, 8
Hyperbolic partial differential equation, 212
HYTRAN, 176, 181, 183, 186, 189,

IBM, 9–10, 153, 155, 160, 391
Impulses, as control perturbations, 326–327
Incremental analog-digital converters, 37–39
Influence coefficients, 266, see Sensitivity functions
Initial conditions, effect on optimization, 284–285
Initial-value problem, 49
Integral equations, 422–433
  errors in hybrid solution, 426–432
  examples of hybrid solution, 426–430
  Fredholm type, 422
  solution by Fisher method, 424–426
  solution by Neumann method, 423–424
  Volterra type, 432
International Business Machines, Inc., see IBM
Interrupt, 161, 205
INTRACOM, 165, 168
Iterative differential analyzers, 8

JOSS, 184

Known-slope method, 134

Ladder network digital-analog converter, 29–30
Lancer Electronic Corp., 34
Laplace's equation, 211, 223–228
Latter-Day Saints Hospital, 405
Learning model, 264
Leg movements, simulation of, 417–419
Linear programming, 260–262
Linear systems, identification of, 356–358
  mean-squared response of, 358–360
  time-varying, 359
  with random inputs, 355–360
Lockheed Aircraft Co., 165, 204
Locomotion, 415–420

SUBJECT INDEX

McDonnell Aircraft Corp., 391
Man-machine systems, 18, 390–403
Matrix-type multiplexer, 45–47
Maxima and minima of functions, 248–249
Mean square, 337–341
Mean value, 337
  confidence limits in estimation of, 341–342
  from ensemble averages, 339
  from EWP averages, 340
  from sampled data, 341
  from time averages, 339
  of biological signals, 407
Michigan, University of, 156–157
MIDAS, 10
Milne integration method, 71–74, 200
MIMIC, 10
MINCC, 181, 197–199
Model-reference adaptive system, 263
Modified open-shop, 156–158
Monitor program for hybrid system, 181, 188–189
Monte-Carlo method, 17
  digital, 223–228
  hybrid, 239–242, 360
Motivation for hybridization, 13–16
Multiplexers, 21, 42–47
  electromechanical, 43, 47
  errors, 127–129
Multiverter, 155

Navigation Computer Corp., 34
Newton-Raphson method, 273
  modified, 281
Noise generators, analog, 286, 350–352
  shift-register, 352–353
Nominal trajectory, 324
Non-self-starting integration methods, 51

Objective function, 261
Open-shop operation, 156–157
Operating cost of hybrid facilities, 159–160
Optimal control problems, 301–330
  admissible controls for, 304
  canonical equations, 306
  computational considerations, 307–310
  cost functional, 302
  free right-end problem, 304

Optimal control problems (*continued*)
  linear minimum-time, 310–316
  solution by Neustadt's method, 312
  trial-and-error solution of, 311
  two-point boundary values, 308
Optimization by steepest descent, 256
Optimization problems, classification, 247
  in the chemical industry, 320–321
  mathematical considerations, 247

Packard-Bell, Inc., 155
Parabolic partial differential equations, 211–242
Parameter optimization, 244–300
  applications of, 245
  continuous, 264
  cyclical, 280
  discrete, 270–280
  geometrical interpretation, 253
  in the presence of noise, 297–298
  neighboring grid method, 283
  one parameter search, 282
  random search, 284–297
Parameter space, 251, 255, 323
Parseval's theorem, 358
Pastoriza Electronics Corp., 34
PDP computer, 155
Performance criteria, 244
Per-step errors in hybrid systems, 117
Perturbation equations, 324–326
Perturbation functions, 317, 324
Perturbation of cost functional, 327
Phase errors, in sample-hold circuits, 130–133
Planning, for hybrid computation, 162–164
Poisson's equation, Monte-Carlo method, 227
Polynomials, roots of, 246
Pontryagin Maximum Principle, 305–307
Power spectral density, 347–348, *see* Spectral analysis
Predictor-corrector method, 52, 70–74, 200
Predictor-modifier-corrector method, 52, 74–83
Preston Scientific Corp., 34
Probability density functions, 333–337
Probability distribution functions, 332–337

Pseudo-random sequences, 286, 354
Pyramid-type multiplexer, 45

Quantization grain 126–127
Quantizing errors, 125–127, 146
Quasi-linearization, 298

Ramo-Wooldridge Corp., 15, 154
Rand Corp., 354
Random noise in hybrid systems, 350–355
Random process studies, 17, 331–362
Random search optimization, 256, 285–289
  absolute biasing of parameter increments in, 287
  convergence of, 288–290
  example, 290–297
  global optimum algorithm, 288–289
  in optimal control problems, 311–312
  local optimum algorithm, 287–288
  of satellite acquisition system, 290–297
Random walk, 224–227, 240–242
Real-time computation, 179
Rectangular integration, 100–103, 107–111
Redcor Corp., 34
Reed relay, 43, 44, 127
Reentrant subroutines, 190, 200
Relaxation method, 256
Repeatability, of analog solutions, 293
Round-off errors, 50
Runge-Kutta methods, 52, 58–67, 200
  fifth order, 61
  fourth order, 61–62
  systems of equations, 63–67
  third order, 60
  z-transform analysis, 105–106

Sampled data, in random process studies, 335, 341, 345–346
Sampled-data systems simulation, 16
Sample-hold, 23
  errors, 124–125, 129–135, 146
  for analog-digital conversion, 31–32
Sampler errors, 115, 124–125
Sampling rate, sensitivity analysis of, 147–148
Sampling theorem, 124–125, 133
Scientific Data Systems, Inc., 9–10, 34, 153, 155

Selection pyramids for multiplexers, 44–45
Self-starting integration methods, 51
Sense lines, 190
Sensitivity analysis, 135–148
  of analog circuits, 137–138
  of hybrid systems, 139–143
  of sampling rate, 147–148
  time-varying perturbations, 138
Sensitivity equations, 266
  for adjoint system, 308
Sensitivity functions, in optimal control problems, 316–318
Sensitivity matrix, 317
Shannon's sampling theorem, 124–125, 133
Simpson's rule, 200
Simultaneous analog-digital converter, 35
Skewing errors, 129
Slewing errors, 128
Sloan-Kettering Institute, 405
Space vehicle simulation, 372–378, 391–394
Spectral analysis, 347–350
  analog, 347–348
  fast Fourier transform method, 348–349
  window problems, 349
Spectral density functions, shaping of, 356, see Spectral analysis
Spectrum of hybrid computer techniques, 6–10
Staff of computer facilities, 158–159
State transition matrix, 312
State variables, 250, 284, 302, 304
Steepest descent methods, approximate, 262, 265
  continuous, 254–255
  difficulties with, 264–265
  in flight simulation, 378
  in functional optimization, 326–327
  in optimal control problems, 318–321
Subranging analog-digital converter, 39–41
Subsystem, input-output errors, 117–135
Successive approximation analog-digital converter, 37–39
System Engineering Laboratories, 34
System identification, 356–358

Target set, in optimal control problems, 304

Taylor series methods, in perturbation analysis, 316–317
  integration formulas based on, 51, 54–58
  z-transform analysis of, 94–96
Technical Measurements Corp., 405
Texas Instruments, Inc., 34
Threshold gate, 418
Time-delay, generation of variable, 204–206
Time-delay errors, 117–124, 142–143
  compensation of, 120–124, 142–144, 382
  in multiplexers, 128
Time-interval analog-digital converter, 41–42
Time-optimal control problems, 310–311
  solution by continuous steepest descent, 319–320
  solution by Neustadt's method, 312–316
  solution by random search, 311–312
T-locus, 100–101
Trapezoidal integration method, z-transform analysis of, 95–96, 103–111

True hybrids, 10
Truncation errors, 50
  control of, 78–87
  in Milne method, 71–73
  in Taylor series, 58
TRW Systems, 154, 392

Univac, 155
Utility library, 181, 197–206

Van der Pol's equation, 128, 142–144
Vector differential equation, 304

Weighted-resistor digital-analog converter, 26–30
Weighting function, 358–359
Wiener-Khinchine equations, 347
Wiener-Lee relations, 355

Zero-order hold, 118, 129–135
z-transform techniques, 90–104, 381–385